I0066102

Sustainable Irrigation Management

Sustainable Irrigation Management

Edited by **Davis Twomey**

R CALLISTO
REFERENCE

New York

Published by Callisto Reference,
106 Park Avenue, Suite 200,
New York, NY 10016, USA
www.callistoreference.com

Sustainable Irrigation Management
Edited by Davis Twomey

© 2016 Callisto Reference

International Standard Book Number: 978-1-63239-765-2 (Hardback)

This book contains information obtained from authentic and highly regarded sources. Copyright for all individual chapters remain with the respective authors as indicated. All chapters are published with permission under the Creative Commons Attribution License or equivalent. A wide variety of references are listed. Permission and sources are indicated; for detailed attributions, please refer to the permissions page and list of contributors. Reasonable efforts have been made to publish reliable data and information, but the authors, editors and publisher cannot assume any responsibility for the validity of all materials or the consequences of their use.

The publisher's policy is to use permanent paper from mills that operate a sustainable forestry policy. Furthermore, the publisher ensures that the text paper and cover boards used have met acceptable environmental accreditation standards.

Trademark Notice: Registered trademark of products or corporate names are used only for explanation and identification without intent to infringe.

Printed in the United States of America.

Contents

Preface

Every book is initially just a concept; it takes months of research and hard work to give it the final shape in which the readers receive it. In its early stages, this book also went through rigorous reviewing. The notable contributions made by experts from across the globe were first molded into patterned chapters and then arranged in a sensibly sequential manner to bring out the best results.

This book explores all the important aspects of irrigation management in the present day scenario. It strives to provide a fair idea about this discipline and to help develop a better understanding of the latest advances within this field. Irrigation management refers to the practice of using water while farming in an efficient and optimum way. It deals with minimizing the wastage of water and managing it in such a manner that it is available in dry areas and when rainfall is scarce. This book will discuss the various types and branches of irrigation management. It unfolds some innovative aspects of irrigation which will be crucial for the progress of agriculture in the future. Most of the topics introduced in this book cover new techniques and the applications of this area. This text will serve as a reference to a broad spectrum of readers.

It has been my immense pleasure to be a part of this project and to contribute my years of learning in such a meaningful form. I would like to take this opportunity to thank all the people who have been associated with the completion of this book at any step.

Editor

Acinetobacter baumannii in Birds' Feces: A Public Health Threat to Vegetables and Irrigation Farmers

M. Dahiru[1]*, O. I. Enabulele[2]

[1]Federal University, Kashere, Gombe State, Nigeria
[2]University of Benin, Benin, Nigeria
Email: *musahanifa@yahoo.com

Abstract

The rising trend of resistance in *Acintobacter baumannii* had in recent days become a public health care concern with most literature reported from samples collected from hospital environment. This research therefore, wishes to determine the occurrence of multidrug-resistant *A. baumannii* in birds' droppings, associated with irrigated farms vegetables, for epidemiological update and future clinical forecast. Forty eight birds fecal samples were collected and processed for isolation and identification of *A. baumannii* on MacConkey agar and Microbact 24E (Oxoid), and tested against 10 commonly used antibiotics (quinolones, fluoroquinolones, aminoglycosides, sulfonamides). *A. baumannii* was isolated from 31.25% of samples and had shown more resistant to ceporex (100.00%) and to streptomycin with 80.00% and 90.00% for Jakara and Sharada farms' fecal samples respectively; isolates were however sensitive to co-trimoxazole. Forty eight (46.67%) of the isolates were resistant to at least 6 drugs, with strong correlation between some drugs. By this result, wild birds' fecal materials demonstrate high potential of *A. baumannii* carrying capacity and dissemination, and thus pose risk of contaminating vegetables, infecting human and transmitting resistance phenotype to other non-multidrug-resistant bacteria—a situation quite challenging to health care management and public health. And thus it further suggests for screening of additional probable contributing factors, so as to develop possible detailed transmission pathway and control strategies.

Keywords

Wild Birds, Vegetables, Public Health, Ceporex, Co-Trimoxazole

---

*Corresponding author.

1. Introduction

Acinetobacter species are usually commensal organisms, but they occasionally cause infections, predominantly in susceptible patients in hospitals. They are opportunistic pathogens that may cause urinary tract infections, pneumonia, bacteraemia, secondary meningitis and wound infections. These diseases are predisposed by factors such as malignancy, burns, major surgery and weakened immune systems, in neonates and elderly individuals. There is no evidence of gastrointestinal infection through ingestion of *Acinetobacter* spp. The emergence and rapid spread of multidrug resistant *A. calcoaceticus baumannii* complex, causing nosocomial infections, are of concern in health care facilities. *A. baumannii* is a prevalent species that causes epidemic outbreaks of nosocomial *Acinetobacter* infections [1]. *A. baumannii* is occasionally isolated from environmental samples such as soil and water.

The exact natural habitat of many of the *Acinetobacter* species is yet to be fully understood and may require intense efforts to identify [2]. Thus, overall diversity of habitat, predilection to accumulate antimicrobial resistance, resistant to desiccation, ability to form biofilm, and propensity to cause hospital infection outbreaks make *Acinetobacter* a remarkable microorganism. *A. baumannii* strains are generally more resistant than other species of this genus and often express a multi-drugs resistant (MDR) phenotype. Therefore, treatment of nosocomial infections caused by *A. baumannii* has become complicated because of the widespread antimicrobial resistance among these organisms [3]. The rising trend of resistance in *A. baumannii* strains, particularly to newer antimicrobial agents, has therefore become a public health care concern. The organism expresses multiple mechanisms of antibiotic resistant that likely leads to the development of multiply resistant or even "pan-resistant" strains.

Some authors have suggested hospital environment as the major reservoir of the multidrug resistant *A. baumannii* [4] [5]. It still remained apparent that *A. baumannii* has been isolated from quit a number of other non-hospital sources. For example, Byrne-Bailey *et al.* in [6] had isolated multidrug resistant *A. baumannii* in England from soil fertilized with pig manure, and similarly Zhang and his colleagues in [7] had also isolated multi-drugs resistant *A. baumannii* from livestock in China. Many bacteria were reported as part of the normal intestinal flora of birds. Bird species associating with potentially contaminated environments, such as refuse dumps, sewage treatment facilities, agricultural sites, and bird feeders, are likely to harbor pathogenic bacteria in their intestinal tracts [8]. For example, Craven *et al.* [9] implied that house sparrows and European starlings may be responsible for transmitting *Salmonella* spp., *Campylobacter jejuni*, and *Clostridium perfringens* to chickens on poultry farms. Ahmed *et al.* [10] have isolated non-multi-drug resistant *A. baumannii* from birds' feces kept in Zoo, Japan. Thus, birds may inadvertently ingest bacteria in their environment, and the bacteria pass through the intestinal tract with no adverse effects to the carrier bird; yet in such cases the birds may aid in dispersal of the bacteria within the environment. The research therefore aims at determining the prevalence of multidrug resistant *A. baumannii* in birds' droppings (carrier rate), especially those associated with vegetables that are minimally process and consumed, in Kano State, and thus, accesses the potential role of wild birds in transmission of *A. baumannii* for public health importance. In Africa, there is paucity of data on the prevalence of *A. baumannii* carried in the fecal samples of domestics or wild birds feeding in irrigation farms in Nigeria; most data on multidrug resistant *A. baumannii* were from Europe and Asia. Therefore an attempt to document these in Nigeria will surely complement the efforts made, for epidemiological and clinical forecast.

2. Material and Methods

The study was carried out in 2 irrigation sites, with the first site located along Jakara wastewater canal irrigation farms and the second site located along Sharada canal wastewater irrigation farms, both located in Kano City. All samples were initially processed to separate the non-fermenters from other Gram negative bacilli on Mac-Conkey agar and 5% sheep blood agar at 37°C for 24 hours. There after identification was done to confirm presence of *Acinetobacter* spp. Samples were sub cultured from primary isolation media and grown further on nutrient agar (NA), from which colonies on NA were Gram stain, other biochemical tests conducted include oxidase, catalase gelatin liquefaction, motility and other sugar fermentation were undertaken. These were done in accordance with Microbact 24E (Oxoid) for the identification of unknown oxidase negative bacteria.

Species differentiation on Microbact 24E was done on basis of 24 reactions, and inoculates at 37°C and 44°C. Antimicrobial susceptibility tests using disc diffusion method was carried out in accordance with [11] using the following antibiotics, Ofloxacin (OFX), Pefloxacin (PEF), Ciproflox (CPX), Amoxicillin-clavulanic acid (AU), Gentamycin (CN), Streptomycin (ST), Ceporex (CEP), Nalidixic acid (NA), Co-trimoxazole (SXT), Ampicilin

(PN). The results were interpreted in accordance with Clinical and Laboratory Standards Institute (CLSI) criteria [12].

Spearman correlation analysis was used to compare the similarities in response to each antibiotic tested on *A. baumannii* isolated

3. Results

From a total of 48 fecal samples, 24 from each site, 5 (20.83%) *Acinetobactaer baumannii* isolates were obtained from fecal samples collected along Jakara irrigation farms, while 10 (41.67%) were obtained from fecal samples collected from Sharada irrigation farms. The overall distribution of antibiotic resistant, for *Acinetobacter* isolates, detected from each site are shown in **Table 1**. Isolates from all sources were more resistant to ceporex (100.00%) followed by streptomycin with 80.00% and 90.00% resistance phenotype in isolates from Jakara farms and Sharada farms fecal samples respectively. All isolates were less resistant to co-trimoxazole with 20.00% and 50.00% in fecal samples from Jakara and Sharada farms respectively. Multi dug resistant of two or more antibiotics was observed among all isolates, with the highest seen in isolates from fecal sample collected in Jakara, for example number JK16 was resistant to all drugs tested. Impact, forty six (46.67%) of the isolates were resistant to at least 6 antibiotics, as showed in **Table 2**. Although, the isolates showed considerable multi drugs resistance, there were however few correlation in the patterns of resistant demonstrated by isolates on each antibiotic. For example only three positive values were observed, which were CEP to PN (67.8%, $p < 0.01$), CPX to

Table 1. Percentage distribution of *Acinetobacter baumannii* occurrence in birds' feces and antibiotic resistant profiles.

Sources	No. sampled	No. Isolated	% Occurrence	Percentage Resistant of Antibiotic									
				ST	PN	CEP	OFX	NA	PEF	CN	AU	CPX	SXT
Jakara Farms	24	5	20.83	80:00	80:00	100:00	80:00	60:00	100:00	60:00	80:00	40:00	20:00
Sharada Farms	24	10	41.67	90:00	60:00	100:00	70:00	40:00	70:00	60:00	70:00	80:00	50:00

Key: No = Total number, % = percentage, ST = Streptomycin, PN = Ampicilin, CEP = Ceporex, OFX = Ofloxacin, NA = Nalidixic acid, PEF = Pefloxacin, CN = Gentamycin, AU = Amoxicillin-clavulanic acid, CPX = Ciproflox, SXT = Co-trimoxazole.

Table 2. Distribution of *Acinetobacter baumannii* isolate from birds' feces and multidrug resistant profiles.

S/N	Sample number	Source	Name of antibiotic										% Multidrug resistance
			ST	PN	CEP	OFX	NA	PEF	CN	AU	CPX	SXT	
1	JK 5	Jakara farms	R	R	R	R	S	R	S	R	S	S	60:00
2	JK 8		S	S	R	R	S	R	R	R	S	S	50:00
3	JK 9		R	R	R	R	R	R	R	R	R	S	90:00
4	JK 16		R	R	R	R	R	R	R	R	R	R	100:00
5	JK 20		R	R	R	S	R	R	S	S	S	S	50:00
6	SH 11	Sharada farms	R	S	R	S	S	R	R	R	R	S	60:00
7	SH 12		S	S	R	S	R	R	R	R	S	R	60:00
8	SH 13		R	R	R	R	S	R	S	S	R	S	60:00
9	SH 14		R	R	R	R	R	R	R	R	S	R	90:00
10	SH 17		R	S	R	R	R	S	R	S	R	S	60:00
11	SH 18		R	R	R	S	S	R	S	R	R	R	70:00
12	SH 19		R	R	R	R	S	R	R	R	R	S	80:00
13	SH 20		R	R	R	R	R	R	S	R	R	R	90:00
14	SH 21		R	S	R	R	S	S	S	R	R	R	60:00
15	SH 23		R	R	R	R	S	S	R	S	R	S	60:00
	Column Total		86:67	66:67	100:00	73:33	46:67	80:00	60:00	73:33	66:67	40:00	

Key: OFX = % = percentage, ST = Streptomycin, PN = Ampicilin, CEP = Ceporex, OFX = Ofloxacin, NA = Nalidixic acid, PEF = Pefloxacin, CN = Gentamycin, AU = Amoxicillin-clavulanic acid, CPX = Ciproflox, SXT = Co-trimoxazole, S = Sensitive, R = Resistant.

ST (57.6%, p < 0.05), AU to SXT and (55.1%, p < 0.05) as shown in **Table 3**.

4. Discussion

There have been limited data on the occurrence of *Acinetobatcer baumannii* especially in association with agricultural produce. *A. baumannii* was reported to multiply not only on human and animal skin, but also in soil and water and thus has a diversity of reservoirs [1]. Our present finding is clearly in support of this hypothesis, as demonstrated by the high occurrence of *Acinetobatcer baumannii* isolated from birds' fecal samples. Although other researchers [6] [7] have demonstrated the isolation of *Acinetobatcer* species from some livestock, there was no report of isolation from wild birds' fecal sample. Previously, report has demonstrated bacteria as part of the normal intestinal flora of birds, and therefore bird species associating with potentially contaminated environments, such as refuse dumps, sewage treatment facilities, agricultural sites, and bird feeders, are likely to harbor pathogenic bacteria in their intestinal tracts [8].

The report of Ahmed *et al.* [10] on isolation of non-multi-drug resistant *A. baumannii* from birds' feces kept in Zoo, Japan, does not really demonstrate isolation from wild bird; since the birds were in captivity, this report is more of prevalence or occurrence of *A. baumannii* in wild birds, whose date were previously scarce for *A. baumannii*. Previously birds have been implicated in transmission of pathogenic bacteria; for example, Craven *et al.* [9] implicated house sparrows and European starlings for transmission of *Salmonella* spp., *Campylobacter jejuni*, and *Clostridium perfringens* to chickens on poultry farms. This suggests that wild birds' fecal samples at farms environment could disseminate multi-drug resistance *A. baumannii* and the possible risk of contaminating and infecting human population is likely, through the consumption of minimally processed fresh vegetables. This could be supported by previously researches, for example Solomon and his colleagues [13] had demonstrated the ability of *E. coli* 0157:H7 to enter lettuce plant through the root system and migrate throughout the edible portion of the plant and thus *A. baumannii* could be expected to have this ability as bacterium; however a more specific research is required to confirm that. Enterobacteriacea were isolated with increasing frequency from fresh produce, including beans, sprouts, cantaloupes, apples, lettuce, [13] and carrot, [14]. In the area of this study Dahiru *et al.* [15] isolated *E. coli* Ol57:H7 in cabbage and lettuce leaves; Uzeh and Adepoju [16] had also reported isolation of *Escherichia coli* O157:H7 and *Listeria monocytogenes* from different salad vegetables: cucumber, cabbage, carrot, and lettuce. A number of reports on contamination of vegetables and other agents of transmission have been documented in [13] [14]; as a whole, leafy green vegetables were cited as a source of 26% of the food-borne outbreaks in United States, between 1998 and 1999 [17].

Table 3. Antibiotic resistant similarities of isolates to the antibiotics tested.

	ST	PN	CEP	OFX	NA	PEF	CN	AU	CPX	SXT
S										
PN	0.342									
CEP	0.310	0.678**								
OFX	0.318	0.059	0.261							
NA	0.234	0.213	0.193	−0.252						
PEF	0.086	0.397	0.026	−0.102	0.020					
CN	0.074	0.068	0.236	0.354	0.006	−0.513				
AU	−0.309	0.086	0.100	−0.233	0.170	0.388	−0.001			
CPX	0.576*	−0.034	−0.077	0.258	−0.241	−0.152	0.337	−0.218		
SXT	−0.040	0.275	0.344	−0.326	0.469	0.165	0.069	0.551*	−0.389	

Key: * = (p) 0.05, ** = (p) 0.01, ST = Streptomycin, PN = Ampicilin, CEP = Ceporex, OFX = Ofloxacin, NA = Nalidixic acid, PEF = Pefloxacin, CN = Gentamycin, AU = Amoxicillin-clavulanic acid, CPX = Ciproflox, SXT = Co-trimoxazole.

While birds scavenging in irrigation farms poses risk of contaminating the environment and even the vegetables with bacteria, *A. baumannii* remained to be a challenge to public health, especially due to its pronounce multidrug-resistance mechanisms. In this research birds' fecal samples have demonstrated a high resistance profile to most antibiotics commonly used, not only resistance per say, but a multi-drug or pan drug resistance as demonstrated by some species isolated. Most alarmingly all isolates were resistance to more than one drug, with quite an amount of resistant to at least five drugs. Although this phenomenon was not new in *A. baumannii*, the isolation from birds' fecal sample called for close monitoring and surveillance, so as to address the possible occurrence and rapid transmission, of multi-drug resistance phenotype not only within the genus but also among other Enterobacteriaceae in farm environments. The use of fecal materials or dung of goats, sheep, donkey, birds and many other domestic and wild animals as a source of nutrients to crops is well established practice in Nigeria; this habit contributes in directly amplification of the spread of multi-drug resistance *A. baumannii* and other pathogenic bacteria, even to environments which were never reported to inhabit.

The observation of high resistance by *A. baumannii* isolated from this work to ceporex (β-lactam drug) and streptomycin (aminoglycoside drug), is in harmony with what was previously reported on *Acinetobacter* species, which were shown to exhibit different mechanisms of resistance to antibiotics, including the capability to produce enzymes modifying aminoglycosides, β-lactamases of a wide spectrum of activity, carbapenemases, as well as mechanisms resulting from the changes in outer membrane proteins, in penicillin binding proteins and in topoisomerases [18]. These lead to the formation of multidrug resistant (MDR) strains, and even the pandrug-resistant (PDR) strains which are resistant to all available drugs [19].

The correlation analysis of the resistance phenotype by *A. baumannii* on the ten antibiotics has demonstrated high positive relationship between ceporex and penicillin (both β-lactam); however cotrimothazole which was more sensitive than other drugs has also demonstrated more than fifty percent positive correlation with amoxicillin-clavulanate. These were in agreement with Bonomo and Szabo, who reported resistance mechanisms that are expressed frequently by *Acinetobacter* including β-lactamases, alterations in cell-wall channels (poring), and efflux pumps. *A. baumannii* can become resistant to quinolones through mutations in the genes *gyrA* and *parC* and can become resistant to aminoglycosides by expressing aminoglycoside-modifying enzymes [20], and thus the exhibition of MDR by greater percentage of *A. baumannii* is isolated on this work. The presence study marks the beginning of understanding the importance of multi-drug resistance phenotype by *A. baumannii* in agricultural produce. The occurrence or spillover of multi-drug resistance by wild birds suggests a rather new pathway of transmission that may demand further research input, so as to properly suggest best ways and practice in the control and prevention of diseases caused by *A. baumannii* and other *Acinetobacter* species.

References

[1] Joshi, S.G. and Litake, G.M. (2013) *Acinetobacter baumannii*: An Emerging Pathogenic Threat to Public Health. *World Journal of Clinical Infectious Diseases*, **3**, 25-36. http://dx.doi.org/10.5495/wjcid.v3.i3.25

[2] Visca, P., Seifert, H. and Towner, K.J. (2001) Acinetobacter Infection—An Emerging Threat to Human Health. *IUBMB Life*, **63**, 1048-1054. http://dx.doi.org/10.1002/iub.534

[3] Seifert, H., Baginski, R., Schulze, A. and Pulverer, G. (1993) Antimicrobial Susceptibility of *Acinetobacter* Species. *Antimicrobial Agents and Chemotherapy*, **37**, 750-753. http://dx.doi.org/10.1128/AAC.37.4.750

[4] Dijkshoorn, L., Nemec, A. and Seifert, H. (2007) An Increasing Threat in Hospitals: Multidrug-Resistant *Acinetobacter baumannii*. *Nature Reviews Microbiology*, **5**, 939-951. http://dx.doi.org/10.1038/nrmicro1789

[5] Gootz, D.T. and Marra, A. (2008) *Acinetobacter baumannii*: An Emerging Multidrug-Resistant Treat. *Expert Review of Anti-Infective Therapy*, **6**, 309-325. http://dx.doi.org/10.1586/14787210.6.3.309

[6] Byrne-Bailey, K.G., Gaze, W.H., Kay, P., Boxal, A.B., Hawkey, P.M. and Wellington, E.M. (2009) Prevalence of Sulfonamide Resistance Genes in Bacterial Isolates from Manured Agricultural Soil and Pigs Slurry in the Unoited Kingdom. *Antimicrobial Agents and Chemotherapy*, **53**, 696-702. http://dx.doi.org/10.1128/AAC.00652-07

[7] Zhang, W.J., Lu, Z., Schwartz, S., Zhang, R.M., Wang, X.M., Si, W., Yu, S., Chen, L., and Liu, S. (2013) Complete Sequence of Bla (NDM-1) Carrying Plasmid pNDM-AB from *Acinetobacter baumannii* of Food Animal Origin. *Journal of Antimicrobial Chemotherapy*, **68**, 1681-1682. http://dx.doi.org/10.1093/jac/dkt066

[8] Casanovas, L., Desinion, M., Ferrer, M.D., Arques, J. and Monzon, G. (1995) Intestinal Carriage of Campylobacters, Salmonellas, Yersinias and Listerias in Pigeons in the City of Barcelona. *Journal of Applied Bacteriology*, **78**, 11-13. http://dx.doi.org/10.1111/j.1365-2672.1995.tb01666.x

[9] Craven, S.E., Stern, N.J., Line, E., Bailey, J.S., Cox, A. and Fedorka-Cray, P. (2000) Determination of the Incidence of

Salmonella spp. *Campylobacter jejuni* and *Clostridium perfringens* in Wild Birds near Broiler Chicken Houses by Sampling Intestinal Droppings. *Avian Diseases*, **44**, 715-720. http://dx.doi.org/10.2307/1593118

[10] Ahmed, A.M., Motoi, Y., Sato, M., Maruyama, A., Watanebe, H., Fukumotom, Y. and Shimamoto, T. (2007) Zoo Animals Reservoir of Gram-Negative Bacteria Harboring Integrons and Antibiotics Resistance Genes. *Applied and Environmental Microbiology*, **73**, 6686-6690. http://dx.doi.org/10.1128/AEM.01054-07

[11] Cheesbrough, M. (2005) District Laboratory Practice for Tropical Countries, Part 2. Cambridge University Press, Cambridge, 426 p. http://dx.doi.org/10.1017/CBO9780511581304

[12] Clinical and Laboratory Standards Institute (2007) Performance Standards for Antimicrobial Susceptibility Testing; Seventeenth Informational Supplement. CLSI Document M100-S17. Clinical and Laboratory Standards Institute, Wayne, Pennsylvania, USA.

[13] Solomon, E.B., Yaron, S. and Matthews, K.R. (2002) Transmissino of *Escherichia coli* O157:H7 from Contaminated Manure and Irrigation Water to Lettuce Plant Tissue and Its Subsequent Internalization. *Applied and Environmental Microbiology*, **68**, 397-400. http://dx.doi.org/10.1128/AEM.68.1.397-400.2002

[14] Beuchat, L.R. (1999) Survival of Enterohemorrhagic *Escherichia coli* O157:H7 in Bovine Feces Applied to Lettuce and the Effectiveness of Chlorinated Water as a Disinfectant. *Journal of Food Protection*, **62**, 845-849.

[15] Dahiru, M., Uraih, N., Enabulele, S.A. and Kawa, A.H. (2008) Prevalence of *E. coli* O157:H7 in Some Vegetables in Kano City, Nigeria. *BEST*, **5**, 221-224.

[16] Uzeh, R.E. and Adepoju, A. (2013) Incidence and Survival of *Escherichia coli* O157:H7 and *Listeria monocytogenes* on Salad Vegetables. *International Food Research Journal*, **20**, 1921-1925.

[17] Codex Alimentarius Commission (2003) Risk Profile for Enterohemorrhagic *E. coli*, Including the Identification of Commodities of Concern, Including Sprouts, Ground Beef and Pork. Codex Alinorm 03/13A: Report of Codex Committee for Food Hygiene 2003, Joint Food and Agriculture Organization (FAO)/World Health Organization (WHO) Food Standards Programme, FAO, Rome, 60-64.

[18] Slama, T.G. (2008) Gram-Nagative Antibiotic Resistance: There Is a Price to Pay. *Critical Care*, **12**, 11-17. http://dx.doi.org/10.1186/cc6820

[19] Wang, S.H., Sheng, W.H., Chang, Y.Y., Wang, L.H., Lin, H.C., Chen, M.L., Pan, H.J., Ko, W.J., Chang, S.C. and Lin, F.Y. (2003) Healthcare-Associated Outbreak due to Pan-Drug Resistant *Acinetobacter baumannii* in a Surgical Intensive Care Unit. *Journal of Hospital Infection*, **53**, 97-102. http://dx.doi.org/10.1053/jhin.2002.1348

[20] Bonomo, R.A. and Szabo, D. (2006) Mechanisms of Multidrug Resistance in *Acinetobacter* Species and *Pseudomonas aeruginosa*. *Clinical Infectious Diseases*, **43**, S49-S56. http://dx.doi.org/10.1086/504477

2

Carbohydrates, Growth and Production of "Roxo de Valinhos" Fig Tree in Initial Development under Irrigation Management

Manoel Euzébio de Souza[1], Sarita Leonel[2], Andréa Carvalho da Silva[3],
Adilson Pacheco de Souza[3], Rafaela Lopes Martin[2], Adriana Aki Tanaka[3]

[1]Department of Agronomy, University of the Mato Grosso State, Nova Xavantina, Brazil
[2]Department of Plant Production, School of Agronomic Sciences, State University of Sao Paulo, Botucatu, Brazil
[3]Institute of Agricultural and Environmental Sciences, Federal University of Mato Grosso, Sinop, Brazil
Email: mseuzebio@gmail.com, sarinel@fca.unesp.br, acarvalho@ufmt.br, adilsonpacheco@ufmt.br

Abstract

The carbohydrates translocation and consequently growth and production of fig tree (*Ficuscarica* L.) vary according to the different management on cultivation conditions. The aim of this study was to evaluate the changes in the levels and total carbohydrates accumulation together with growth and "Roxo de Valinhos" fig trees production onimplementation of orchards in initial phase, cultivated with and without irrigation. We adopted a factorial arrangement (2 x 7) with four repetitions distributed in installments (with and without irrigation) subdivided in time (collect time). Destructive analyzes were performed at 40, 80, 120, 160, 200, 240 and 280 days after pruning (DAP) and are measured: stem diameter and branch, stem length and branch, number of leaves, internodes and fruit. Subsequently, the plant parts were sectioned to obtain the leaf area, length and roots volume, fresh and dry matter weight. The number, weight and total productivity of fruits were evaluated. The media of all growth attributes and production characteristics were higher in treatments with water irrigation. The total carbohydrate content was higher at 120 and 160 DAP and the carbohydrates accumulation was increasing for most institutions over the plants development, except for the leaves that showed a decrease in the levels at 160 DAP. The fruits showed greater carbohydrates accumulation in relation to the other evaluated organs.

Keywords

Ficus carica L., Water Management, Growth Analysis

1. Introduction

The fig tree culture in Brazil has presented a significant advancement in recent years, increasing at 45.72% in the produced volume, 116.16% and 404.03% on the amount in export value [1]. The main importing countries of the Brazilian figs are the Netherlands, France, Germany and England [2]. In commercial terms, the "Roxo de Valinhos" fig tree is the only variety grown in Brazil. This variety excels due to high economic value, hardiness, high vigor and productivity, good adaptation to drastic pruning and with fruits can be used for both fresh consumption and industry [3] [4].

In Brazilian conditions, according to the cultural techniques of the fig tree production, the annual pruning to canopy training and/or fruiting is carried out during the winter, followed by thinning of shoots in the autumn, setting so the number of branches in each plant per production cycle [5]. Subsequent to the drastic pruning starts an intense assimilates translocation to root system for the new branches and fruits formation, however, environmental changes and phenological characteristics provides variations on carbohydrates translocation dynamics to the different partitions of plant. Among the various climatic factors, the availability of air temperature and water promotes the greatest influences in carbohydrates levels on plant, and furthermore, can cause reductions in their growth by affecting the development and production [6] [7].

The carbohydrate storage by plant is critical to sustain development in stress periods, vegetative dormancy, emission and formation of branches and fruiting [8]. The metabolic interactions study involving the carbohydrates interconversion provides an understanding of assimilates flow and consumption both in whole plants as in separate plant organs. The determination of carbohydrates availability in storage structures has great importance for planning the time of pruning, defoliation, breaking the dormancy control, vegetative growth, fertilization management and production [9].

Together with the evaluation of carbohydrates, dynamics is necessary to relate it with the fig tree growth characteristics. According to Larcher [7] and Benincasa [10], the growth analysis is expressed to morphophysiological conditions of plants and then quantifies at production derived from the photosynthetic process, that is, which is the result of assimilatory system performance. Thus, each stage of development and plant growth is strongly limited by the environment through edaphic and climatic factors, plus the cultural techniques that can have major effects on survival and plants productivity [11].

Given that most research on the carbohydrates flow in fruit only restricts the seedlings stage and/or isolated parts of adult plants are fundamental studies that report the dynamics of carbohydrates and growth characteristics during a cycle full of perennial crops, mainly for the different scenarios of Brazilian fruit production. Therefore, this study aimed to evaluate the changes in the levels and total carbohydrates accumulation in different phenological stages (plant partitions) of "Roxo de Valinhos" fig tree with and without irrigation, at soil and weather conditions of Botucatu, São Paulo State, Brazil.

2. Material e Methods

The research was conducted at Fruit Cultivation Experimental Sector (Orchard) of the Department of Horticulture, Faculty of Agricultural Sciences (FCA), São Paulo State University (UNESP), Botucatu, Brazil. The local geographic coordinates are 22°52'47"S, 48°25'12"W and altitude of 810 m above sea level. The predominant climate type on region, based on the Köppen Classification System is CFa, characterized as warm temperate (mesothermal) with rainfall in summer and dryin winter; precipitation and annual temperature of 1530 mm and 21°C, respectively [12]. The soil in this area was classified as Red Nitossol [13], whose the results of soil chemical analysis on the initial experimentare presented in **Table 1**.

During the experimental period, the rainfall variation and average, minimum and maximum air temperature in data daily were monitored by the Weather Station of Natural Resources Department in the Faculty of Agricultural Sciences, UNESP, as shown in **Figure 1**. The monthly average air temperatures ranged from 16.40 (June) and 24.63°C (January) for the year 2011. I n 2011 the rainfall was 1984 mm, with 37.61% accumulated only in January, however, between May and September of 2011 were registered only 98.2 mm, thus indicating the importance of irrigation in the regional context.

The seedlings transplanting of "Roxo de Valinhos" fig tree was conducted in December 2010 when the seedlings had heithg of 0.30 m. The seedlings were obtained with the Coordination of Integral Technical Assistance (CATI)-unit of Botucatu-SP, and in general, the propagating cuttings are from certified producers of the center of origin at Valinhos-SP. We adopted a spacing of 2.5 m between planting rows and 2.5 m between plants. The

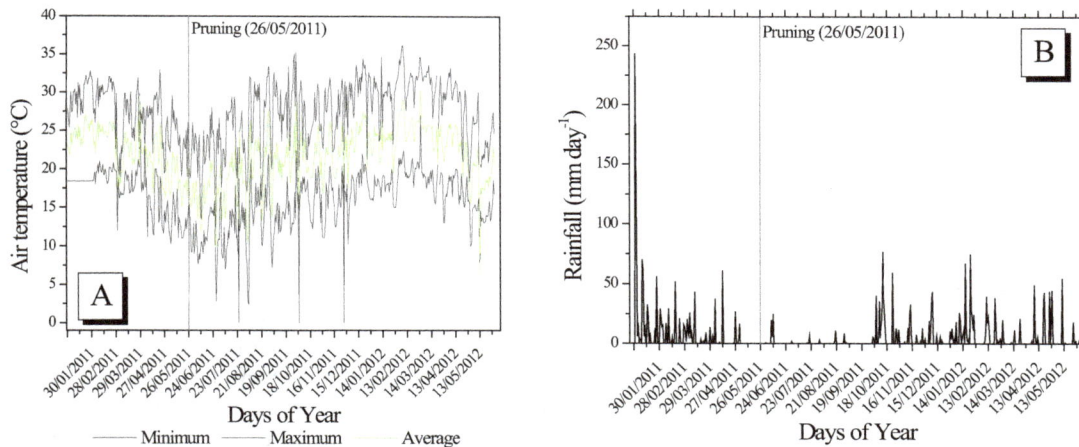

Figure 1. Daily values of air temperature (a) and rainfall (b) between 01/01/2011 and 25/05/2012, at Botucatu, São Paulo State, Brazil.

Table 1. Chemical soil characterization before transplantation of "Roxo de Valinhos" fig tree seedlings, on 15/11/2010.

Sampling Depth (cm)	pH	Organic Matter	Aluminum Al^{3+}	H + Al	Potassium	Calcium	Magnesium	Total Base	Cation Exchange Capacity	Base Saturation (%)
	CaCl₂	g/dm³				-------- mmol_c/dm³ --------				
0 - 20	5.5	40	0	24	2.0	28	11	41	65	64
20 - 40	5.1	40	0	29	2.8	29	10	42	71	59

	P_resin	Sulfur	Boron	Copper	Iron	Manganese	Zinc
			-------------- mg/dm³ --------------				
0 - 20	17	3.0	0.22	7.8	77	30.2	1.5
20 - 40	14	3.0	0.19	7.3	61	25.1	1.1

Source: Soil Fertility Laboratory. Department of Soil Science, Faculty of Agricultural Sciences, UNESP, Botucatu, São Paulo State, Brazil.

furrows for planting had 0.50 m deep was previously fertilized with 1 L of cattle manure, 0.5 kg of dolomitic calcareous, 0.5 kg of magnesium thermo-phosphate (containing 0.1% boron and 0.25% zinc). The coverage fertilizations were based on the recommendation of Campo Dall'Orto et al. [14] calling for the fertilizers application according to the chemical analysis results of soil and plant age. Thus, in 2011 and early 2012 were applied 0.045 and 0.035 kg of urea and potassium chloride per plant every two months.

The fig plants were pruned on 28/06/2011 to conform the plant canopy with three productive branches in the first year, and if necessary were performed desprouts. For *Cerotelium fici* (rust) fungus control was used the products with cooper and others applications of Tebuconazole (Folicur®) and Thiophanate methyl (Cercobin®) fungicides were performed whenever necessary. The weed control between plant rows was done through periodic mechanical mowing and the plants crowning was performed by hand weeding.

The water retention curves were obtained experimentally with undisturbed soil samples colected by volumetric rings (0.05 × 0.05 m for height and diameter) and the relations between matric potential values and soil moisture was obtaining by the Richards porous plate method [15]. The soil density obtained was 1.4822 g cm⁻³ of 0 to 20 cm and 1.3593 g cm⁻³ of 20 to 40 cm. The water retention equations were obtained by the squared deviations minimization method were adjusted using the Excel Solver optimization tool and had determination coefficients (r^2) of 0.9974 and 0.9930, to the depths of 0 - 20 cm and 20 - 40 cm, respectively. The adopted drip irrigation system with two emitters was at flow rate of 1.5 L·h⁻¹ per plant. The regime was based on the permanence of matrix potential between the field capacity and the maximum value of 60 kPa, monitored by tensiometers.

$$\theta_{20} = 0.169 + \frac{0.437}{\left[1 + \left(2.8111\psi_m\right)^{1.363}\right]^{0.2663}} \tag{1}$$

$$\theta_{40} = 0.210 + \frac{0.641}{\left[1 + \left(2.77\psi_m\right)^{1.5044}\right]^{0.3353}} \tag{2}$$

where: θ—volumetric water content ($cm^3 \cdot cm^{-3}$); θ_{20}—volumetric water content at 20 cm depth ($cm^3 \cdot cm^{-3}$); θ_{40}—volumetric water content at 40 cm depth ($cm^3\ cm^{-3}$); ψ_m—matric water potential of soil monitored by tensiometers (kPa) [15].

The irrigation management was based on tensiometers technique, with batteries of two mercury tensiometers installed in each treatment and repetition. The first tensiometer mounted at 20cm depth (relative to the center of the porous capsule) was regarded as a "decision", as based on their readings were taken irrigations. The second tensiometer was considered "control" was installed at 40 cm depth (with respect to the center of the porous capsule) for drainage control. In general, tensiometers have reliable operation up to the range of −80 kPa [15], with the variations in readings when increasing the potential increases. The management was based on the permanence of the matric potential next of −30 kPa or water content equal to 0.2988 cm^3 water cm^{-3}_{soil} (reference content). With ψ_m rates assessed in reading, was found that the level of water content by difference between current reading and reference content. Thus, the water volume applied still depended on the volume used by the plant root system, which in this case was monitored by the collection of root systems for destructive evaluations on growth analysis. The rainfall and irrigation levels distribution that have been applied during the trial period are shown in **Table 2**.

The growth analyzes were performed every 40 days after pruning (DAP) in a total of 7 collections between June 2011 to April 2012 (07/08; 17/09; 25/10; 04/12/11 and 13/01; 24/02; 03/04/2012) corresponding to 40, 80, 120, 160, 200, 240 and 280 DAP. These assessments were obtained the following measures in plants: stem di-

Table 2. Monthly accumulated values of effective rainfal and irrigated levels of "Roxo de Valinhos" fig trees in initial development, in Botucatu, São Paulo State, Brazil.

Monthly	Rainfall (mm)	Irrigation Levels (L·m⁻²) CI	Irrigation Levels (L·m⁻²) SI	Received Total Water (L·m⁻²) CI	Received Total Water (L·m⁻²) SI
Jan/11	712.25	−	−	712.25	712.25
Fev/11	188.13	−	−	188.13	188.13
Mar/11	163.50	−	−	163.50	163.50
Abr/11	126.50	−	−	126.50	126.50
Mai/11	16.50	−	−	16.50	16.50
Jun/11	49.90	−	−	49.90	49.90
Jul/11	7.00	16.34	−	23.34	7.00
Ago/11	24.75	14.8	−	39.55	24.75
Set/11	0.00	11.98	−	11.98	0.00
Out/11	359.58	47.85	−	407.43	359.58
Nov/11	102.50	18.06	−	120.56	102.50
Dez/11	143.38	25.27	−	168.65	143.38
Jan/12	357.25	23.28	−	380.53	357.25
Fev/12	166.75	32.99	−	199.74	166.75
Mar/12	58.88	21.93	−	80.81	58.88
Sum	2476.85	212.5	−	2689.35	2476.85

ameter (SD), branch diameter (BD), branch length (BL), plant height (PH), number of leaves (NL), internodes number (IN) and number of fruits (NF) per plant. For the characteristics of BD, BL and IN was considered the average of three branches, while for NL, NF (ripe and green) and leaf area per plant was given by the sum of the three branches.

To adopt as a reference was held the first collection before pruning to characterize the plants before treatment application. In each collection were uprooted eight plants (four per treatment) with the aid of a backhoe. After removal, the plants were cut and separated in roots, stems, branches, leaves and fruit. Each plant part was washed with water and detergent and packaged to obtain fresh weight. Subsequently, to obtain the dry mass of the plant material was subjected to air drying oven with forced circulation at a temperature of 65°C for 72 hours. The root system volume was measured by the Becker method, in which the roots are inserted in the becker with a known water volume, and after the increase indicates the root contribution (cm^3). The leaf area was obtained with the aid of leaf area integralizator model Licor 3000 (cm^{-2}).

The total carbohydrates were determined by the phenol-sulfuric method [16]. The extraction was performed in water, treating 10 mg of sample with 10 ml of distilled water for 40 min at 40°C in a water bath and centrifuged for 30 minutes at 5000 rpm. Sample aliquots of 0.5 ml were used with addition of 0.5 ml phenol and 2.5 ml of concentrated sulfuric acid. The reading was performed in spectrophotometer by absorbance 490 nm. We calculated the total carbohydrate content in mg g^{-1} dry matter of plant material and carbohydrate accumulation was obtained by the product content with the dry matter (in mg·kg^{-1}).

The ripe fruit harvest was carried out between January and March 2012. The number, weight and total productivity of fruits were obtained by the sum within each collection, considering green and ripe fruit. The experimental design adopted was randomized block, in factorial arrangement (2 × 7) with 4 replicates distributed in installments (with and without irrigation) subdivided in time (collection time). For growth analysis and total carbohydrate was not considered the first collection.

When significant, regressions (with independent variable as the evaluation time—DAP) were adjusted by Sisvar statistical package and graphic representations made in Origin 6.0. The differences between means were subjected to variance analysis by F test and compared by Tukey test at 5% significance, using Sisvar program [17].

3. Results and Discussion

3.1. Growth Analysis

In the plants, growth dynamics was observed that the highest mean of leaves number (NL) and leaf area (LA) for two treatments were performed at 200 DAP (**Figure 2**) and that water supplementation provided mainly differences in leaf area (**Figure 1(B)**). The leaf area importance in this culture is reflected mainly on productivity, because the photosynthetic process depends on the interception of light energy and its conversion in to chemical energy [18]. McCree & Fenández [19] claimed that the response called most prominent of plants to drought, constitutes in decrease on LA production, closing of the stomata, senescence acceleration and leaves abscission.

After 200 DAP there was a significant decrease in the number of leaves due to the leaves for natural senescence, and also because this period coincided with the maximum fruit production (considered preferred drains). These results corroborate the results found by Silva [20], which researching the assimilates allocation labeled with carbon 13 (^{13}C) and the source-sink relationship in fig trees of this variety, found that in plants that had fruit and shoots, the isotopic signals were higher in fruits, indicating a higher assimilates translocation to these organs. Hopkins [21] explains that the relative importance of source-sink change over the plant cycle and according to the spatial distribution. During the growing season, the apical and root meristems are important but in the fruiting stages occur changes between these demands.

For most growth variables in the different plant organs were adjusted polynomial regression (**Table 3**), except for root length (linear) (**Table 1**). For Benincasa [10], the sigmoidal curve is usually better to plant growth, which presents initial pattern of slow growth followed by a fast phase. However, polynomial or linear models are important because they present evaluation possibilities of maximum and minimum points when they are subjected to numerical derivation.

In general, higher values of growth for all plant organs were observed for the treatments with irrigation, thus indicating the water importance to the plant growth. Kerbauy [22] reported that among there sources that the plant needs to growth, the water is the primary and most limiting. Thus, the vegetation distribution on the sur-

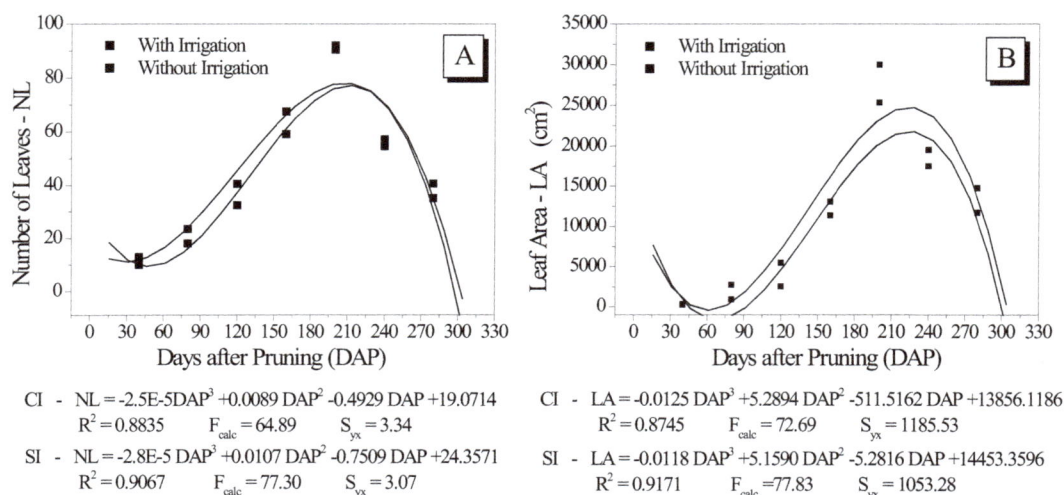

CI - $NL = -2.5E-5DAP^3 +0.0089 DAP^2 -0.4929 DAP +19.0714$
$R^2 = 0.8835$ $F_{calc} = 64.89$ $S_{yx} = 3.34$

SI - $NL = -2.8E-5 DAP^3 +0.0107 DAP^2 -0.7509 DAP +24.3571$
$R^2 = 0.9067$ $F_{calc} = 77.30$ $S_{yx} = 3.07$

CI - $LA = -0.0125 DAP^3 +5.2894 DAP^2 -511.5162 DAP +13856.1186$
$R^2 = 0.8745$ $F_{calc} = 72.69$ $S_{yx} = 1185.53$

SI - $LA = -0.0118 DAP^3 +5.1590 DAP^2 -5.2816 DAP +14453.3596$
$R^2 = 0.9171$ $F_{calc} = 77.83$ $S_{yx} = 1053.28$

Figure 2. Regressions of number of leaves and leaf areaof "Roxo de Valinhos" fig tree grown inirrigated (I) and non-irrigated (NI) system in Botucatu, São Paulo State, Brazil, 2013.

face as to agricultural productivity is mainly controlled by the water availability.

The diameter, length, fresh weight and dry matter of the stem were moderately higher in treatments with water irrigation. The increase in branch diameter is due to the vascular activity, which is strongly influenced by factors such as photoperiod, rainfall and availability of physical space [23]. These results differ from those found by Silva [24] for the Roxo de Valinhos fig tree grown in the presence or absence of mulch and irrigated and non-irrigated systems, as the initial stage of development was not observed differences on stem diameter between plants irrigated and non-irrigated.

The growth attributes evaluated for the branch (length, diameter, number of internodes, fresh weight and dry matter) showed maximum values at 270 DAP. The irrigated system resulted in higher means when compared with the non-irrigated systems because of the constant water supply throughout the plants cycle. Benincasa [10] reports that one of the essential water roles is to maintain turgor, which is essential for cell growth and elongation. Taiz & Zeiger [18] confirmed this information when reporting that the largest component of the plant cell growth is governed by expanding turgor pressure, which is controlled by the water.

The root system had linear response for the length, both the irrigation system and the non-irrigated system. On the other hand, the volume, fresh weight and dry matter of roots was adjusted by polynomial model, reaching maximum values at 240 DAP. The highest rates were observed for the treatments with water irrigation, mainly for the volume root had maximum values of (935 cm³) in relation (679.5 cm³) non-irrigated plants. This result shows that the water availability favors the root volume increase.

In Botucatu-SP, Silva [11] evaluating the explored volume and the depth reached by the root system of fig Roxo de Valinhos subjected to treatments with the presence or absence of mulch (sugarcane bagasse) and the supply or no supplemental irrigation, found that in treatments with irrigation, the root system reached greated depths and explored greater volume of soil.

In general, the maximum growth values for the roots, branches, stems and leaves coincided with the highest daily average of temperature and rainfall (January, February and March) (**Figure 1**). This reinforces the idea that the plant expresses the greatest potential for growth when is in ideal conditions of temperature and water availability.

3.2. Evaluation of Production

Figure 3 shows the fig tree "Roxo de Valinhos" irrigated. The harvest for the two cultivation systems occurred in the samples 5, 6, 7 and 8, which correspond to 160, 200, 240 and 280 DAP, respectively (**Table 4**). The average number of fruits, production and productivity were lower in non-irrigated system to 160 DAP, due to low rainfall occurred in the months before fruiting have promoted delay in fruit formation. This also highlights the irrigation importance to the antecipation of the production plant cycle. There was no difference between the

Table 3. Coefficients of the polynomial regression [$Y = a_3DAP^3 + a_2DAP^2 + a_1DAP + a_0$] adjusted for different growth variables of "Roxo de Valinhos" fig trees in initial stage of development, irrigated and non-irrigated systems, in Botucatu, Brazil.

Growth variable	Regression coefficient				R^2	$F_{calculado}$	S_{yx}
	a_3	a_2	a_1	a_0			
Irrigation system							
Stem diameter (mm)	−8.0E−5	0.00473	−0.5488	38.0714	0.9835	49.73	2.78
Stem length (cm)	−1.0E−6	0.00519	−0.3314	38.2957	0.9989	1.70	1.89
Branch diameter (mm)	−3.0E−6	0.00158	−0.1064	7.6343	0.9919	159.78	0.91
Branch length (cm)	−4.6E−5	0.02499	−2.7512	96.1536	0.9938	184.30	7.75
Number of internodes	−6.0E−6	0.00341	−0.3098	13.6089	0.9980	397.14	1.08
Root system volume (cm^3)	−4.62E−4	0.1811	−15.0727	414.6429	0.9246	37.82	60.53
Root length (cm)	----	----	45.8929	0.1507	0.9621	6.17	5.84
Fresh stem mass (g)	−4.39E−4	0.24007	−30.3237	1158.2857	0.8766	70.16	91.73
Dry stem mass (g)	−2.46E−4	0.13231	−16.1277	568.6786	0.8886	57.05	39.60
Fresh branch mass (g)	−1.01E−4	0.03992	−8.2515	335.2475	0.9985	59.87	165.08
Dry branch mass (g)	−8.8E−5	0.06685	−9.2528	311.5807	0.9689	64.32	51.93
Fresh root system mass (g)	−9.807E−5	0.05406	−4.6441	196.5761	0.9353	34.19	131.04
Dry root system mass (g)	−0.000061	0.03121	−3.6544	143.5961	0.9087	27.49	40.45
Fresh leaves mass (g)	−6.33E−4	0.27757	−28.7161	805.6564	0.9089	68.93	60.14
Dry leaves mass (g)	−1.63E−4	0.07132	−7.4638	215.2939	0.9212	2923	22.10
Non-irrigation system							
Stem diameter (mm)	−4.1E−4	0.00505	−0.5590	34.5357	0.9923	42.95	2.18
Stem length (cm)	−3.0E−5	0.02336	−0.4480	37.0357	0.9519	0.33	2. 17
Branch diameter (mm)	−7.0E−6	0.00327	−0.3417	15.70	0.9843	70.81	1.21
Branch length (cm)	−5.5E−5	0.02851	−3.1879	99.3571	0.9874	140.73	8.57
Number of internodes	−9.0E−6	0.00489	−0.5271	20.4271	0.9876	169.03	1.55
Root system volume (cm^3)	−2.69E−4	0.12844	−14.6956	556.71	0.9521	23.43	54.75
Root length (cm)	--	--	41.0357	0.1424	0.9666	2.75	7.55
Fresh stem mass (g)	−3.63E−4	0.21515	−27.3129	1090.7071	0.9236	56.00	82.86
Dry stem mass (g)	−1.38E−4	0.73123	−9.1107	3.58	0.9486	21.57	44.28
Fresh branch mass (g)	−3.35E−5	0.21468	−28.7136	925.2143	0.9643	23.75	204.92
Dry branch mass (g)	−8.7E−5	0.05992	−8.2035	267.0714	0.9759	32.13	54.84
Fresh root system mass (g)	−2.23E−4	0.10771	−12.0448	469.35	0.9660	22.97	54.69
Dry root system mass (g)	−0.000036	0.01917	−2.15811	0.9731	0.0731	29.56	16.49
Fresh leaves mass (g)	−6.45E−4	0.28880	−31.8142	922.6096	0.9220	33.08	78.21
Dry leaves mass (g)	−2.01E−4	0.08898	−9.8550	286.2218	0.8421	32.29	23.38

Table 4. Number of fruits, total production and productivity of Roxo de Valinhos fig tree in the early stage of development, under irrigation and non-irrigation system, in Botucatu-SP, Brazil, 2013.

Systems of Management	160 DAP	200 DAP	240 DAP	280 DAP
Fruit Number				
Irrigated	24.00 A	45.25 A	52.25 A	51.25 A
Non-Irrigated	0.50 B	44.00A	51.50 A	53.25 A
CV%	7.45	24.87	34.57	25.88
Fruit Weight (g)				
Irrigated	120.30 A	539.25 A	1.655.00 A	2.104.50 A
Non-Irrigated	4.87 B	534.75 A	909.75 B	1.522.00 B
CV%	12.11	26.47	28.60	41.39
Productivity (kg ha^{-1})				
Irrigated	192.50 A	863.25 A	2.648.25 A	3.367.50 A
Non-Irrigated	7.75 B	855.50 A	1.455.25 B	2.435.50 B
CV%	12.24	26.50	28.62	21.38

Means followed by uppercase letters in the column do not differ by Tukey test t 5% probability.

Figure 3. Pictures of "Roxo de Valinhos" plants employed in the research, in Botucatu, São Paulo State, Brazil, 2013.

systems to 200 DAP in any production variables assessed, however, after 240 DAP the average fruit weight and plants productivity subjected to water supplementation were higher.

Addressing the irrigation influence on the yield characteristics of Roxo de Valinhos fig tree, Silva [24] found that fig plants in the early development stages (first production cycle) treated with irrigation showed an increase of 350 kg·ha^{-1} in total productivity when compared with trees that non-irrigated. Also in this sense, Leonel & Damatto Junior [25] evaluated the effect of cattle manure doses in nutrition on Roxo de Valinhos fig production under irrigation, found that after four years of cultivation, the fig trees showed a productivity of 6.3t ha^{-1}, higher than that found in this work.

3.3. Accumulation and Levels of Total Carbohydrates

The total carbohydrates levels to each plant organ were present in **Table 5**. It had no significant difference to carbohydrates levels among irrigation systems in most of all collection dates and evaluated organs, except for the branch partition, where there was a higher concentration in the irrigation system, at 120 DAP. In simple terms, the carbohydrates levels present in plant reserves bodies before pruning and treatments (with and without irrigation) were enough to keep the plants metabolic activities that did not receive water irrigation. These results corroborate with Cruz [26], which did not detect differences in starch content in acid lime "Tahiti" treated with different doses of paclobutrazol in irrigated plants and plants subjected to water stress.

All organs were detected higher levels of total carbohydrates at 120 and 160 DAP, regardless of the water regime (**Table 5**). This increase may be due to high air temperature observed during the same period (September and October 2011) (**Figure 1(A)**), and also by the fact that the plants were in full vegetative growth in the two water management systems. Temperature is one of the climatic elements which is directly connected with the carbohydrates mobilization present in woody plant organs [27]. Rodrigues [28] elucidate that the temperature is

Table 5. Total carbohydrates level on dry matter mass (mg·g^{-1}) in roots, stems, branches, leaves and fruits of Roxo de Valinhos fig plants under irrigated and non irrigated system in Botucatu-SP, 2013.

	40	80	120	160	200	240	280
				Roots			
I	41.33 Ab	78.00 Aab	86.67 Aa	89.00 Aa	68.33 Aab	41.30 Ab	59.67 Aab
NI	52.67 Ab	98.67 Aa	70.67 Aab	77.00 Aab	63.67 Aab	53.67 Ab	67.00 Aab
				Stems			
I	51.66 Abc	47.00 abc	74.67 Aab	80.33 Aa	36.67 Ac	29.33 Ac	42.67 Ac
NI	53.33 Aabc	57.67 Aab	78.33 Aa	63.33 Aab	38.67 Abc	27.00 Ac	47.33 Abc
				Branches			
I	61.33 Aab	81.00 Aab	96.33 Aa	62.66 Aabc	54.33 Abc	32.67 Ac	48.00 Abc
NI	73.67 Aa	93.00 Aa	63.33 Bab	82.33 Aa	64.67 Aab	29.67 Ab	64.33 Aab
				Leaves			
I	61.33 Ac	81.00 Aabc	111.0 Aab	108.33 Aab	120.33 Aa	73.00 Abc	69.00 Abc
NI	73.67 Aab	93.00 Aab	101.33 Aab	93.33 Aab	115.00 Aa	62.67 Ab	68.33 Ab
				Fruits			
I	*nd	Nd	Nd	124.00 Ab	281.33 Aab	323.33 Aa	280.33 Aab
NI	Nd	Nd	Nd	11.00 Bb	290.33 Aa	325.33 Aa	271.00 Aab

Means followed by lowercase in line and uppercase letters in the column do not differ by Tukey test t 5% probability. *nd: not determined.

an important factor on adaptation of temperate climate fruits plants. Larcher [7] also reports that the intensity of mobilization of carbohydrates influences the branch growth, flowering and fruit production.

The results of this study are in agreement with those found by Corsato [29], which evaluated the carbohydrates variation in the root and branch persimmon variety Rama Fort, observed higher total soluble carbohydrates in roots and branches in the months of September and October (spring) due to the air temperature increase.

Rodrigues *et al.* [28], studying the changes in starch and soluble sugars in two pear cultivars gems from February to September, found higher levels of total soluble sugars in September. The authors explain that the elevated total sugars levels occurred during this period due to starch degradation process of soluble carbohydrates (sorbitol, sucrose, glucose and fructose) which have been translocated from bud basis for developing floral structure.

It is worth emphasizing that for the two water regimes, most of the evaluated plant organs showed increasing trend for the total carbohydrate content up to 160 DAP, showing intense carbohydrates partition to the organs growing. However, the fruit was noted that the increase on total carbohydrates increased up to 240 DAP, and the body more carbohydrate accumulated in this period, indicating a possible translocation of these reserves in mature leaves (sources) for fruit (drains). This can be confirmed when analyzing the carbohydrates levels in leaves, which decreases while the increase in the fruit content. The mature leaves are photosynthetically active, and the carbohydrate production of these plant organs is greater than the need for maintenance and growth [30]. Therefore, the surplus is translocated to the organs with little or no photosynthetic activity, as young leaves, branches, buds, flowers and fruits.

The total carbohydrate accumulation varied among plant organs during the growing season, and increasing for most organs both two water regimes (**Table 6**). The plants received water supplemented accumulated more dry matter, which may be the result of increased CO^2 assimilation and conversion of that molecule photosynthate. At 280 days after pruning, end of the experiment, the irrigated plants fruits accumulated more carbohydrates, followed by branches, stem, root and leaves.

4. Conclusions

The water supplementation promoted greater growth and increased productivity by 27.68%, which corresponded to 932 kg·ha^{-1} fruit more compared to non-irrigated system.

The higher carbohydrate content in all evaluated organs were observed at 120 and 160 days after pruning, coinciding with the fruiting onset, the fruit being the organs that had the highest total carbohydrates.

Table 6. Values of dry matter mass and carbohydrate (mg·kg^{-1}) accumulated over 280 days after Roxo de Valinhos fig plants pruning under irrigated and non-irrigated system in Botucatu, 2013.

	Roots				Stems				Branches				Leaves				Fruits			
	MS (g)		CT (mg·kg⁻¹)		MS (g)		CT (mg·kg⁻¹)		MS (g)		CT (mg·kg⁻¹)		MS (g)		CT (mg·kg⁻¹)		MS (g)		CT (mg·kg⁻¹)	
DAP	I	NI	I	NI	I	NI	I	NI	I	NI	I	NI	I	NI	I	NI	I	NI	I	NI
40	59.0	32.3	2.5	1.9	65.67	103.0	3.5	5.5	11.7	1.7	0.7	0.1	11.7	4.7	0.7	0.3	-	-	-	-
80	38.7	30.7	3.0	3.4	87.33	72.7	4.2	4.1	14.7	3.3	1.2	0.3	25.3	11.0	2.0	0.9	-	-	-	-
120	38.7	31.7	3.4	2.4	111.67	60.3	8.4	4.8	44.7	5.7	4.5	0.4	47.0	22.0	5.3	2.2	-	-	-	-
160	59.3	80.3	5.3	6.6	155.00	162.0	12.6	9.9	93.7	56.0	5.8	4.8	145.0	106.0	15.7	9.7	15.3	0.7	1.9	0.01
200	218.7	148.0	14.6	9.5	456.00	267.0	17.2	10.1	355.0	310.3	20.1	20.1	292.3	361.0	34.9	41.9	90.3	75.0	24.5	19.5
240	224.3	163.3	9.1	8.6	834.67	479.0	24.4	12.9	777.0	484.3	24.9	14.1	247.3	224.0	17.9	13.7	198.0	96.7	64.1	26.7
280	224.0	178.3	13.5	12.3	592.67	484.7	25.5	21.9	1053.0	703.7	49.6	45.0	128.7	99.7	8.9	6.7	253.7	126.0	75.2	22.9

The total carbohydrates accumulation at 280 days after pruning was higher in the fruits under irrigation system presenting 75.20 mg·kg^{-1}.

References

[1]   Fachinello, J.C., Nachtigal, J.C. and Kersten, E. (1996) Fruit Growing: Fundamentals and Practice. Editora Universitária, Universidade Federal de Pelotas, Pelotas, 31.

[2]   FAO.FAOSTAT (2011) Production-Crops. http://faostat.fao.org/site/567/DesktopDefault. aspx?PageID=567#ancor

[3]   Maiorano, J.A., Antunes, L.E.C., Regina, M.A., Abrahão, E. and Pereira, A.F. (1997) Botany and Characterization of Fig Cultivars. *Informe Agropecuário*, **18**, 22-24.

[4]   Penteado, S.R. (1999) The Fig Tree Cultivation in Brazil and in the World. In: Corrêa, L.S. and Boliani, A.C., Eds., *Fig Culture-from Planting to Marketing*, FAPESP, Ilha Solteira, 1-16.

[5]   Silva, A.C. (2011) Growth, Productivity and Allocation of the Fig Tree Reserves Indifferent Conditions. Thesis (PhD in Agronomy/Horticulture), Faculty of Agricultural Sciences, State University of São Paulo, Botucatu, 126f.

[6]   Gupta, U.C. (2001) Micronutrients and Toxic Elements in Plants and Animals. In: Ferreira, M.E., Pessôa, M.C., Raij, B.V. and Abreu, C.A.A., Eds., *Micronutrients and Toxic Elements in Agriculture*, POTAFOS, CNPq, FAPESP, Jaboticabal, 13-41.

[7]   Larcher, W. (2000) Plant Ecophysiology. Rima, São Carlos, 531.

[8]   Borba, M.R.C., ScarpareFilho, J.A. and Kluge, R.A. (2005) Levels of Carbohydrates in Peaches Submitted to Different Intensity of Green Pruning in Tropical Climate. *Revista Brasileira de Fruticultura*, *Jaboticabal*, **27**, 68-72. http://dx.doi.org/10.1590/S0100-29452005000100019

[9]   Gonçalves, B.H.L. (2014) Carbohydrate Content in Peach Grown in Subtropical. Thesis (PhD in Agronomy/Horticulture). Faculty of Agricultural Sciences, State University of São Paulo, Botucatu, 70f.

[10]   Benincasa, M.M.P. (2003) Plant Growth Analysis: Basic Knowledge. 2nd Edition, Funep, Jaboticabal, 41 p.

[11]   Flore, J.A. (1994) Stone Fruit. In: Schaffer, B. and Andersen, P.C., Eds., *Handbook of Environmental Physiology of Fruit Crops*, CRC Press, Boca Raton, 467.

[12]   Cunha, A.R., *et al.* (1999) Climate Classificationfor the City of Botucatu, SP, According to Köppen. In: FEPAF, *Simpósio em Energia na Agricultura*, 1999, *Botucatu, SP. Anais...* Faculdade de Ciências Agronômicas, Botucatu, 490-491.

[13]   EMBRAPA, Centro Nacional de Pesquisa de Solos (2006) Brazilian System of Soil Classification. 2nd Edition, EMBRAPA SOLOS, Rio de Janeiro, 306.

[14]   Campo-Dall'Orto, F.A., *et al.* (1996) Temperate Fruits: II. Fig, Apple, Quince, Pearand Peachorchard System. In: Van Raij, B., *et al.*, Eds., *Recommendations Fertilization and Liming to the State of Sao Paulo*, 2nd Edition, Instituto Agronômico, Fundação, Instituto Agronômico de Campinas, Campinas, 139-140.

[15]   Klar, A.E. (1988) The Water in the System Soil-Plant-Atmosphere. 2nd Edition, Nobel, São Paulo.

[16]   Dubois, M., Gilles, K.A., Hamilton, J.K., Rebers, P.A. and Smith, F. (1956) Colorimetric Method for Determination of Sugars and Related Substances. *Analytical Chemistry*, **28**, 350-356. http://dx.doi.org/10.1021/ac60111a017

[17]   Pimentel Gomes, F. (2009) Experimental Statistics Course. 15th Edition, Fealq, Piracicaba, 451.

[18]   Taiz, L. and Zeiger, E. (2004) Plant Physiology. Artmed, Porto Alegre, 719.

[19]   McCree, K.J. and Fernández, C.J. (1989) Simulation Model for Studying Physiological Water Stress Responses of Whole Plants. *Crop Science*, **29**, 353-360. http://dx.doi.org/10.2135/cropsci1989.0011183X002900020025x

[20]   Silva, A.C., Souza, A.P., Ducatti, C. and Leonel, S. (2011) Carbon-13 Turnover in Fig Trees "Roxo de Valinhos". *Revista Brasileira de Fruticultura*, **33**, 660-665.

[21]   Hopkins, W.G. (1995) Introduction to Plant Physiology. John Wiley e Sons, New York, 464.

[22]   Kerbauy, G.B. (2004) Plant Physiology. Guanabara Koogan S.A, Rio de Janeiro, 452.

[23]   Lojan, L. (1968) Tendências del crecimiento radial de 23 especies forestales del trópico. *Turrialba*, **18**, 275-281.

[24]   Silva, A.C., Leonel, S., Souza, A.P., Souza, M.E. and Tanaka, A.A. (2011) Fig Tree Growth under Different Crop Conditions. *Pesquisa Agropecuária Tropical*, **41**, 539-551.

[25]   Leonel, S. and Damatto Junior, E.R. (2008) Effects of Coat Manure in the Soil Fertility, Plants Nutrition and Yield of Fig Orchards. *Revista Brasileira de Fruticultura*, **30**, 534-539. http://dx.doi.org/10.1590/S0100-29452008000200046

[26]   Cruz, M.C.M., Siqueira, D.L., Salomão, L.C.C., Cecon, P.R. and Santos, D. (2007) Levels of Carbohydrates in Acid

Lime Tree "Tahiti" Treated with Paclobutrazol. *Revista Brasileira de Fruticultura*, **29**, 42-47.

[27] Herter, F.G., Veríssimo, V., Camelatto, D., Gardin, J.P. and Trevisan, R. (2011) Flower Bud Abortion of Pear in Brazil. In: *Seminário de Fruticultura de Clima Temperado no Brasil*, Anais, Florianópolis, 106-114.

[28] Rodrigues, A.C., Herter, F.G., Veríssimo, V., Campos, A.D., Leite, G.B. and Silva, J.B. (2006) Balance of Carbohydrates in Flower Bud of Two Pear Tree Genotypes under Mild Winter Conditions. *Revista Brasileira de Fruticultura*, **28**, 1-4. http://dx.doi.org/10.1590/S0100-29452006000100003

[29] Corsato, C.E., Scarpare Filho, J.A. and Junquiera de Sales, E.C. (2008) Carbohydrate Content in Persimmon Tree Woody Organs in Tropical Climate. *Revista Brasileira de Fruticultura*, **30**, 414-418. http://dx.doi.org/10.1590/S0100-29452008000200025

[30] Dantas, B.F., Ribeiro, L.S. and Pereira, M.S. (2007) Soluble and Insoluble Sugars Content in cv. Syrah Grapevine Leaves in Different Positions of the Branch and Seasons. *Revista Brasileira de Fruticultura*, **29**, 42-47.

3

Assessment of the Irrigation Capacity during the Dry Season Using Remote Sensing and Geographical Information (Case Study in the Binh Thuan Province, Vietnam)

Hien Thi Thu Le[1*], Thang Nguyen Ngoc[1], Luc Hens[2]

[1]Institute of Geography, Vietnam Academy of Science and Technology (IG/VAST), Hanoi, Vietnam
[2]Flemish Institute for Technological Research (VITO), Antwerp, Belgium
Email: [*]hientuanphuong@yahoo.com, nnthang0101@yahoo.com, luchens51@gmail.com

Abstract

Today satellite system provides a main instrument supplying regular data on land inventory and monitoring of land use/land cover changes, in a timely manner. These data are keys to many applications in different sectors: environment, forestry, hydrology, irrigation, agriculture, geology, resource management and planning. Using Landsat image and change detection, this paper presents a method to extract changes of agricultural land, as the basis for the assessment and development of irrigation systems, which enhance production and protect land resources. During the period 1996-2014, the agricultural land in Binh Thuan decreased from 43.5% (in 1996) to 40.1% (February, 2014) of the total land surface in the province. However, the land area under cultivation tends to decrease rapidly, from 25.7% in 1996 to 14.0% in 2014. Combining the results of land use change and assessing the capacity of the irrigation systems show which areas are frequently irrigated versus those are not. This allows proposing irrigation development needs contributing to more production while protecting land resources.

Keywords

Land Use Change, Irrigation, Landsat, Vietnam

1. Introduction

Already during the Stockholm Conference on the Human Environment in 1972, "land use and land cover change"

---

[*]Corresponding author.

(LULC) was recognized as a major problem of global environmental change. This was confirmed by the International Conference on Environment and Development (UNCED) in 1992. The International Geospatial and Biosphere Program (IGBP) stimulated research on LULC. The quantification of the location, extent and trend of the change is an important task for scientists. Nowadays, satellite images provide frequent and timely data on land inventories and allow monitoring changes. Long time observations are necessary to establish a database serving the important goals of development trends [1].

Remote sensing data on vegetation, land-use status and its change offer keys for applications in environment, forestry, hydrology, agriculture, and geology. The rationality in natural resource management, planning and monitoring depends on the accuracy and timeliness of the information provided. Methods of monitoring changes in vegetation by remotely sensed data proved the reusefulness and cost-effectivity [2]-[4]. Historical data of LULC changes of the Earth's surface were ultimate importance for any kind of sustainable development programs, including irrigation development. The study of LULC from which solutions for irrigation development orientation were proposed has been carried out since 1970 at different scales: global, regional and local [5].

Local by often, one or more irrigation basins or areas are studied. The limited surface allows investigators to deepen their knowledge of the study sites. As often remote sensing studies demonstrate the application of remote sensing rather than an operational assessment. Moreover, irrigation practices vary with the size of the study area. Consequently methods developed at a particular moment for one place may not be useful for other locations and periods. Two mapping methods that have been widely used in local studies are: visual interpretation and digital image classification. Early work with satellite imagery at the local scale relied on visual interpretation to identify irrigated areas [6] [7]. Landsat images were used drawing the boundaries of irrigated fields by hand. The early visual interpretation of satellite data was later on replaced by integrated automated procedures [8] [9]. These methods used the separation of irrigated fields from harvested and fallow land in the visible and near-infrared portions of the electromagnetic spectrum. Although visual interpretation is still useful, more recent studies focus on digital image classification, of which the analysis time for analysis is shorter and the mapping costs are lower. Multi-stage classification, unsupervised clustering and decision tree classification are widely used in this context.

Binh Thuan is a province of the South Central Coast of Vietnam. It stretches from 10°33'42" to 11°33'18" north, and from 107°23'41" to 108°52'42" east. The province covers 781.282 ha, including 40.06% agricultural land, 46.10% forested land, and 10.08% non-agricultural land, while other uses apply to the rest of the land [10]. However, in reality, the land area used for agricultural production during the dry season is limited as a result of water shortage. In particular these lands are prone to degradation and desertification. Therefore, these regions need to be monitored and appropriate measures for the reuse and protection are imperative.

This article presents an approach using Landsat imagery to assess the land use change. The focus is on practical aspects of farming during the dry season, assessing the hydraulic capacity, inducting how irrigation system should be developed, enhancing and protecting the arable land.

2. Materials and Methodology

2.1. Satellite Data

Satellite images allow a fast and reliable review of the vegetation and the basic land characteristics. They equally allow repetitive direct observation of the surface, mapping and monitoring surface objects and monitoring, and assessing changes over time and space. Landsat images with a ground resolution of 30 m × 30 m, allow mapping at large and medium scale. An image Database has been built since 1972 (up to present) is now freely accessible in the US Geological Survey (USGS) store, the images are suitable for LULC research and monitoring. For this work, remotely sensed data of the Binh Thuan province were downloaded from USGS indicated in **Table 1** [11].

Determining whether the agricultural land is cultivated or not during the dry season in the Binh Thuan province. Using Landsat images captured during the dry season (January to April are the months of dry season). The images are spread over: 1996; 2001; 2005; 2010 and 2014. 1996 is the first year providing good quality data; 2005 and 2010 are important milestones in orienting and planning the country's development, and 2014 is the most recent year under study.

2.2. Image Classification

The maximum likelihood is the algorithm used for supervised classification is. The sample selection was carried out using a histogram in combination with field sampling of all types of land use: cultivated agricultural land; non-cultivated agricultural land; dense forest; seasonal forest; sparse forest, shrub, settlement, fallow land; water surfaces. The reflectance characteristics and digital number values on Landsat data, agricultural land is hard to recognize, cultivated agricultural land is often confused with sparse forest, or shrub; agricultural area (nonvegetation) with dry soil is often confused with fallow land. Therefore, it is necessary to use ancillary data: field surveys complete the current land use map.

To monitor the change in land use, this paper applies detection method with separately-classified images [12]. The flowchart of processing Landsat images to monitor agricultural area changes are shown in **Figure 1**. Landsat images have been calibrated, enhanced and classified independently. Land use changes were used in pair analysis between the years: 1996-2001; 2001-2005; 2005-2010 and 2010-2014. The long and continuous observation period allows identifying the cultivated and non-cultivated agricultural area, this further provide basis for proper use of resources and irrigation development.

The classification results were assessed using the Kappa coefficient: Kappa allows assessing the accuracy of the classification based on a random method [13].

Table 1. Landsat data of Binh Thuan province from 1996-2014.

No	Satellite	Path/row	Date	No	Satellite	Path/row	Date
1.	Landsat-5	123-52	23/2/1996	9	Landsat-7	124-53	3/4/2005
2.	Landsat-5	124-52	1/3/1996	10	Landsat-7	124-52	5/2/2010
3.	Landsat-5	124-53	1/3/1996	11	Landsat-7	124-52	12/2/2010
4.	Landsat-7	123-52	17/4/2001	12	Landsat-7	124-53	16/3/2010
5	Landsat-7	124-52	8/4/2001	13	Landsat-8	123-52	23/1/2014
6	Landsat-7	124-53	8/4/2001	14	Landsat-8	124-52	15/2/2014
7	Landsat-7	124-52	11/3/2005	15	Landsat-8	124-53	15/2/2014
8	Landsat-7	124-52	2/3/2005				

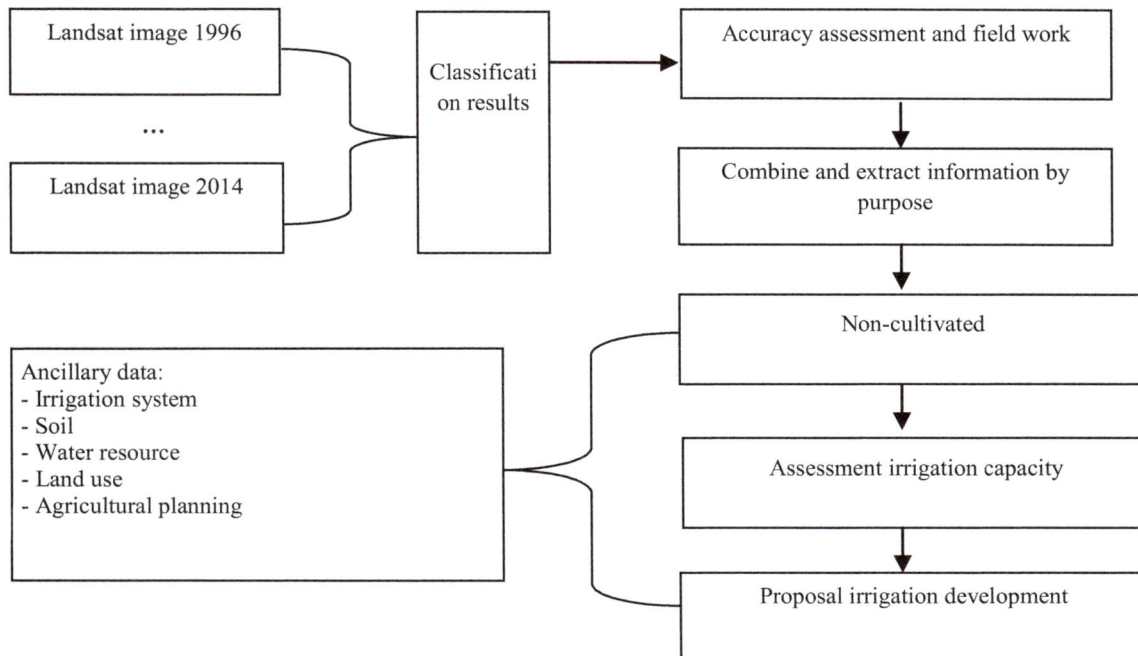

Figure 1. Flowchart of the study rationale: processing data analysis and proposal of irrigation development.

$$Kappa = A/B$$

A = correct by classified pixels − errorously classified pixels.

B = total number of classified pixels.

Kappa is a measure of this difference, standardized to lie on a −1 to 1 scale, where 1 is perfect agreement; 0.81 - 0.99: almost perfect; 0.61 - 0.80: substantial; 0.41 - 0.6 moderate; 0.21 - 0.40: fair; 0.01 - 0.20: slight; 0 is poor agreement; and <0 less than chance agreement [14].

Pixels are selected based on a random method, verified using land use map combined with field observations The accuracy assessment of the classification results are presented in **Table 2**.

Table 2 shows the accuracy of image interpretation results for land use in the study area. All the Kappa coefficients ranging from 0.73 to 0.93; so the results are interpreted on the substantial or almost perfect agreements.

2.3. Assessment of the Irrigation Capacity

Agricultural production is influenced by many factors, but the supply of water or irrigation capacity is the important one. This study approach towards evaluated the capacity of the irrigation system using of satellite images.

Satellite image interpretation is one of the methods establishing the capacity of the irrigation system. The approach is supported by data on spatial distribution of the reservoir; dams; canals; the pumping station; and the irrigated area.

Classification results from 1996-2014 show the distribution of unused agricultural land (non-planted during the dry season). Analysis and comparison of the irrigated and non-irrigated areas reveal the capacity of the irrigation system.

The following analytical methods were used: density; statistical and comparison methods.

3. Results and Discussion

3.1. Change in Cultivated Agricultural Land during the Period 1996-2014

89% of the land in Binh Thuan is used for agriculture: 40% of that area is for annual cropping. The analysis of land use change using satellite data showed minimal changes in this areas coved (annual crops): from 43% (in 1996) to 40% (in 2014) (**Table 3**). In contrast "specialized" land (homestead, industrial land, transportation land) more often changes use. This corresponds with other result on land use and land cover change in Binh Thuan Province [15].

However, a major concern is that the surface of land cultivated during the dry season only accounts for a small portion and is significantly decline. In other words: the agricultural area that is left unused during the dry season is increases (**Figure 2** and **Table 3**). These results were the same as the data collected from field trip by local agencies [16]. This is explained by the change in the climate: the weather becomes more extreme during the dry season and the current irrigation capability can not meet the production demand.

Meteorological data show the change in precipitation in the region: the amount of rainfall reduces during the dry season and increased during the rainy season. Climate change and drought scenarios for the periods 2015-2100 forecast that the even suffering from severe drought will expand to the North and the East of the province, occupying the entire Phan Thiet city and half of the districts Bac Binh, Tuy Phong, Ham Thuan Bac and Ham Thuan Nam [17]. Water shortage is becoming more and more severe in these areas.

Table 2. Accuracy assessment of the classification results using the Kappa coefficient.

Land use type	Year				
	1996	2001	2005	2010	2014
Cultivated land	0.73	0.88	0.89	0.91	0.81
Non-cultivated land	0.88	0.93	0.74	0.85	0.88
Others	0.80	0.89	0.83	0.87	0.86
Average Kappa coefficient	*0.80*	*0.86*	*0.86*	*0.85*	*0.88*

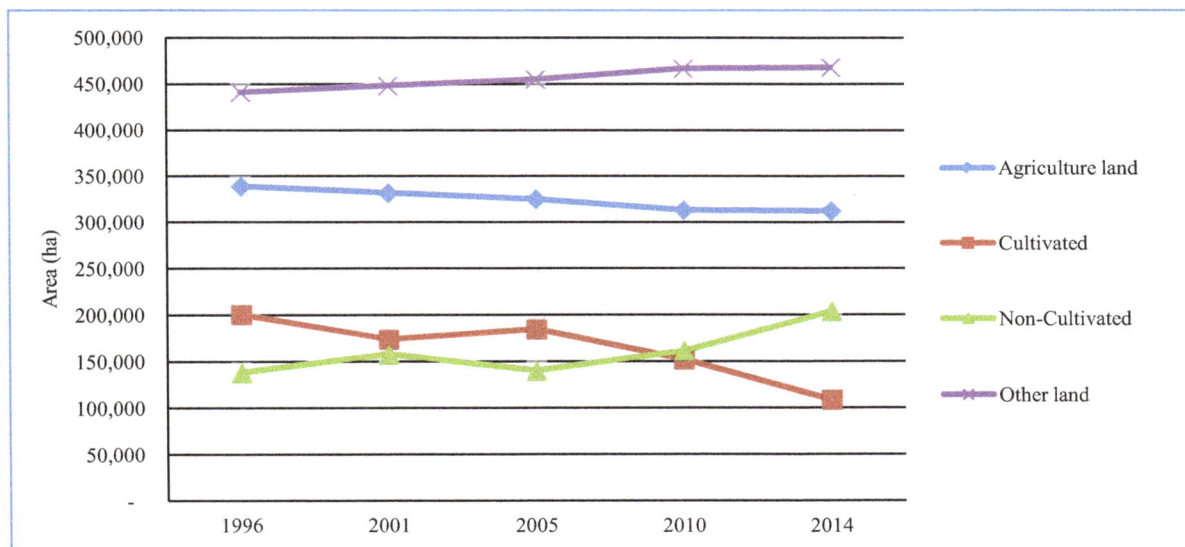

Figure 2. Land use change in Binh Thuan during the period 1996-2014.

Table 3. Change in agricultural area in dry season during the period 1996-2014.

Land use types	Year 1996		Year 2001		Year 2005		Year 2010		Year 2014	
	Area (ha)	%	Area(ha)	%	Area(ha)	%	Area(ha)	%	Area(ha)	%
Agricultural land:	339,582	43.5	332,426	42.5	325,593	41.7	314,023	40.2	312,967	40.1
-Cultivated	200,880	25.7	174,287	22.3	185,153	23.7	152,495	19.5	108,997	14.0
-Non-cultivated	138,701	17.8	158,139	20.2	140,439	18.0	161,528	20.7	203,970	26.1
Others	441,701	56.5	448,857	57.5	455,690	58.3	467,260	59.8	468,315	59.9

The fast development of the irrigation capacity in Binh Thuan over the past 10 years, shows that lakes have been built, while the canal system is still limited. Consequently even when sufficient amounts of water are available in the lakes, the crops are still suffering drought.

3.2. Irrigation Capacity

Table 4 shows the area of agricultural land is cultivated frequently during the dry season in both areas (have irrigation systems and non-irrigation) which accounts for a very small areas (here taken as the high number of cultivate times in 5 years study from 1996-2014). And the large of the agricultural land areas are unstable cultivated and quite big of the agricultural land areas has not been cultivated in a long time. This suggests that irrigation systems perform poorly during the dry season.

Figure 3 shows 3 areas which are still cultivated during the dry season: the Luy River Basin (Tuy Phong district); the Quao River Basin (located in Ham Thuan Bac) and the La Nga River Basin (Tanh Linh district). These areas are intensively cultivated which is possible because of the dams; and pumping stations.

3.3. Irrigation Development

Developing a more comprehensive irrigation system should be based on the combined results of change detection analysis and the assessment irrigation capacity. Change detection analysis shows that the status of agricultural land and irrigation of Binh Thuan during the dry season, is different for: 1) without irrigation; 2) unstable irrigation (occasional watering); 3) and stable irrigation (frequent watering). This allows proposing different amounts of irrigation water demand, both to secure the irrigation sources for the crops and to maintain soil moisture, preventing the soil from degradation. Details entail:

Table 4. Cultivated areas in the dry season during period 1996-2014.

The number of cultivated times	In areas with built irrigation		In areas without built irrigation	
	Ha	% of the total agriculture land in 2014	Ha	% of the total agriculture land in 2014
1	39,553.65	0.13	114,308.1	0.37
2	37,823.40	0.12	87,382.2	0.28
3	24,364.26	0.08	45,168.1	0.14
4	13,490.19	0.04	18,039.2	0.06
5	5,199.84	0.02	4,549.0	0.01
Total	120,431.34	0.38	269,446.6	0.86

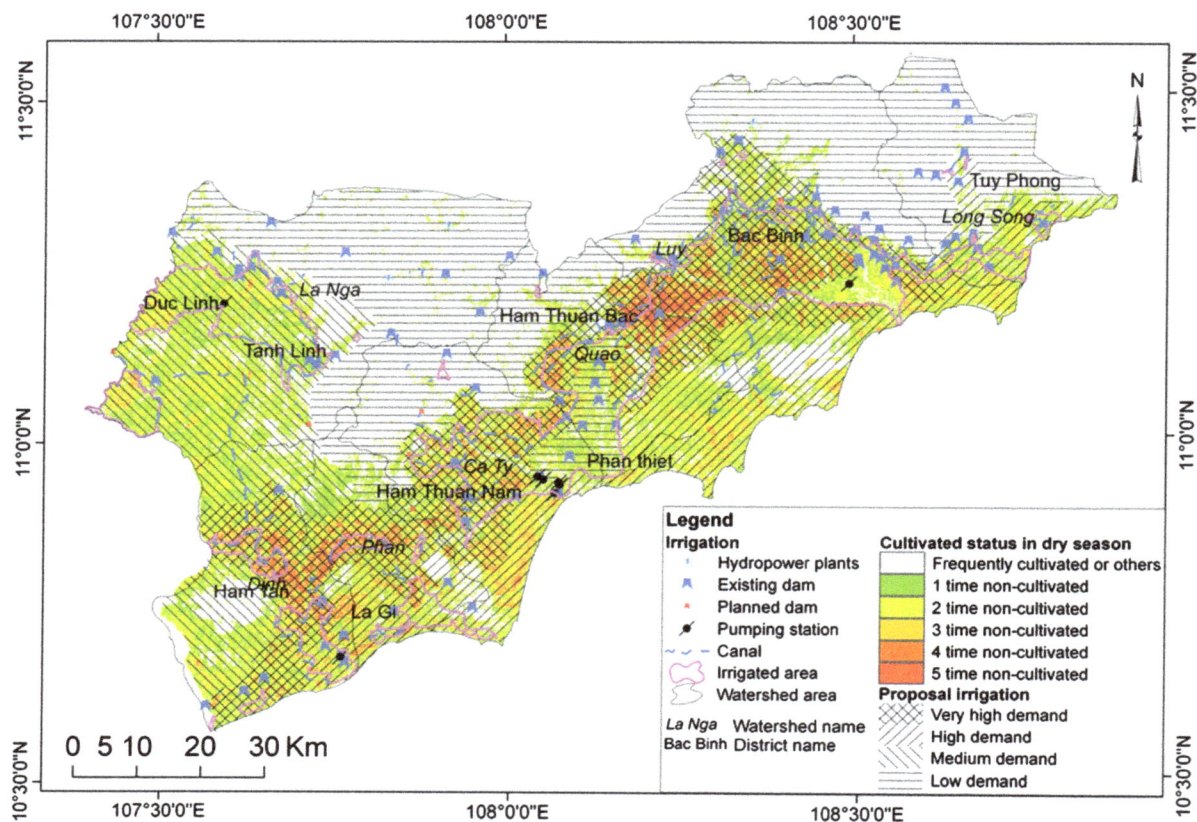

Figure 3. Current agricultural land use status and irrigation.

- Areas for agricultural production are often not cultivated during the dry season (in the period 1996-2014) due to the lack of irrigation in the foothill plains of the Bac Binh, Ham Thuan Bac, Ham Thuan Nam and Ham Tan districts. These areas are during the dry season, are often rarely used, dry and consequently shows signs of degradation. This region needs to be equipped with irrigation systems to improve the soil moisture during the dry season, and to reduce the risk of degradation. The need for an irrigation system is very high in these districts.

- Agricultural land is not cultivated all over the dry season as a result of unreliable irrigation water distribution mainly from Tuy Phong to Ham Tam. Here sandy soils dominate huge sand dunes and poor agricultural situations. In many places, sand encroaches the fields. These areas are in a high need of irrigation and protective forest to combat desertification.

- The agricultural land which already has stable irrigation (in Duc Linh, Tanh Linh and some other parts of Ham Thuan Bac and Bac Binh) should have an irrigation system at medium level.

4. Conclusions

Landsat satellite images allow analyzing land use changes at large and medium scale. The analysis locates and quantifies the area of change, and identifies the change of land use over time. The method provides recent data and allows updated and cost-effective assessments.

The agricultural land in Binh Thuan has declined in a limited way, from 43% (in 1996) to 40% (in 2014). But the area of agricultural land during the dry season only accounts for a small fraction and significantly decreases. This is related to changes in the weather and an insufficient irrigation system.

Combining land use changes and the irrigation system assessment, allows recommending for the region: frequently irrigated or non-irrigated areas, as the basis of an irrigation system providing sufficient water, fulfill the needs for production and land resource protection.

Acknowledgements

This work was a part of the research project code VAST.DLT 13/13-14 and funded by Vietnam Academy of Science and Technology (VAST).

References

[1]   NASA (2014) Land Cover/Land Use Change Program. http://lcluc.umd.edu/program_information.php?tab=1

[2]   Prakasam, C. (2010) Land Use and Land Cover Change Detection through Remote Sensing Approach: A Case Study of Kodaikanal Taluk, Tamil Nadu. *International Journal of Geomatics and Geosciences*, **1**, 150-158.

[3]   Ryngnga, P.K., Bring, B. and Ryntathiang, L. (2013) Dynamics of Land Use Land Cover for Sustainability: A Case of Shillong, Meghalaya. *International Journal of Scientific & Technology Research*, **2**, 235-239.

[4]   SIC (2014) Satellite Imagery and GIS for Land Cover and Change Detection. http://www.satimagingcorp.com/applications/environmental-impact-studies/land_cover_and_change_detection/

[5]   Ozdogan, M., Yang, Y., Allez, G. and Cervantes, C. (2010) Remote Sensing of Irrigated Agriculture: Opportunities and Challenges. *Remote Sensing*, **2**, 2274-2304. http://dx.doi.org/10.3390/rs2092274

[6]   Robert, C.N. (1997) An Inventory of Irrigated Lands for Selected Counties within the State of California Based on LANDSAT and Supporting Aircraft Data. University of California, Berkeley, Space Sciences Laboratory. http://ntrs.nasa.gov/archive/nasa/casi.ntrs.nasa.gov/19770024625.pdf

[7]   Heller, R.C. and Johnson, K.A. (1979) Estimating Irrigated Land Acreage from Landsat Imagery. *Photogrammetric Engineering & Remote Sensing*, **45**, 1379-1386.

[8]   Haack, B., Wolf, J. and English, R. (1998) Remote Sensing Change Detection of Irrigated Agriculture in Afghanistan. *Geocarto International*, **13**, 65-75. http://dx.doi.org/10.1080/10106049809354643

[9]   Thiruvengadachari, S. (1981) Satellite Sensing of Irrigation Patterns in Semiarid Areas: An Indian Study [Tamil Nadu State, India]. *Journal of Photogrammetric Engineering and Remote Sensing*, **47**, 1493-1499.

[10]   Binh Thuan Statistical Office (2014) Statistical Yearbook Binh Thuan Province. http://cucthongke.vn/ngtk/ngtk2013/index.html#/8

[11]   USGS (2014) Earth Explorer. http://earthexplorer.usgs.gov/

[12]   Mas, J.F. (1999) Monitoring Land-Cover Changes: A Comparison of Change Detection Techniques. *International Journal of Remote Sensing*, **20**, 139-152. http://dx.doi.org/10.1080/014311699213659

[13]   Congalton, R.G. (1991) A Review of Assessing the Accuracy of Classifications of Remotely Sensed Data. *Remote Sensing of Environment*, **37**, 35-46. http://dx.doi.org/10.1016/0034-4257(91)90048-B

[14]   Viera, A.J. and Garrett, J.M. (2005) Understanding Interobserver Agreement: The Kappa Statistic. *Family Medicine*, **37**, 360-363.

[15]   Hountondji, Y. and Ozer, P. (2011) Land Use and Land Cover Change Analysis 1990-2002 in Binh Thuan Province, South Central Vietnam. *1st International Conference on Energy, Environment and Climate Change*.

[16]   Binh Thuan Province's Committee (2005) Report on Drought and Others Remedial Measure to Support Farming. Department of Agriculture and Rural Development, Binh Thuan.

[17]   Huong, P.T.T. (2010) Scenarios of Climate Change and Drought in Binh Thuan Province. Report on the Project: Research on the Effects of Climate Change and Desertification South Central Region (Pilot Study Binh Thuan Province), Institute of Geography (VAST), Vietnam.

4

Growth and Yield of Sesame (*Sesamum indicum* L.) under the Influence of Planting Geometry and Irrigation Regimes

Asif Nadeem[1*], Shahabudin Kashani[2], Nazeer Ahmed[2], Mahmooda Buriro[3], Zahid Saeed[1], Fateh Mohammad[2], Shafeeque Ahmed[2]

[1]Agriculture Extension, Panjgoor, Pakistan
[2]Livestock Research Institute, Turbat, Pakistan
[3]Sindh Agriculture University, Tandojam, Pakistan
Email: [*]Shahabkashani64@gmail.com

Abstract

A field study to evaluate the "growth and yield of sesame (*Sesamum indicum* L.) under the influence of planting geometry and irrigation regimes" was carried out at Oilseeds Section, Agriculture Research Institute, Tandojam located at (25°25'60"N, 68°31'60"E) during Kharif 2013. The experiment was laid out in a three replicated randomized complete block design (RCBD) factorial, having net plot size 3 × 3 m (9 m²). The treatments comprised three planting geometry (30 × 20 cm, 45 × 15 cm and 60 × 10 cm) and three irrigation regimes (2 irrigations at 20 and 40 DAS, 3 irrigations at 20, 40 and 60 DAS and 4 irrigations at 20, 40, 60 and 80 DAS). The analysis of variance showed that all the planting geometry and irrigation regimes significantly (P < 0.05) affected growth and yield of sesame. Planting geometry of 45 × 15 cm resulted in maximum branches plant⁻¹ (15.67), capsules plant⁻¹ (38.00), seeds capsule⁻¹ (51.44), seed weight plant⁻¹ (31.89 g), seed index (2.83 g), biological yield (2301.23 kg·ha⁻¹), seed yield (742.33 kg·ha⁻¹) and harvest index (30.44%), followed by planting geometry of 60 × 10 cm in all the parameters. However, minimum growth and yield traits of sesame were recorded under planting geometry of 30 × 20 cm. Moreover, plant height was maximum (99.89 cm) under planting geometry of 30 × 20 cm. Among irrigation regimes, four irrigations (20, 40, 60 and 80 DAS) recorded maximum plant height (103.33 cm), branches plant⁻¹ (16.44), capsules plant⁻¹ (41.22), seeds capsule⁻¹ (54.56), seed weight plant⁻¹ (33.22 g), seed index (2.92 g), biological yield (2321.21 kg·ha⁻¹), seed yield (748.78 kg·ha⁻¹) and harvest index (31.00%), followed by three irrigations (20, 40 and 60 DAS) almost in all the traits. However, minimum growth and yield traits of sesame were recorded when crop was applied two irrigations (20 and 40 DAS). In case of interactive effects, the interaction of planting geometry of 45 × 15 cm × four irrigations (20, 40, 60 and 80 DAS) resulted in maximum attributes,

---

[*]Corresponding author.

particularly seed yield (765.00 kg·ha⁻¹), whereas minimum traits were noted in the interaction of planting geometry of 30 × 20 cm × two irrigations (20 and 40 DAS). Furthermore, the results of four and three irrigations had non-significant differences with each other means not sharing the same letter differ significantly at 0.05 probability level. Hence, it is concluded that interaction of 45 × 15 cm planting geometry and three irrigations (20, 40 and 60 DAS) is conducive to produce maximum seed yield (kg·ha⁻¹) of sesame.

Keywords

Sesame, Planting Geometry, Irrigation Regimes, High Yield

---

1. Introduction

Sesame (*Sesamum indicum* L.) is the most conventional oilseed crop cultivated for its edible oil in the sub-continent. Sesame is known as the king of oil seeds due to the high oil content (50% - 60%) of its seed [1]. Sesame oil is used as foods (cooking and salad), medicine and soap manufacturing etc. Its seeds and young leaves are eaten as stews and soaps in Asia [2]. Til oil-cake is good feed for poultry, goat, sheep, fish and cattle [3].

Sesame crop is considered as a drought tolerant crop [4]. It was found in Africa and a small number in the Indo-Pakistan regions. According to archeological evidence, it was cultivated at Harappa in the Indus Valley between 2250 and 1750 BC, and a more recent charred sesame seeds had been found in MiriQalat and Shahi-Tump in the Makran region of Pakistan [5].

Pakistan is facing acute shortage of edible oil. Demand of edible oil in Pakistan is increasing with increase in population but production of edible oil is decreasing every year. The local production estimated at 0.680 million tons, meets only 24% of domestic requirement of edible oil while the remaining 76% is met through imports. Total availability from all sources is provisionally estimated at 1.749 million tons. Higher production of sesame may contribute towards edible oil and to meet country's requirement which may help in reducing the import [6].

The area under sesame cultivation in Sindh, Punjab, KPK and Balochistan during the year 2006 was 3.4, 75.1, 0.1 and 3.4 thousand hectares with the production of 1.4, 31.6, 0.1 and 2.0 thousand tons. The yield per hectare of sesame in Pakistan is low as compared to many other sesame growing countries of the world. The causes behind this setback are varied, which may include use of marginal land impure seed, low yielding varieties; improper sowing time, irrigation frequencies, fertilizer; and other cultural practices [7].

In Pakistan, sesame is grown in Kharif season (June-July). Low yield of sesame in Pakistan may be attributed to the lesser availability of good quality seed, sowing method (broadcast method), sowing time (early or late sowing) and less or over plant population [8]. Differential response of varieties to sowing dates showed that the yield of sesame was decreased with delay in sowing beyond third week of July [9]. To get high yield of a crop it is necessary to understand the interaction of crops with weather it plays important role in plant growth. Proper inter row and intra row spacing is of primary importance as it determines the proper plant population in the field. All these factors are also affected by the planting geometry of the crop. The environmental factors and management practices influence sesame productivity [10]. Suitable environmental conditions are necessary for optimum growth of crop and in field the crop yield also depends upon the row spacing between plants. Planting geometry is one of the major factors exploring the yield potential of a particular variety of oilseed crop [11]. The space available for individual plant growing in a community affects the yield and quality of produce. If space is more there may be less competition between plants and the chance of weed growth is definitely high. Whereas, space is less there may be high competition for growth factor like light, carbon-dioxide, moisture and nutrients among crop plants [12]. The establishment of an adequate and uniform crop stand is critical to achieve high seed yield. Seed yield of oilseed crop is a function of population density pods plant⁻¹, seeds pod⁻¹ and seed weight [13].

Due to decrease or increase in row space plant population of the crop varies and without optimum plant population one cannot get good crop yield. the higher yield of sesame was obtained when row to row distance was 30 cm and yield was decreased at 45 cm [14]. On reducing row space the plant height was increased but capsule plant⁻¹ decreased [15]. The recommended range of planting density for branched type is from 200 to 400 thousand plants ha⁻¹. Five cultivation patterns of 30 × 10, 30 × 15, 30 × 20, 30 × 25 and 30 × 30 cm for branched sesame varieties were evaluated. The results concluded that a seed yield increased from 417, 552 and 561 kg·ha^{-1} while a row space increase from 10 to 20 cm and a wider row space increase to 25 and 30 cm caused a seed yield decrease to 510 and 395 kg·ha^{-1} [16]. The maximum plant height (158.9 cm), plants m^{-2} (60.33), capsules plant^{-1} (24.90), biological (4.003 Mg·ha^{-1}) and seed yield (0.857 Mg·ha^{-1}), harvest index (21.42%) and oil yield (0.371 Mg·ha^{-1}) of sesame were recorded in plots where sesame was sown with row spacing of 15 cm [17].

Irrigation optimization is a very important practice used in crop management, which could reduce irrigation water losses and maintain high yield. With increasing costs associated with irrigation, there is a need to ensure the maximum return from each unit of input applied. The need of crops for water is related to moisture sensitive periods. If moisture sensitive periods could be identified for wheat crop under field conditions, it would have important implications for irrigation practice [18]. Irrigation scheduling is important for achieving the higher crop yields. Irrigation frequency is the decision of when and how much water to apply to a field. Its purpose is to maximize irrigation efficiencies by applying the exact amount of water needed to replenish the soil moisture to the desired level. Irrigation scheduling saves water and energy, and all irrigation scheduling procedures consist of monitoring indicators that determine the need for irrigation. The purpose of irrigation scheduling is to determine the exact amount of water to apply to the field and the exact timing for application. The amount of water applied is determined by using a criterion to determine irrigation need and a strategy to prescribe how much water to apply in any situation [19].

Water depletion is estimated from root zone and effect on sesame yield and water consumptive use. In a recent study, the predictions for moisture stress suggested applying five irrigations to sesame [20]. Applying six irrigations gave the highest values of yield and its attributes, whereas the lowest values were recorded from applying five irrigations and skipping one at the beginning of flowering [21]. Whereas, Davut *et al.* (2007) achieved highest sesame yield under 6 days interval of irrigation and lowest when irrigation applied at 24 days interval [22]. Keeping in view the facts stated above, a study was designed under the agro-ecological conditions of Tandojam.

2. Materials and Methods

The field study was carried out at Oilseeds Section, Agriculture Research Institute, Tandojam (25°25'60"N, 68°31'60"E) during Kharif 2013. The experiment was laid out into a three replicated randomized complete block design (RCBD) factorial, having net plot size 3 m × 3 m (9 m^2). Planting geometry and irrigation regimes were maintained as per treatments, whereas all other inputs and cultural operations were adopted as per the recommendations of Oilseeds Section, Agriculture Research Institute, Tandojam. The treatments comprised of three planting geometry (30 × 20 cm, 45 × 15 cm and 60 × 10 cm) and three irrigation regimes (2 irrigations at 20 and 40 DAS, 3 irrigations at 20, 40 and 60 DAS and 4 irrigations at 20, 40, 60 and 80 DAS).

2.1. Experimental Soil

Soil condition of experimental area was clay loam in texture, non-saline (EC 0.14 to 0.68 dS·m^{-1}) in nature, slightly alkaline in reaction (pH 7.24 to 7.75), calcareous (CaCO$_3$ 9.5%), low in organic content (0.62% - 0.86%), total nitrogen content varied from 0.03% - 0.06% and available phosphorus ranged from 6.8 to 9.2 mg·kg^{-1}. However, soil of the experimental area was marginal in extractable potassium (60 - 110 mg·kg^{-1}) content.

2.2. Land Preparation

For preparation of fine seedbed, disc harrow was run to open and pulverize the soil, and the land was leveled and planked. After soaking dose, when the land came in condition, the cultivator was used, followed by rotavator. The treatments were managed in such a way to separate each treatments and replications easily, and the channels and bunds were developed to facilitate the irrigation water application and interculturing.

2.3. Sowing Time and Method

The seed of sesame variety S-17 was grown on June 2013 with the help of single row hand drill. The row to row spacing of 75 cm and plant to plant distance 22.5 cm was maintained apart. Seed rate 5 kg/hac.

2.4. Fertilizer

Nitrogen and phosphorus fertilizers were applied at recommended rate of Full dose of phosphorus in the form of

DAP, whereas half dose of Nitrogen in the form of Urea were applied at the time of sowing. Remaining half dose of Nitrogen was applied at the time of first irrigation.

2.5. Weeding

The weeds were controlled by interculturing at 1st and 2nd irrigations.

2.6. Procedure for Recording Observations

1) Plant height (cm): Plant height was recorded at maturity of crop using measurement tape from bottom to tip of the randomly selected five plants in each plot and was averaged in (cm).

2) Branches plant^{-1}: Branches in randomly selected five plants were counted and accordingly average plant^{-1} was worked out in each treatment.

3) Capsules plant^{-1}: The capsules randomly selected five plants were counted in each plot and averaged plant^{-1} basis.

4) Seeds capsule^{-1}: At maturity, the capsules from the randomly selected five plants were threshed and number of seeds in each capsule was counted and averaged.

5) Seed weight plant^{-1} (g): At maturity, the seed obtained from all the randomly selected five plants were weighed and calculated on the basis of number of plants threshed to get average weight of seeds plant^{-1} (g).

6) Seed index (1000 seeds weight, g): One thousand seeds from the seed lot threshed in each plot was collected and weighed to obtain seed index in (g).

7) Biological yield (kg·ha^{-1}): At maturity, the sesame crop in each plot was harvested; weighed and biological yield was converted to kg·ha^{-1}.

8) Seed yield (kg·ha^{-1}): At maturity, the sesame crop in each plot was harvested and threshed, and yield ha^{-1} was calculated by the following formula:

$$\text{Seed yield ha}^{-1}(\text{kg}) = \frac{\text{Seed yield plot}^{-1}(\text{kg})}{\text{Plot size}(\text{m}^2)} \times 10000$$

9) Harvest index (%): Harvest index is the ratio between seed yield and biological yield. Harvest index (%) was calculated by the following formula.

$$\text{Harvest index}(\%) = \frac{\text{Seed yield}(\text{kg}\cdot\text{ha}^{-1})}{\text{Biological yield}(\text{kg}\cdot\text{ha}^{-1})} \times 100$$

2.7. Statistical Analysis

The collected data were subjected to statistical analysis using MSTAT-C. The differences among the treatments means were compared by the least significant difference (LSD), where necessary (Russel and Eisensmith, 1983).

3. Result and Discussions

The results of this study showed that all the levels of planting geometry and irrigation regimes affected significantly ($P < 0.05$) the growth and yield of sesame. Planting geometry of 45 × 15 cm resulted in maximum branches plant^{-1} (15.67), capsules plant^{-1} (38.00), seeds capsule^{-1} (51.44), seed weight plant^{-1} (31.89 g), seed index (2.83 g), biological yield (2301.23 kg·ha^{-1}), seed yield (742.33 kg·ha^{-1}) and harvest index (30.44%), followed by planting geometry of 60 × 10 cm with branches plant^{-1} (12.22), capsules plant^{-1} (35.22), seeds capsule^{-1} (47.00), seed weight plant^{-1} (27.11 g), seed index (2.36 g), biological yield (2208.58 kg·ha^{-1}), seed yield (712.44 kg·ha^{-1}) and harvest index (28.33%). However, minimum growth and yield traits of sesame were recorded under planting geometry of 30 × 20 cm (**Table 1**). Moreover, plant height was maximum (99.89 cm) under planting geometry of 30 × 20 cm. Planting geometry affected the plant height of sesame significantly. The crop which was sown at planting geometry of 30 × 20 cm produced significantly taller plants than at other planting geometry. The increase in plant height in case of 30 × 20 cm planting geometry may be due to narrow row spacing, when the number of plants m^{-2} increases then the competition for light increases and plant grows taller to intercept maximum light. These results are in accordance with the findings of Caliskan et al. (2004) who reported taller plants with increase in plant population. The crop which was at planting geometry of 60 × 10 cm produced significantly shortest plants [23]. The crop which was sown at planting geometry of 45 × 15 cm produced significantly maximum branches plant^{-1}, capsules plant^{-1}, seeds capsule^{-1}, seed weight plant^{-1}, seed index (g), biological yield, seed yield and harvest index. This may be due to optimum plant population and had more chance to get nutrients. These results are in line with those of El-Naim *et al.* (2010) who demonstrated that increasing plant density significantly decreased the number of branches plant^{-1}, the number of capsules plant^{-1} and seed yield plant^{-1}, decreased the number of branches plant^{-1}, the number of capsules plant^{-1} and seed yield plant^{-1}, but led to increased seed yield per unit area [24]. Similarly, Tahir *et al.* (2012) revealed that maximum plant height, plants m^{-2}, capsules plant^{-1}, biological, and seed yield, harvest index and oil yield were recorded in plots where sesame was sown at 15th June with row spacing of 15 cm [17]. In another study, Roy *et al.* (2009) reported that higher yield of sesame was obtained when row to row distance was 30 cm and yield was decreased at 45 cm [14].

As it is shown in **Table 2**, Four irrigations (20, 40, 60 and 80 DAS) recorded maximum plant height (103.33 cm), branches plant^{-1} (16.44), capsules plant^{-1} (41.22), seeds capsule^{-1} (54.56), seed weight plant^{-1} (33.22 g), seed index (2.92 g), biological yield (2321.21 kg·ha^{-1}), seed yield (748.78 kg·ha^{-1}) and harvest index (31.00%), followed by three irrigations (20, 40 and 60 DAS) with 101.78 cm plant height, 15.33 branches plant^{-1}, 40.22 capsules plant^{-1}, 52.78 seeds capsule^{-1}, 32.11 g seed weight plant^{-1}, 2.77 g seed index, biological yield (2292.62 kg·ha^{-1}), seed yield (739.56 kg·ha^{-1}) and harvest index (30.11%). However, minimum growth and yield traits of sesame were recorded when crop was applied two irrigations (20 and 40 DAS) [25].

Table 3 shown the analysis of variance regarding plant height and branches plant^{-1} highly significant (P < 0.01) effect of planting geometry irrigation regimes. The result regarding capsules plant^{-1}, seeds capsule^{-1}, weight plant^{-1}, seed index, biological yield, seed yield and harvest index their analysis of variance showed highly significant (P < 0.01) effect of planting geometry and irrigation regimes, whereas non-significant (P > 0.05) effect of their interaction [25].

Table 1. Means of growth and yield traits of sesame under planting geometry.

Planting geometry	Plant height (cm)	Branches plant^{-1}	Capsules plant^{-1}	Seed capsules^{-1}	Seed weight plant^{-1} (g)	Seed index (g)	Biological yield (kg/ha^{-1})	Seed yield (kg·ha^{-1})	Harvest index (%)
30 × 20 cm	99.89 a	12.22 c	35.22 c	47.00 b	27.11 c	2.36 c	2208.58 b	712.44 b	28.33 c
45 × 15 cm	96.56 b	15.67 a	38.00 a	51.44 a	31.89 a	2.83 a	2301.23 a	742.33 a	30.449 a
60 × 10 cm	92.44 c	13.67 b	36.44 b	49.56ab	30.78 a	2.61 b	2242.68 b	723.44 b	29.44 b

Note: Means with same letter in each column are not significantly different at a 5% probability level.

Table 2. Means of growth and yield traits of sesame under different irrigation regimes.

Irrigation regimes	Plant height (cm)	Branches plant^{-1}	Capsules plant^{-1}	Seed capsules^{-1}	Seed weight plant^{-1} (g)	Seed index (g)	Biological yield (kg·ha^{-1})	Seed yield (kg·ha^{-1})	Harvest index (%)
2 irrigations (20 & 40 DAS)	83.78 c	9.78 c	28.22 c	40.67 b	24.44 b	2.11c	2138.66 b	689.89 b	27.11c
3 irrigations (20, 40 & 60 DAS)	101.78 b	15.33 b	40.22 a	52.78 a	32.11 a	2.77 a	2292.62 a	739.56 a	30.11 b
4 irrigations (20, 40, 60 & 80 DAS)	103.33 a	16.44 a	41.22 a	54.56 a	33.22 a	2.92 a	2321.21 a	748.78 a	31.00 a

Note: Means with same letter in each column are not significantly different at a 5% probability level.

Table 3. Analysis of variance for growth and yield parameters.

Source of variation	DF	Plant height	Branches plant^{-1}	Capsules plant^{-1}	Seed capsules^{-1}	Seed weight plant^{-1}	Seed index	Biological yield	Seed yield	HI
Planting geometry (P)	2	**	**	NS	**	*	**	**	**	**
Irrigation regimes (I)	2	**	**	**	**	**	**	**	**	**
P × I	4	**	*	NS	NS	NS	NS	NS	NS	NS

Notes: NS, non-significant; * and **, significant at the 0.05 and 0.01 levels of probability, respectively.

4. Conclusion

The results concluded that growth and yield attributes particularly seed yield of sesame was significantly ($P <$ 0.05) affected by all the levels of planting geometry and irrigation regimes. Planting geometry of 45 × 15 cm produced maximum parameters, particularly seed yield (742.33 kg·ha^{-1}). Among irrigation regimes, four irrigations (20, 40, 60 and 80 DAS) recorded highest traits, particularly seed yield (748.78 kg·ha^{-1}). In case of interaction, planting geometry 45 × 15 cm × four irrigations (20, 40, 60 and 80 DAS) resulted in greatest values, particularly seed yield (765.00 kg·ha^{-1}).

Acknowledgements

The author (A N) is highly grateful to Professor, Dr. Shamsuddin Tunio, Sindh Agriculture University, Tandojam and Shahab-u-din, Scientific officer, Livestock Research Institute Turbat for their technical support throughout the research work.

References

[1] Toan, D.P., Thuy-Duong, T.N.A., Carlsson, S. and Bui, T.M. (2010) Morphological Evaluation of Sesame (*Sesamum indicum* L.) Varieties from Different Origins. *Australian Journal of Crop Science*, **4**, 498-504.

[2] Pakissan.com (2010) Sesame Production Practices in Pakistan.
http://www.pakissan.com/english/allabout/crop/sesame.shtml

[3] Khan, M.H.A., Sultana, N.A., Islam, M.N. and Zaman, M.H. (2009) Yield and Yield Contributing Characters of Sesame as Affected by Different Management Practices. *American-Eurasian Journal of Scientific Research*, **4**, 195-197.

[4] Jefferson, T. (2003) Sesame a High Value Oil Seed. *Growing Sesame Production Tips, Economics and Mare. Htm*, 1-4.

[5] Wikipedia (2007) Sesame: From Wikipedia, the Free Encyclopedia Wikimedia Foundation, Inc.

[6] GOP (2010) Government of Pakistan, Ministry of Food, Agriculture and Livestock, Finance Division, Economic Advisor Wing, Islamabad Pakistan. 23.

[7] Saleem, M.F., Ma, B.L., Malik, M.A., Cheema, M.A. and Wahid, M.A. (2008) Yield and Quality Response of Sesame (*Sesamum indicum* L.) to Irrigation Frequencies and Planting Patterns. *Canadian Journal of Plant Science*, **88**, 101-109.
http://dx.doi.org/10.4141/CJPS07052

[8] Ashri, A. (1998) Sesame Breeding. *Plant Breeding Reviews*, **16**, 179-228.
http://dx.doi.org/10.1002/9780470650110.ch5

[9] Mahdi, A., Amin, S.E.M. and Ahmed, F.G. (2007) Effect of Sowing Date on the Performance of Sesame (*Sesamum indicum* L.) Genotypes under Irrigation Conditions in Northern Sudan. *African Crop Sciences Conference Proceedings*, **8**, 1943-1946.

[10] Adebisi, M.A. (2004) Variation, Stability and Correlation Studies in Seed Quality and Yield of Sesame (*Sesamum indicum* L.). Ph.D. Thesis, University of Agriculture, Abeokuta.

[11] Mahan, R.K.S., Singh, U.P. and Verma, N.K. (2008) Effect of Planting Geometries in Relation to Fertilizer Combinations on Growth and Yield of Mustard (*Brassica juncea* coss) var. "Varuna" under Bundelkhand Region of Uttar Pradesh.
http://www.indianjournals.com/ijor.aspx?target=ijor:asd&volume=28&issue=2&article=024

[12] Singh, K., Dhaka, R.S. and Fageria, M.S. (2004) Response of Cauliflower (*Brassica oleracea* var. *botrytis* L.) Cultivars to Row Spacing and Nitrogen Fertilization. *Progressive Horticulture*, **36**, 171-173.

[13] Wysocki, D. and Sirovatka, N. (2010) Effect of Row Spacing and Seeding Rate on Winter Canola in Semiarid Oregon. *Journal of Science*, **85**, 444-446.

[14] Roy, N., Abdullah, S.M. and Jahan, M.S. (2009) Yield Performance of Sesame (*Sesamum indicum* L.) Varieties at Varying Levels of Row Spacing. *Research Journal of Agriculture & Biological Sciences*, **5**, 823-827.

[15] Rahnama, A. and Bakhshandeh, A. (2006) Determination of Optimum Row Spacing and Plant Density for Uni-Branched Sesame in Khuzestan Province. *Journal of Agricultural Science and Technology*, **8**, 25-33.

[16] Tiwari, K.P., Jain, P. and Raghuwanshi, S. (1990) Effect of Sowing Date and Plant Densities on Seed Yield of Sesame Cultivars. *Crop Research*, **8**, 404-406.

[17] Tahir, M., Saeed, U., Ali, A., Hassan, I., Naeem, M., Ibrahim, M., *et al.* (2012) Optimizing Sowing Date and Row Spacing for Newly Evolved Sesame (*Sesamum indicum* L.) Variety TH-6. *Pakistan Journal of Life & Social Sciences*, **10**, 1-4.

[18] Maqsood, M., Ali, A., Aslam, Z., Saeed, M. and Ahmad, S. (2002) Effect of Irrigation and Nitrogen Levels on Grain

Yield and Quality of Sesame. *International Journal of Agriculture & Biology*, **4**, 164-165.

[19] Broner, I. (2005) Irrigation Scheduling for Sesame Production. Colorado State University Cooperative Extension, Fort Collins, 1-4.

[20] Hassanzadeh, M., Ebadi, A., Panahyan-e-Kivi, M., Jamaati-e-Somarin, Sh., Sacidi, M. and Gholipouri, A. (2009) Investigation of Water Stress on Yield and Yield Components of Sesame, *Sesamum indicum* L. in Moghan Region. *Research Journal of Environmental Sciences*, **3**, 239-244. http://dx.doi.org/10.3923/rjes.2009.239.244

[21] Detphirattanamongkhon, S. (2002) Influence of Different Water Regimes and Irrigation Intervals on the Growth and Yield of Sesames. http://agris.fao.org/agris-search/search.do?recordID=TH2001002924

[22] Karaaslan, D., Boydak, E., Gercek, S. and Simsek, M. (2007) Influence of Irrigation Intervals and Row Spacing on Some Yield Components of Sesame Grown in Harran Region. *Asian Journal of Plant Sciences*, **6**, 623-627. http://dx.doi.org/10.3923/ajps.2007.623.627

[23] Caliskan, S., Arslan, M., Arioglu, H. and Isler, N. (2004) Effect of Planting Method and Plant Population on Growth and Yield of Sesame (*Sesamum indicum* L.) in a Mediterranean Type of Environment. *Asian Journal of Plant Sciences*, **3**, 610-614.

[24] El-Naim, A.M., Elday, E.M. and Ahmed, A.A. (2010) Effect of Plant Density on the Performance of Some Sesame (*Sesamum indicum* L.) Cultivars under Rain Fed. *Research Journal of Agriculture and Biological Sciences*, **6**, 498-504.

[25] Moghadam, P.R., Poor, G.N., Nabati, J. and Abadi, A.A.M. (2008) Effects of Different Irrigation Intervals and Plant Density on Morphological Characteristics, Grain and Oil Yields of Sesame, *Sesamum indicum* (L.). *Iranian Journal of Field Crops Research*, **6**, 112-116.

5

Fluoride Contamination of Groundwater and Health Hazard in Central India

Nohar Singh Dahariya, Keshaw Prakash Rajhans, Ankit Yadav, Shobhana Ramteke, Bharat Lal Sahu, Khageshwar Singh Patel*

School of Studies in Chemistry/Environmental Science, Pt Ravishankar Shukla University, Raipur, India
Email: *patelkhageshwarsingh@gmail.com

Abstract

The basic bed rocks of central India are contaminated with fluorite minerals. The overuse of groundwater for irrigation causes increased mineralization of F- in the groundwater. This contaminated groundwater is widely used for drinking and other household purposes. The excess F- is excreted through urine of animals. In this work, the exposure of contaminated groundwater in domestic animals of Dongargarh city, Chhattisgarh, India is studied. The symptoms of fluorosis diseases in the domestic animals *i.e.* cattle and buffalo are surveyed. The quality and sources of the contaminants of the groundwater are discussed.

Keywords

Groundwater, Fluoride, Exposure, Fluorosis

1. Introduction

Abnormal levels of F^- in the groundwater are common in India due to weathering of the fractured hard rock pegmatite veins composing of minerals viz. topaz, fluorite fluorapatite, villuamite, cryolite, ferro magnesium silicate, etc. [1]. The F^- contamination of groundwater in several states of the country was reported [2]-[21]. Millions of people and animals were exposed to excessive amount of F^- through drinking water contaminated from geogenic and anthropogenic sources, suffering with various types of fluoride diseases [22]-[31]. The goal of this work is to study F^- contamination of the groundwater and its exposure in domestic animals *i.e.* cattle and buffalo of Dongargarh city, India.

---

*Corresponding author.

2. Materials and Methods

2.1. Study Area

Dongargarh (21.18842°N and 80.75875°E) is a tourist city in central India with population of 0.1 million inclusive of neighboring villages. The town was settled near majestic mountains. The contaminated groundwater is widely used for drinking, cooking, washing and agricultural purposes. Four minerals *i.e.* oligoclase, rectorite, kaolinite and feldspar have been identified in the studied area. Feldspar is one of the most dominant mineral constituents of all the above-mentioned rocks, which is highly susceptible to chemical weathering and produces various types of clay minerals [32].

2.2. Sample Collection

Forty eight groundwater samples were collected from the tube wells of the Dongargarh city from ≈ 100 km^{-2} area in the post monsoon (January) and pre monsoon (May) period, 2014 by using established method, **Figure 1**

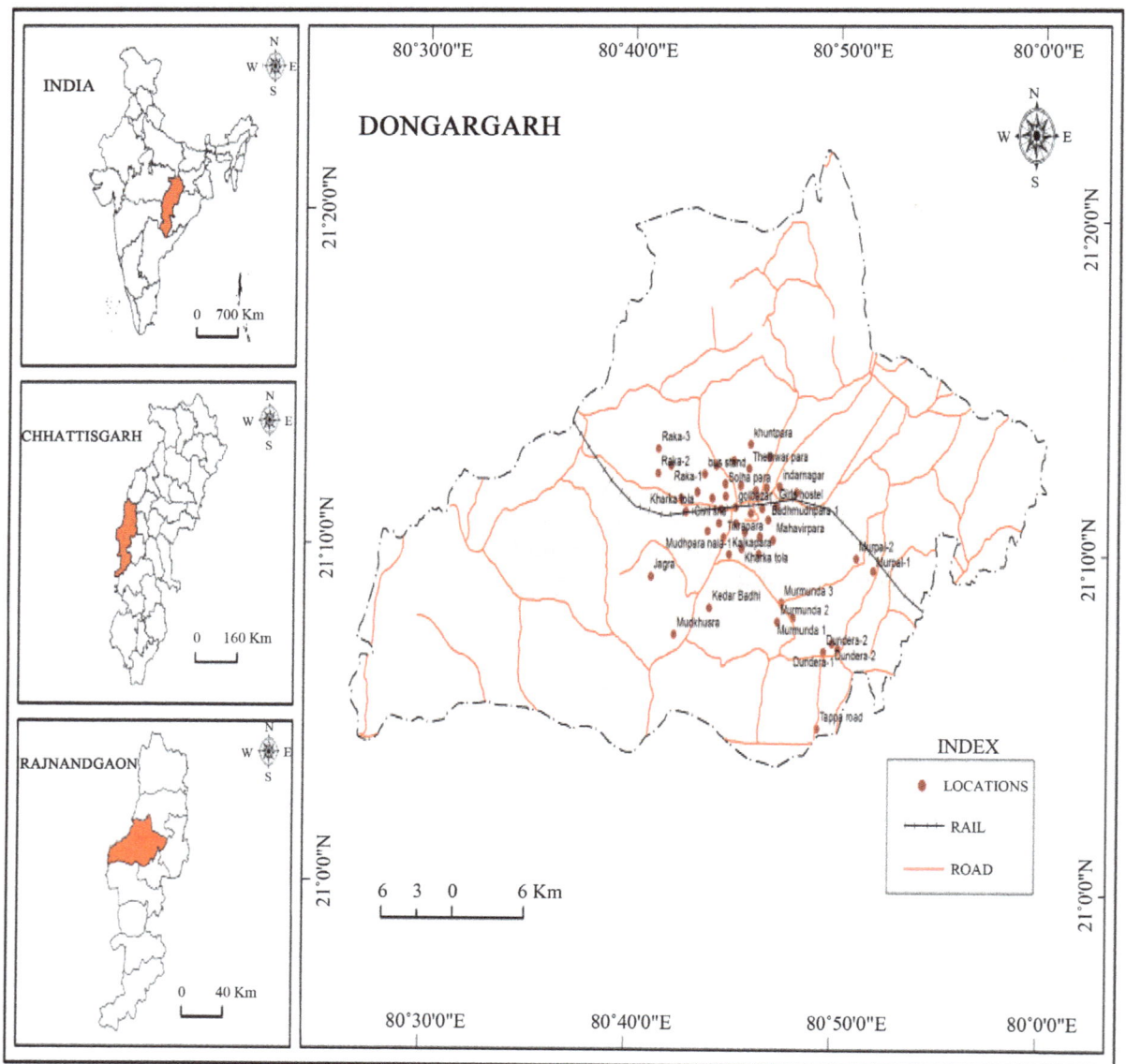

Figure 1. Representation of sampling locations in Dongargarh area, Chhattisgarh, India.

[33]. The groundwater sample was stored in 1-L cleaned polyethylene bottle. The physical parameters *i.e.* pH, temperature (T), electrical conductivity (EC), dissolved oxygen (DO) and reduction potential (RP) of the water were analyzed at the spot. The water samples were dispatched to the laboratory and preserved in the deep freezer.

The first morning urine sample (100 ml) was collected in plastic bottles containing 0.2 g EDTA. Total 40 urine samples of cattle and buffalo were collected in January, 2014. They were shipped to the laboratory in insulated container at about 4°C and refrigerated at –20°C until use.

2.3. Analysis

The total dissolved solid (TDS) value was determined by evaporation of the filtered water sample (through glass fiber filter) by drying at constant weight. The total hardness (TH) and total alkalinity (TA) values were analyzed by titration methods [34]. The F^- content was analyzed by using Metrohm ion meter-781 in the presence of 1:1 total ion strength adjustment buffer (TISAB). The buffer was prepared by adding 58 g NaCl + 5 g CDTA (trans-1, 2, NNNN, cyclodiamine tetra acetic acid) +57 ml glacial acetic acid and deionized water by adjusting pH value to 5.5 with 8 N NaOH in 1-L volumetric flask.

The ion (*i.e.* Cl^-, NO_3^-, SO_4^{2-}, NH_4^+, Na^+, K^+, Mg^{2+} and Ca^{2+}) content of the water was analyzed by Dionex-1100 ion chromatography equipped with the anion and cation columns.

The water quality index (WQI) of the groundwater was computed by using the weighed arithmetic method. The value of 6 parameters *i.e.* pH, DO, EC, TDS, TA and NO_3^- was used in calculation of the WQI with the help of following expression.

$$WQI = \sum q_n W_n / \sum W_n$$

where:

$$q_n = 100 \left(V_o - V_{io} \right) / \left(V_s - V_{io} \right)$$

q_n = Quality rating of the nth water quality parameter;

V_n = Estimated value of the nth parameter of a given water;

S_n = Standard permissible value of the nth parameter;

V_{io} = Ideal value of the nth parameter of pure water (*i.e.*, 0 for all other parameters except pH and dissolved oxygen (7.0 and 14.6 mg/L, respectively);

W_n = Unit weight for the nth parameter;

K = Proportionality constant.

Multivariate statistical analysis such as factor analysis (FA) was employed for the source apportionment. The windows statistical software Statistica-7.1 was used for the multivariate statistical calculation.

3. Results and Discussion

3.1. Geological Characteristics of Tube Well

The geological characteristics of the tube wells is summarized in **Table 1**. The tube wells of the studied area lie in the deeper zone, ranging from 45 - 110 m. The life was ranged from 2 - 50 Yr old. Tube wells are recharged by rain and runoff water during rainy season. The water table is varied from 20 - 50 m, depending on seasons and water uses. The higher T value for deeper tube well was observed due to geothermal energy. In turn, the higher DO value of shallow tube wells was marked.

3.2. Physical Characteristics of Groundwater

The physical characteristics of the groundwater in the post monsoon period is shown in **Table 1**. The value of pH, DO,T, RP, EC, TDS, TA and TH of groundwater located in 48 tube wells was ranged from 6.0 - 8.1, 8.4 - 9.2 mg/L, 20.0°C - 25.0°C, 237 - 330 mV, 221 - 1938 μS/cm, 342 - 2598 mg/L, 128 - 659 mg/L and 99 - 687 mg/L with mean value of 7.2 ± 0.1, 8.9 ± 0.1, 23.0 ± 0.3, 285 ± 5 mV, 861 ± 96 μS/cm, 1138 ± 121 mg/L, 383 ± 47 mg/L and 344 ± 37 mg/L, respectively. Two ions *i.e.* Na^+ and SO_4^{2-} were found to be responsible for contributing the EC value of the water. The DO, EC, TDS, TA and TH value of water was found to be higher than recommended value of 4.0 mg/L, 300 μS/cm, 500 mg/L, 120 mg/L and 300 mg/L, respectively [35] [36]. The

Table 1. Characteristics of tube well and groundwater in the post monsoon period, 2014.

S. No.	Location	Age, Yr	Depth, m	T°C	pH	EC, μS/cm	RP, mV	DO, mg/L
1	Mahavir Para	20	75	21	7.3	831	283	8.8
2	Goal Bajar	1	82	21	6.1	1017	310	8.9
3	Bhagat Shih Ward	3	75	21	8.1	967	237	8.6
4	Civil Line	7	69	22	6.2	991	311	8.8
5	Shubhash Ward	1	90	22	6.5	1013	319	8.9
6	Thana Chowk	7	69	22	6.4	1139	313	8.9
7	Kachari Chowk	12	75	21	6.6	603	304	8.8
8	Thethwar Para	12	82	21	6.5	344	309	8.7
9	BUS Stand	25	75	22	6.0	609	319	8.8
10	Solh Para	2	105	22	6.1	982	330	8.8
11	Ek-batti Char Rasta	10	66	23	8.0	733	265	9.1
12	Kumhar Para	15	54	22	6.5	385	269	9.0
13	School	10	69	22	6.3	276	297	9.2
14	Kumhar Para	1	75	22	6.9	349	298	8.8
15	Ambedkar Ward	1	90	23	7.3	613	254	8.9
16	Station	3	92	23	7.6	888	246	8.8
17	Bheem Nagar	4	105	22	7.3	1451	254	8.7
18	Utkarsh Nagar	8	110	22	7.3	1329	290	8.9
19	Kalka Para	10	75	22	7.2	1331	293	8.7
20	Indra Nagar	15	60	23	7.4	969	277	9.0
21	Rajiv Nagar	6	62	21	7.1	1178	304	8.7
22	Sanjay Nagar	10	66	22	7.2	1062	285	9.1
23	Khuta Para	12	72	22	7.2	951	302	8.4
24	Danteshwari Para	10	75	23	7.4	670	290	8.8
25	Kharka Tola	10	69	22	7.9	1033	282	9.0
26	Tikra Para	4	84	22	7.7	858	276	8.9
27	Kedar Badhi	23	75	21	7.6	1014	286	8.8
28	Badhiya Tola	30	82	23	7.1	1199	288	9.0
29	Raka Panchayat	24	90	22	7.2	637	275	9.0
30	Raka-2	50	75	21	7.2	322	291	9.1
31	Raka-3	3	92	21	7.1	1202	291	8.9
32	Murmunda-1	4	84	22	7.2	817	292	8.8
33	Murmunda-2	5	75	22	7.2	687	285	9.0
34	Murmunda-3	3	69	22	7.3	735	283	9.1
35	Murpal-1	10	45	22	7.3	539	276	8.9
36	Murpal-2	4	56	23	7.2	625	299	9.0
37	Murpal-3	6	64	23	7.4	1938	277	8.7
38	Jagra	2	75	22	7.4	221	272	8.9
39	Badhmudh Par-1	5	72	22	7.6	623	278	8.6
40	Badhmudhpar-2	21	72	21	7.5	707	276	8.7
41	Mudhpara Nala-1	7	75	22	7.8	1053	287	9.0
42	Mudhpara Nala-2	12	75	21	7.8	1057	272	8.9
43	Dundera-1	5	66	23	7.7	895	275	9.0
44	Dundera-2	20	69	22	7.6	1057	280	8.9
45	SBI Bank	10	90	22	7.2	605	294	8.9
46	Tappa Road	20	75	22	7.8	785	262	9.0
47	Mudhkhusra	10	66	21	7.3	681	269	9.1
48	Shivnikunj	10	75	22	7.7	1373	272	9.0

RP value of water was marked to be just of a half of recommended value of 600 mV.

3.3. Chemical Characteristics of Groundwater

The chemical characteristics of the groundwater in the post monsoon period is shown in **Table 2**. The concentration of ions *i.e.* F^-, Cl^-, NO_3^-, SO_4^{2-}, NH_4^+, Na^+, K^+, Mg^{2+} and Ca^{2+} in the groundwater of 48 tube wells was ranged from 2.0 - 10.3, 43 - 408, 48 - 152, 12 - 161, 10 - 144, 8.0 - 75, 3.0 - 25, 9.0 - 57 and 24 - 172 mg/L with mean value of 4.9 ± 0.5, 121 ± 22, 75 ± 8, 47 ± 9, 55 ± 11, 41 ± 5, 6.1 ± 1.5, 28 ± 3 and 87 ± 9 mg/L, respectively. The F^- and NO_3^- crossed the recommended limit of 1.5 and 45 mg/L, respectively in the water of all tube wells [35] [36]. However, Mg and Ca concentration was above recommended limit of 30 and 75 mg/L in the 33% and 67% tube wells [35] [36].

3.4. Seasonal Variation and Sources of Fluoride

The chemical data for the pre monsoon period, 2014 is presented in **Table 3**. The variation of physical and chemical parameters of the water in the pre monsoon period (May 2014) is presented in **Figure 2**. The value of pH, EC, TDS, TA, TH, F, Na, Mg and Ca was found to be increased ≥30%, may be due to increase of water temperature, ≈4°C, and deduction of water level up to 50 m. The F^- with the metals *i.e.* Na^+, Mg^{2+} and Ca^{2+} had good correlation (r = 0.78 - 0.85), indicating origin from the rock weathering, **Figure 3**. Other ions (*i.e.* Na^+, Mg^{2+}, Ca^{2+}, Cl^- and SO_4^{2-}) among themselves had fare correlation, suggesting origin from multiple sources, **Table 4**. However, two ions *i.e.* NH_4^+ and NO_3^- had good correlation, originating from similar anthropogenic sources.

3.5. Factor Analysis

The factor analysis of data has extracted six factors which explained 77.25% of the variance in the data set. The loadings of variables, eigenvalues and cumulative variance for each factor are shown in **Table 5**. Factor-1 accounted for 32.66% of the total variance with high positive loadings of Ca^{2+}, Mg^{2+}, F^- and TH. This factor suggests the role of dissolution/precipitation processes of some minerals such as CaF_2 and $CaCO_3$. Factor-2 accounted for 13.31% of representation with strong positive loading of pH which is negatively correlated with redox potential (RP). This factor suggests occurrence of redox processes which determine the acidic or alkaline nature of groundwater. Factor-3 represents 10.80% of the total variance with strong positive loadings of NH_4^+ and NO_3^-. This factor loadings shows the anthropogenic influences on these parameters. Factor-4 yielded 8.39% of the total variance with strong positive loadings of SO_4^{2-} and Na^+ suggesting mineral weathering. Factor-5 accounted for 6.32% of the total variance with high positive loading of K^+. In groundwater, K could proceed from fertilizers or weathering of K-feldspar. Factor-6 accounted for 5.66% of the total variance with a high positive value of temperature (T). The T value is a variable which controls many reactions.

3.6. Water Quality Index

The WQI of the water in the post monsoon period was ranged from 22 - 226 with mean value of 97 ± 12. The value of TDS, TA, TH, F^- and NO_3^- in the water of all tube wells was found above permissible limits of 500, 120, 300, 1.5 and 45 mg/L, respectively [35] [36]. However, in the pre monsoon period, the value of F^-, Mg^{2+} and Ca^{2+} crossed significantly the prescribed permissible limit of 1.5, 30 and 75, making water unsafe for drinking purposes.

3.7. Fluoride Toxicities

Chronic ingestion of fluoride water in endemic areas leads to development of fluorosis in the animal e.g. dental discoloration, difficulty in mastication, bony lesions, lameness, disability and mortality [37]. In lower age group, the lesion of teeth, skin, hair and nails were frequently observed. Fluoride enters the animal body mainly through the intake of water and quickly absorbed in the gastrointestinal tract. The excess F^- is excreted largely through the urine. The survey for the fluorosis in domestic animals (3 - 15 Yr) in the Dongargarh area, Rajnandgaon, Chhattisgarh, India was carried out in January 2014. A total of 40 domestic animals were screened for prevalence of various types of fluorosis *i.e.* lesion, dental, horn skin and toe fluorosis. The F^- concentration

Table 2. Chemical characteristics of groundwater in post monsoon period, 2014, mg/L.

S. No.	TDS	TA	TH	F^-	Cl^-	NO_3^-	SO_4^{2-}	NH_4^+	Na^+	K^+	Mg^{2+}	Ca^{2+}
1	1187	293	414	5.5	89	95	50	90	32	3.1	30	108
2	1474	173	327	5.2	195	71	81	64	26	3.2	31	78
3	1411	275	411	6.2	160	66	74	57	34	6.4	29	108
4	1347	128	432	6.1	188	51	41	15	37	14	32	112
5	1308	189	291	4.1	142	107	55	99	28	3.3	25	72
6	1510	165	363	3.4	231	71	37	65	22	12	29	92
7	918	171	297	3.9	85	67	28	63	26	3.2	25	74
8	694	128	111	2.5	60	112	23	94	20	15	11	26
9	933	262	225	2.1	138	53	62	18	25	3.1	19	56
10	1719	177	426	6.1	195	145	63	133	33	3.2	36	106
11	886	275	123	2.2	92	53	68	39	39	12	11	30
12	537	153	171	2.9	60	59	18	34	26	3.4	15	42
13	550	153	99	3.4	56	72	17	69	34	3.5	9	24
14	545	153	120	4.7	59	73	16	66	15	3.6	10	30
15	822	293	246	3.9	89	65	29	52	17	3.7	18	64
16	982	549	297	4.3	85	70	34	61	43	3.8	23	76
17	1876	647	513	7.4	213	58	161	11	61	3.2	41	130
18	1366	634	342	4.3	185	59	65	12	34	3.3	26	88
19	1892	610	447	7.7	220	105	95	91	77	3.1	37	112
20	1204	659	357	7.4	67	89	63	77	73	5.1	33	86
21	1216	561	516	7.3	53	52	110	10	72	3.2	46	126
22	1348	580	360	4.4	174	55	44	49	37	3.2	28	92
23	1710	512	330	3.2	337	54	56	11	22	3.4	26	84
24	825	433	348	4.9	57	51	22	48	36	3.5	28	88
25	729	348	276	2.1	82	53	25	17	23	6.3	18	74
26	968	531	291	3.8	124	53	35	11	45	21	21	76
27	1287	549	330	3.9	149	105	26	94	45	3.2	24	86
28	991	378	438	6.2	56	68	47	57	37	3.3	36	110
29	957	305	282	4.1	75	89	41	85	17	9.1	24	70
30	342	189	126	2.0	58	53	35	12	16	3.4	10	32
31	1646	512	423	5.6	160	111	33	108	26	25	35	106
32	1152	140	300	3.1	114	48	74	42	22	17	26	74
33	1140	464	450	6.1	71	98	41	88	63	3.2	36	114
34	909	397	294	3.1	85	82	12	77	27	3.3	26	72
35	596	427	213	2.7	64	52	22	13	25	3.2	17	54
36	729	366	279	6.1	75	57	41	12	65	3.5	23	70
37	1188	390	435	6.1	75	49	104	11	63	6.2	39	106
38	2598	622	654	9.1	408	88	89	80	58	12	54	164
39	1042	512	534	7.4	43	67	12	56	69	6.3	38	140
40	1014	512	321	3.1	54	145	26	132	52	12	27	80
41	1200	427	315	4.1	107	99	26	86	31	3.7	27	78
42	722	458	285	5.6	71	53	22	10	46	3.1	25	70
43	1137	616	252	6.1	53	152	16	144	65	3.2	20	64
44	1525	470	687	10.3	138	65	48	47	85	3.4	57	172
45	657	299	348	5.1	57	53	15	17	58	3.4	26	90
46	1199	360	342	4.5	124	81	52	72	42	3.5	28	86
47	981	439	555	8.1	78	51	17	15	73	3.2	45	140
48	1638	500	492	5.9	234	53	74	13	53	6.2	42	122

Table 3. Characteristics of groundwater in pre monsoon period, May 2014.

S. No.	pH	EC	F^-	Cl^-	NO_3^-	SO_4^{2-}	NH_4^+	Na^+	K^+	Mg^{2+}	Ca^{2+}
							mg/L				
1	7.7	1064	6.8	71	58	58	58	41	2.9	36	126
2	6.4	1302	6.4	156	43	94	41	33	2.9	37	91
3	8.5	1238	7.6	128	40	86	36	43	5.9	35	126
4	6.5	1268	7.5	150	31	48	10	47	12.9	38	131
5	6.8	1297	5.0	114	65	64	63	36	3.0	30	84
6	6.7	1458	4.2	185	43	43	42	28	11.0	35	108
7	6.9	772	4.8	68	41	32	40	33	2.9	30	87
8	6.8	440	3.1	48	68	27	60	25	13.8	13	30
9	6.3	780	2.6	110	32	72	12	32	2.9	23	66
10	6.4	1257	7.5	156	88	73	85	42	2.9	43	124
11	8.4	938	2.7	74	32	79	25	50	11.0	13	35
12	6.8	493	3.6	48	36	21	22	33	3.1	18	49
13	6.6	353	4.2	45	44	20	44	43	3.2	11	28
14	7.2	447	5.8	47	45	19	42	19	3.3	12	35
15	7.7	785	4.8	71	40	34	33	22	3.4	22	75
16	8.0	1137	5.3	68	43	39	39	55	3.5	28	89
17	7.7	1857	9.1	170	35	187	7	77	2.9	49	152
18	7.7	1701	5.3	148	36	75	8	43	3.0	31	103
19	7.6	1704	9.5	176	64	110	58	98	2.9	44	131
20	7.8	1240	9.1	54	54	73	49	93	4.7	40	101
21	7.5	1508	9.0	42	32	128	6	91	2.9	55	147
22	7.6	1359	5.4	139	34	51	31	47	2.9	34	108
23	7.6	1217	3.9	270	33	65	7	28	3.1	31	98
24	7.8	858	6.0	46	31	26	31	46	3.2	34	103
25	8.3	1322	2.6	66	32	29	11	29	5.8	22	87
26	8.1	1098	4.7	99	32	41	7	57	19.3	25	89
27	8.0	1298	4.8	119	64	30	60	57	2.9	29	101
28	7.5	1535	7.6	45	41	55	36	47	3.0	43	129
29	7.6	815	5.0	60	54	48	54	22	8.4	29	82
30	7.6	412	2.5	46	32	41	8	20	3.1	12	37
31	7.5	1539	6.9	128	68	38	69	33	23.0	42	124
32	7.6	1046	3.8	91	29	86	27	28	15.6	31	87
33	7.6	879	7.5	57	60	48	56	80	2.9	43	133
34	7.7	941	3.8	68	50	14	49	34	3.0	31	84
35	7.7	690	3.3	51	32	26	8	32	2.9	20	63
36	7.6	800	7.5	60	35	48	8	83	3.2	28	82
37	7.8	2481	7.5	60	30	121	7	80	5.7	47	124
38	7.8	283	11.2	326	54	103	51	74	11.0	65	192
39	8.0	797	9.1	34	41	14	36	88	5.8	46	164
40	7.9	905	3.8	43	88	30	84	66	11.0	32	94
41	8.2	1348	5.0	86	60	30	55	39	3.4	32	91
42	8.2	1353	6.9	57	32	26	6	58	2.9	30	82
43	8.1	1146	7.5	42	93	19	92	83	2.9	24	75
44	8.0	1353	12.7	110	40	56	30	108	3.1	68	201
45	7.6	774	6.3	46	32	17	11	74	3.1	31	105
46	8.2	1005	5.5	99	49	60	46	53	3.2	34	101
47	7.7	872	10.0	62	31	20	10	93	2.9	54	164
48	8.1	1757	7.3	187	32	86	8	67	5.7	50	143

Table 4. Correlation matrix of ions.

	F^-	Cl^-	NO_3^-	SO_4^{2-}	NH_4^+	Na^+	K^+	Mg^{2+}	Ca^{2+}
F^-	1								
Cl^-	0.24	1							
NO_3^-	0.08	0.01	1						
SO_4^{2-}	0.38	**0.49**	−0.14	1					
NH_4^+	0.05	−0.02	**0.94**	−0.20	1				
Na^+	**0.78**	−0.01	0.06	0.29	−0.04	1			
K^+	−0.13	0.16	0.09	−0.01	0.11	−0.17	1		
Mg^{2+}	**0.85**	0.42	0.02	**0.48**	0.00	**0.66**	−0.02	1	
Ca^{2+}	**0.83**	0.43	−0.01	**0.41**	−0.03	**0.64**	0.00	**0.97**	1

Table 5. Eigenvalues and factor loadings of Dongargarh groundwater samples.

Variable	Factor-1	Factor-2	Factor-3	Factor-4	Factor-5	Factor-6
Age	−0.18	0.07	−0.06	−0.15	−0.45	−0.66
Depth	−0.05	−0.13	0.19	0.45	0.30	−0.05
T	−0.09	0.17	−0.06	0.00	−0.25	**0.82**
pH	0.07	**0.94**	−0.04	−0.01	0.01	0.00
EC	0.24	0.17	−0.10	0.69	−0.16	0.11
RP	−0.07	**−0.85**	0.12	0.11	−0.09	−0.08
DO	−0.11	0.06	0.05	−0.51	−0.24	0.26
TDS	0.61	−0.04	0.21	0.65	0.23	0.09
TA	0.36	0.66	0.10	0.37	−0.18	0.09
TH	**0.95**	0.12	−0.02	0.25	0.03	−0.05
F^-	**0.90**	0.11	0.06	0.08	−0.11	0.12
Cl^-	0.36	−0.18	−0.04	0.60	0.32	0.12
NO_3^-	0.03	−0.03	**0.98**	−0.02	0.03	−0.02
SO_4^{2-}	0.30	−0.03	−0.24	**0.73**	−0.02	0.06
NH_4^+	0.02	−0.07	**0.98**	−0.03	0.08	0.00
Na^+	0.22	0.36	0.21	**0.71**	−0.21	0.16
K^+	−0.05	0.07	0.06	0.04	**0.82**	−0.06
Ca^{2+}	**0.94**	0.14	−0.03	0.24	0.04	−0.06
Mg^{2+}	**0.94**	0.07	−0.01	0.26	−0.02	−0.02
Eigenvalue	6.21	2.53	2.05	1.59	1.20	1.10
% Total variance	32.66	13.31	10.80	8.39	6.32	5.77
Cumulative %	32.66	45.97	56.77	65.16	71.48	77.25

Significant factor loading in bold >0.7.

in their urine samples were measured and presented in **Table 6**. The concentration of F^- in the buffalo and cattle urines was ranged from 18 - 52 and 26 - 58 mg/L with mean value of 31 ± 4 and 41 ± 4 mg/L, respectively. At least 7 - 10 folds higher F^- content in the urine of animals was marked, may be due intake of higher dose of F^-contaminated water and food. The higher fluorosis prevalence rate was observed in the cattle than buffalo, **Figures 4-6**.

Table 6. Fluoride exposure in animals.

S. No.	Animal	Color	Age, Yr	F⁻, mg/L	Fluorosis
1	Buffalo	B	6	32	DF
2	Buffalo	B	5	35	DF
3	Buffalo	B	13	29	DF
4	Buffalo	Br	8	40	DF
5	Buffalo	B	14	37	DF
6	Buffalo	B	5	35	DF
7	Buffalo	B	5	18	DF
8	Buffalo	B	7	29	DF
9	Buffalo	B	5	22	DF
10	Buffalo	B	8	45	SK
11	Buffalo	B	5	29	SK
12	Buffalo	B	5	21	DF
13	Buffalo	B	6	19	DF
14	Buffalo	B	5	22	DF
15	Buffalo	B	9	52	TF
16	Buffalo	B	7	19	TF
17	Buffalo	B	7	33	TF
18	Buffalo	B	6	35	TF
19	Buffalo	B	8	27	TF
20	Buffalo	B	15	31	TF
21	Cattle	W	10	58	SF
22	Cattle	W	8	40	SF
23	Cattle	Br	8	39	SF
24	Cattle	R	7	34	SF
25	Cattle	Gray	5	32	SF
26	Cattle	WG	6	36	HF
27	Cattle	WG	11	56	HF
28	Cattle	G	8	42	HF
29	Cattle	BW	8	39	HF
30	Cattle	WG	8	56	HF
31	Cattle	WG	9	46	HF
32	Cattle	WG	5	37	HF
33	Cattle	W	8	58	HF
34	Cattle	W	7	37	HF
35	Cattle	BW	6	32	L
36	Cattle	BW	5	26	L
37	Cattle	W	5	34	L
38	Cattle	W	6	29	L
39	Cattle	Br	7	39	L
40	Cattle	Br	9	47	L

DF = Dental fluorosis, SF = Skin fluorosis, TF = Toe fluorosis, HF = Horn fluorosis, L = Lesion.

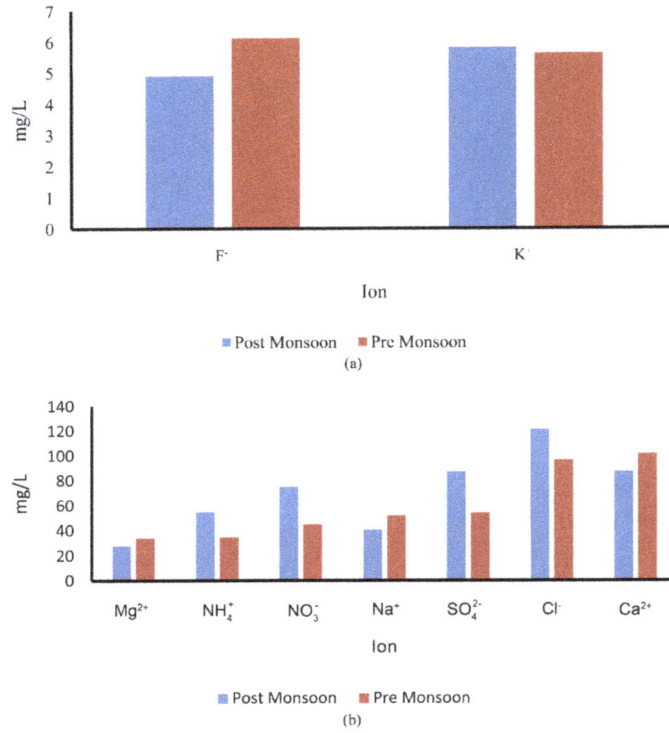

Figure 2. Seasonal variation of ionic concentration in post (a) and pre (b) monsoon period, 2014.

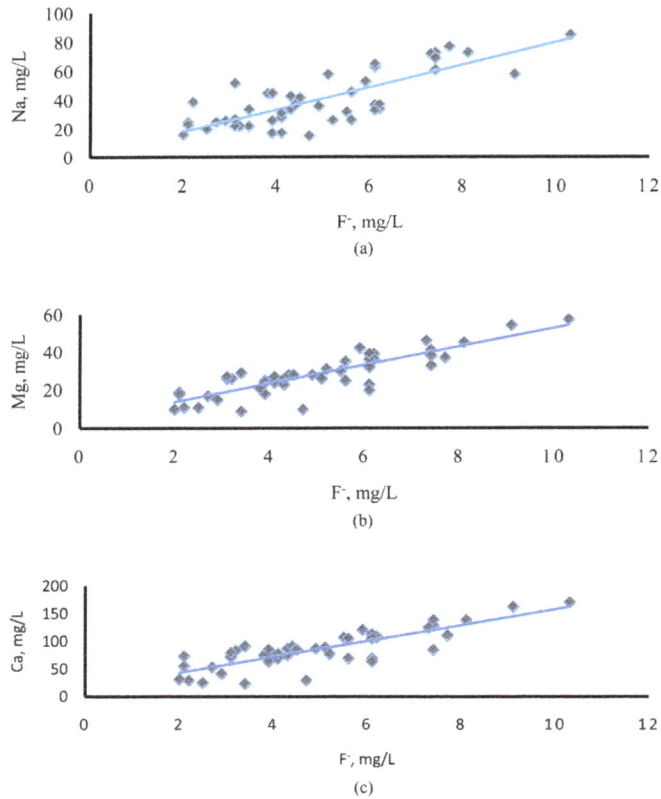

Figure 3. Correlation of F^- with Na^+ (a), Mg^{2+} (b) and Ca^{2+} (c).

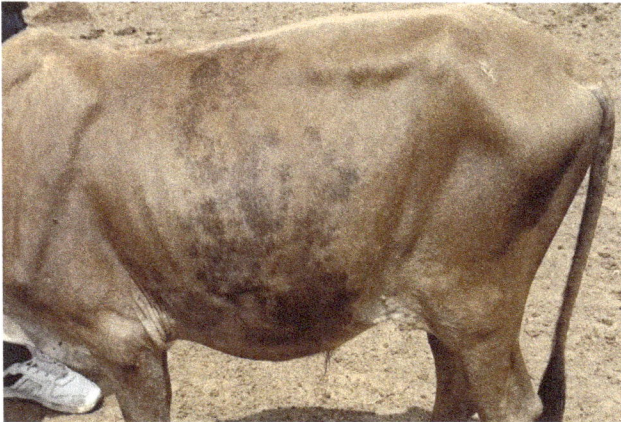

Figure 4. Skin lesson in cattle.

Figure 5. Horn fluorosis in cattle.

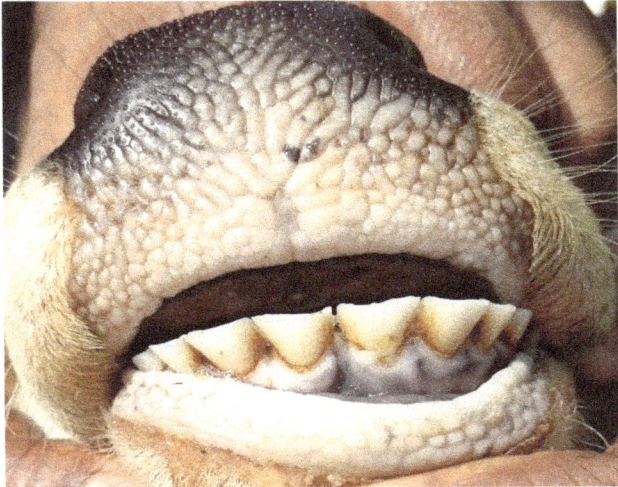

Figure 6. Dental fluorosis in buffalo.

4. Conclusion

The groundwater of Dongargarh is contaminated with F⁻ at dangerous levels due to mineralization of the bed rock F⁻ in the water. The WQI index of water was found to be ≈100, making water unsafe for dinking purposes. The F⁻ levels in urine of cattle and buffalo were found several folds higher than recommended value of 4 mg/L. Around 5% domestic animals of the studied were suffered with different types of fluorosis.

Acknowledgements

We are thankful to the UGC, New Delhi for awarding Rajiv Gandhi Fellowship to one of the author: NSD.

References

[1]   Saxena, V.K. and Ahmad, S. (2001) Dissolution of Fluoride in Groundwater: A Water-Rock Interaction Study. *Environmental Geology*, **40**, 1084-1087. http://dx.doi.org/10.1007/s002540100290

[2]   Datta, A.S., Chakrabortty, A., De Dalal, S.S. and Lahiri, S.C. (2014) Fluoride Contamination of Underground Water in West Bengal, India. *Fluoride*, **47**, 241-248.

[3]   Brindha, K., Rajesh, R., Murugan, R. and Elango, L. (2011) Fluoride Contamination in Groundwater in Parts of Nalgonda District, Andhra Pradesh, India. *Environmental Monitoring and Assessment*, **172**, 481-492. http://dx.doi.org/10.1007/s10661-010-1348-0

[4]   Arveti, N., Sarma, M.R., Aitkenhead-Peterson, J.A. and Sunil, K. (2011) Fluoride Incidence in Groundwater: A Case Study from Talupula, Andhra Pradesh, India. *Environmental Monitoring and Assessment*, **172**, 427-443. http://dx.doi.org/10.1007/s10661-010-1345-3

[5]   Reddy, A.G., Reddy, D.V., Rao, P.N. and Prasad, K.M. (2010) Hydrogeochemical Characterization of Fluoride Rich Groundwater of Wailpalli Watershed, Nalgonda District, Andhra Pradesh, India. *Environmental Monitoring and Assessment*, **171**, 561-577. http://dx.doi.org/10.1007/s10661-009-1300-3

[6]   Rao, N.S. (2009) Fluoride in Groundwater, Varaha River Basin, Visakhapatnam District, Andhra Pradesh, India. *Environmental Monitoring and Assessment*, **152**, 47-60. http://dx.doi.org/10.1007/s10661-008-0295-5

[7]   Suthar, S., Garg, V.K., Jangir, S., Kaur, S., Goswami, N. and Singh, S. (2008) Fluoride Contamination in Drinking Water in Rural Habitations of Northern Rajasthan, India. *Environmental Monitoring and Assessment*, **145**, 1-6. http://dx.doi.org/10.1007/s10661-007-0011-x

[8]   Vikas, C. (2009) Occurence and Distribution of Fluoride in Groundwaters of Central Rajasthan, India. *Journal of Environmental Science and Engineering*, **51**, 169-174.

[9]   Rao, N.S. (2011) High-Fluoride Groundwater. *Environmental Monitoring and Assessment*, **176**, 637-645. http://dx.doi.org/10.1007/s10661-010-1609-y

[10]   Hussain, I., Arif, M. and Hussain, J. (2012) Fluoride Contamination in Drinking Water in Rural Habitations of Central Rajasthan, India. *Environmental Monitoring and Assessment*, **184**, 5151-5158. http://dx.doi.org/10.1007/s10661-011-2329-7

[11]   Kundu, M.C. and Mandal, B. (2009) Assessment of Potential Hazards of Fluoride Contamination in Drinking Groundwater of an Intensively Cultivated District in West Bengal, India. *Environmental Monitoring and Assessment*, **152**, 97-103. http://dx.doi.org/10.1007/s10661-008-0299-1

[12]   Jha, S.K., Nayak, A.K. and Sharma, Y.K. (2010) Potential Fluoride Contamination in the Drinking Water of Marks Nagar, Unnao District, Uttar Pradesh, India. *Environmental Geochemistry and Health*, **32**, 217-226. http://dx.doi.org/10.1007/s10653-009-9277-y

[13]   Srikanth, R., Gautam, A., Jaiswal, S.C. and Singh, P. (2013) Urinary Fluoride as a Monitoring Tool for Assessing Successful Intervention in the Provision of Safe Drinking Water Supply in Five Fluoride-Affected Villages in Dhar District, Madhya Pradesh, India. *Environmental Monitoring and Assessment*, **185**, 2343-2350. http://dx.doi.org/10.1007/s10661-012-2713-y

[14]   Avishek, K., Pathak, G., Nathawat, M.S., Jha, U. and Kumari, N. (2010) Water Quality Assessment of Majhiaon Block of Garwa District in Jharkhand with Special Focus on Fluoride Analysis. *Environmental Monitoring and Assessment*, **167**, 617-623. http://dx.doi.org/10.1007/s10661-009-1077-4

[15]   Singh, C.K. and Mukherjee, S. (2015) Aqueous Geochemistry of Fluoride Enriched Groundwater in Arid Part of Western India. *Environmental Science and Pollution Researches International*, **22**, 2668-2678. http://dx.doi.org/10.1007/s11356-014-3504-5

[16]   Varadarajan, N. and Purandara, B.K. (2008) Fluoride Contamination in Ground Water of Malaprabha Sub Basin. *Journal of Environmental Science and Engineering*, **50**, 121-126.

[17] Sajil Kumar, P.J. (2012) Assessment of Fluoride Contamination in Groundwater as Precursor for Electrocoagulation. *Bulletin of Environmental Contamination and Toxicology*, **89**, 172-175. http://dx.doi.org/10.1007/s00128-012-0638-3

[18] Vijay, R., Khobragade, P. and Mohapatra, P.K. (2011) Assessment of Groundwater Quality in Puri City, India: An Impact of Anthropogenic Activities. *Environmental Monitoring and Assessment*, **177**, 409-418. http://dx.doi.org/10.1007/s10661-010-1643-9

[19] Banerjee, A. (2015) Groundwater Fluoride Contamination: A Reappraisal. *Geoscience Frontiers*, **6**, 277-284. http://dx.doi.org/10.1016/j.gsf.2014.03.003

[20] Patel, K.S., Sharma, R., Sahu, B.L., Patel, R.K. and Matini, L. (2014) Ground Water Quality of Rajnandgaon City. *Asian Journal of Water, Environment and Pollution*, **11**, 31-37.

[21] Dutta, R.K., Saikia, G., Das, B., Bezbaruah, C., Das, H.B. and Dube, S.N. (2006) Fluoride Contamination in Groundwater of Central Assam India. *Asian Journal of Water, Environment and Pollution*, **3**, 93-100.

[22] Arif, M., Husain, I., Hussain, J. and Kumar, S. (2013) Assessment of Fluoride Level in Groundwater and Prevalence of Dental Fluorosis in Didwana Block of Nagaur District, Central Rajasthan, India. *International Journal of Occupational and Environmental Medicine*, **4**, 178-184.

[23] Yadav, J.P., Lata, S., Kataria, S.K. and Kumar, S. (2009) Fluoride Distribution in Groundwater and Survey of Dental Fluorosis among School Children in the Villages of the Jhajjar District of Haryana, India. *Environmental Geochemistry and Health*, **31**, 431-438. http://dx.doi.org/10.1007/s10653-008-9196-3

[24] Nayak, B., Roy, M.M., Das, B., Pal, A., Sengupta, M.K., De, S.P. and Chakraborti, D. (2009) Health Effects of Groundwater Fluoride Contamination. *Clinical Toxicology (Philadelphia)*, **47**, 292-295. http://dx.doi.org/10.1080/15563650802660349

[25] Suthar, S. (2011) Contaminated Drinking Water and Rural Health Perspectives in Rajasthan, India: An Overview of Recent Case Studies. *Environmental Monitoring and Assessment*, **173**, 837-849. http://dx.doi.org/10.1007/s10661-010-1427-2

[26] Nayak, B., Mohan Roy, M., Das, B., Pal, A., Sengupta, M.K., De, S.P. and Chakraborti, D. (2009) Health Effects of Groundwater Fluoride Contamination. *Clinical Toxicology*, **47**, 292-295. http://dx.doi.org/10.1080/15563650802660349

[27] Kotecha, P.V., Patel, S.V., Bhalani, K.D., Shah, D., Shah, V.S. and Mehta, K.G. (2012) Prevalence of Dental Fluorosis & Dental Caries in Association with High Levels of Drinking Water Fluoride Content in a District of Gujarat. *The Indian Journal of Medical Research*, **135**, 873-877.

[28] Marya, C.M., Ashokkumar, B.R., Dhingra, S., Dahiya, V. and Gupta, A. (2014) Exposure to High-Fluoride Drinking Water and Risk of Dental Caries and Dental Fluorosis in Haryana, India. *Asia-Pacific Journal of Public Health*, **26**, 295-303. http://dx.doi.org/10.1177/1010539512460270

[29] Panday, A. (2010) Prevalence of Fluorisis in an Endemic Village in Central India. *Tropical Doctor*, **40**, 217-219. http://dx.doi.org/10.1258/td.2010.100032

[30] Viswanathan, G., Gopalakrishnan, S. and Siva Ilango, S. (2010) Assessment of Water Contribution on Total Fluoride Intake of Various Age Groups of People in Fluoride Endemic and Non-Endemic Areas of Dindigul District, Tamil Nadu, South India. *Water Research*, **44**, 6186-6200. http://dx.doi.org/10.1016/j.watres.2010.07.041

[31] Viswanathan, G., Jaswanth, A., Gopalakrishnan, S., Siva Ilango, S. and Aditya, G. (2009) Determining the Optimal Fluoride Concentration in Drinking Water for Fluoride Endemic Regions in South India. *Science of the Total Environment*, **407**, 5298-5307. http://dx.doi.org/10.1016/j.scitotenv.2009.06.028

[32] Bhattacharya, S., Majumdar, T.J., Rajawat, A.S., Panigrahi, M.K. and Das, P.R. (2012) Utilization of Hyperion Data over Dongargarh, India, for Mapping Altered/Weathered and Clay Minerals along with Field Spectral Measurements. *International Journal of Remote Sensing*, **33**, 5438-5450. http://dx.doi.org/10.1080/01431161.2012.661094

[33] Nielson, D.M. and Nielson, G. (2006) The Essential Handbook of Groundwater Sampling. CRC Press, Boca Raton. http://dx.doi.org/10.1201/9781420042795

[34] Nollet, L.M.L. and De Gelder, L.S.P. (2007) Handbook of Water Analysis. 2nd Edition, CRC Press, Boca Raton.

[35] BIS (2009) Drinking Water—Specification. 2nd Edition, Bureau of Indian Standards, New Delhi. http://bis.org.in/sf/fad/FAD25(2047)C.pdf

[36] WHO (2011) Guidelines for Drinking-Water Quality. 4th Edition, World Health Organization, Geneva. http://www.haceclick.com.uy/documentos/Guia_OMS%202011_4aEd.pdf

[37] Patra, R.C., Dwivedi, S.K., Bhardwaj, B. and Swarup, D. (2000) Industrial Fluorosis in Cattle and Buffalo around Udaipur, India. *Science of the Total Environment*, **253**, 145-150. http://dx.doi.org/10.1016/S0048-9697(00)00426-5

6

Farmers' Ability to Pay for Irrigation Water in the Jordan Valley

Mohammad Tabieh*, Emad Al-Karablieh, Amer Salman, Hussein Al-Qudah, Ahmad Al-Rimawi, Tala Qtaishat

Department of Agricultural Economic and Agribusiness, Faculty of Agriculture, The University of Jordan, Amman, Jordan
Email: *m.tabieh@ju.edu.jo

Abstract

This study aims to analyze the famers' ability to pay for irrigation water in Jordan Valley to investigate the farmer capacity to cope with increasing the water tariff. The residual imputation approach was used based on the enterprise budget for crops cultivated in different geographical locations in Jordan Valley. This methodology deducts the contribution of non-water production inputs, annualized capital cost and fixed costs from the gross output and attributes the remaining value to water. The resulting water value is an indication of the economic efficiency of water and a proxy for the maximum farmer's ability to pay for water. The result shows that cucumber has the highest ability to pay (JD 2.26 m^{-3}); the percentage of water cost to total cost is 1.1%. This low percent does not encourage farmers to save water. The weighted average for maximum farmers' ability to pay for irrigation water in Jordan valley is estimated at JD 0.76 m^{-3}. The result shows the farmer's ability to pay for water used in plastic house is JD 1.34 m^{-3} compared to JD 0.62 m^{-3} for open field. The estimated value of desalinated brackish water is JD 0.59 m^{-3} while the average desalination cost is JD 0.28 m^{-3}. Therefore, the current practice of installing Reverse Osmosis units to irrigate high value cash crops by some farmers is economically rational, since water value is twice the desalination cost of one cubic meter. If farmers have to pay the cost of O&M, they need to pay at least JD 0.065 m^{-3}. Increasing the water prices could encourage more efficient water use, shifts to higher value crops, adoption of plastic houses, and encourage desalination of brackish water.

Keywords

Water Tariff, Water Value, Water Cost, Residual Imputation Approach, Water Resources

---

*Corresponding author.

1. Introduction

In a country facing such a significant imbalance between limited supplies and ever-growing demand, the government must grapple with very difficult policy decisions and trade-offs in order to determine the best ways to allocate water across sectors. This is especially true for the agricultural and industrial sectors which consume significant portions of the national water supply and are central to the Jordanian economy. A critical component of improved resources management is a more informed policy setting process. Jordan's water sector is struggling to keep up with rapid population growth and economic growth. Jordan is one of the most water scarce countries in the world [1], with very limited quantities of renewable water and high costs for providing water to people and businesses. Sustaining public projects for irrigating agricultural activities requires expenditures in a form of capital investment, operations and maintenance, and periodic rehabilitation. Government has historically been the primary sources for financing these projects, but farmer beneficiaries are increasingly expected to contribute substantially toward recovering full costs. Estimates of farmers' ability to pay (ATP), or repayment capacity, for part or all of the costs of irrigation water supply facilities are useful in deciding on how much of these costs farmers can and should pay for irrigation water. Willingness to pay depends at once on individuals' level of income and their perception of risk: the greater a person's aversion to risk, the more he or she will be willing to pay. An individual's ability to pay is the maximum amount that he or she is capable of paying; it is therefore linked to income. Ability to pay is always greater than or equal to willingness to pay, even for persons with a strong aversion to risk. In a context of poverty, however, the levels of ability to pay and willingness to pay are both very low and tend to be indistinguishable from one another [2]. The average tariff billed per cubic meter of water billed to irrigation in 2010 range from JD 0.008 to 0.016 m^{-3}, with an average of JD 0.012 m^{-3}. Based on billed water volume, the average operation and maintenances cost per cubic meter billed are about JD 0.12 per m^3 (0.17 US$ m^{-3}) in the three years (2008-2010), while the average revenue of irrigation water in Jordan Valley (JV) is only JD 0.03 per m^3. Jordan Valley Authority (JVA) is not able to cover its basic operating costs; its revenues fall far short. The decline of JVA's capacity to pay for its operating expenditures has been especially pronounced since 2008. The operating margin is highly negative and shows that currently the total revenues including pumping revenues which were not charged to Water Authority of Jordan (WAJ) do not even cover staff costs. The operating cost coverage ratio is less than 30 percent for all purposes of water use and only 10% of irrigation water [3]. Water in Jordan Valley is charged according to the principle of price discrimination and quota system. In 2004, the JVA revised the quotas system for better supply of water and crop water requirements [4]. The new quotas correspond to 360, 765 and 1255 m^3/du for vegetables, citrus and bananas, respectively, *i.e.*, a cut by about 20 to 25 percent. At a regional scale, this generated total freshwater savings in the northern and middle directorates of approximately 20 mcm. The water saved was subsequently reallocated to domestic use in Amman with about 53 mcm in 2010. In sight the value of water is essential to support policy decision making about 1) investments in water supply system, 2) investments in the water distribution system and the irrigation system, 3) efficient allocation of water with competing sectors, 4) setting water pricing and tariffs, 5) setting cost recovery (O&M and capital recovery) mechanisms, and 6) determining the socio-economic impacts of water management decisions [1]. However, the decision-makers are thus torn between pressures to meet water authorities' demands for expansion and maintenance, and public pressure to restrict water prices, particularly for poor people. Pressures from donor communities to adopt full water cost recovery aggravate the situation. Water pricing is the most important measure in establishing effective demand management to use water efficiently and sustainably. Appropriate and adequate operation and maintenance of water systems are necessary to enable them to meet the current and future requirements for distributing water. Decision makers often lack information on unit value of water to agricultural revenue in different agricultural activities and their relative economic contributions to the local and regional economies. As a result, they cannot adequately assess potential trade-offs amongst different agricultural users under different management schemes. The question that this paper seeks to explore is how much a farmer would pay for water and at what price farmers plan to cultivate the area they have under irrigation. Regardless of the reason for reforming water policies, knowledge of the value of water is essential for efficient allocations of water and when crafting policies to compare the variable impacts of water reform within and across sectors of the economy. However, a major difficulty that policy makers and water resources managers face is over accurately determining net economic value of irrigation water due to a number of economic, political, and physical complexities. Economic gains (or net values of irrigation water) of some agricultural activities are much higher than others but these activities may require major initial investment and take years before economic gains are achieved. Also, the issue of large sunk costs in on-farm infrastructure impacts on the net economic value of irrigation water. Such investment costs have not been incorporated in the previous ISSP water valuation study [5].

2. Materials and Methods

2.1. Review of Literature

The idea of water as an economic good is simple. Water has a value to users who are able to pay for it. Like other goods, consumers will use water as long as the benefit derived from the use of an additional cubic meter exceed the costs so incurred, *i.e.* until the marginal value product of water equals its price. The capacity to pay for water depends directly on farmers' incomes, especially from crop enterprises. According to the economic theory, a farmer will be able and willing to apply water to the crops as long as it generates more income than its per unit cost. If the marginal income from a crop by application of an additional unit of irrigation water is less than the water charge for that crop, the farmer is not able to pay for water. The net income criterion generally serves as a good approximation of farmers' ability to pay for water charges. Net farm income, as a measure to assess the average paying capacity of the water users [6]. Establishing economic value for water is considered to be one of the most discussed and debated issues related to economic efficiency of water use and its allocation [7] [8]. Young stated that: "water valuation presents the economic analyst with a wide range of challenging issues and problems, because water values tend to be quite site-specific, spatial, and temporal, and each case confronts its own unique issues and typically requires its own original valuation" [9]. Effective measurement of water values demands skill and rigor in application of all the tools of the applied economist's trade. These tools include data collection, statistical analysis, optimization models, and research reporting. However, researchers have employed many methods for assessing the value of irrigation water. These methods have been classified into two major groups, namely "inductive techniques" and "deductive techniques" [9]. The inductive techniques (based on observation in market) for valuing irrigation water differ mainly according to the type and source of data and the form of statistical model, if any, used to estimate the productivity relationship. Most commonly used inductive techniques include: (a) direct observations on water entitlement markets, (b) land value method by imputing value of water via land and implementing valuation from land market data, (c) hedonic property (or revealed preference) value method; and (d) econometric valuation of irrigation water from primary and secondary data including stated preference techniques [1] [5]. Ability to pay (ATP) and willingness to pay (WTP) are economic concepts, which aims to determine the amount of money a consumer is able or willing to pay for the supply of additional unit of goods and services and it is water in our case. The consumers' ATP and WTP are becoming increasingly popular and are one of the standard approaches that are used by market researchers and economists to place a value on goods or services for which no market-based pricing mechanism exists [10]. Literature suggests that two approaches are being used to analyses the consumers' ATP/WTP. The direct approach, involves taking a survey through a structured questionnaire of consumers' ATP/WTP specified prices for hypothetical services, also referred as contingent valuation method (CVM). The direct approach used in CVM has been to directly ask survey respondents to state their exact maximum ATP/WTP for the particular use or non-use value of the water. The ATP/WTP is defined as the amount that can be taken away from person's income while keeping his utility constant [11]. The CVM still have serious methodological and theoretical shortcomings when used to assess ATP/WTP for non-market based goods and services, such as format bias, embedding effect, ordering problem, starting bid effects, strategic bias, information bias, non response bias, payment vehicle, free rider problem, warm glow effect [12]-[14]. However, CVM is still useful tool for water resource management in developing countries. The price of water in a water market should reflect water's economic value. Because water is usually supplied by public agencies who price water at its average financial delivery cost rather than its value to producers, water is rarely priced at its marginal economic value [9]. Water can be valued from a supply (*i.e.* depending on the cost of water provision) or demand perspective (value added due to water use in productive activities), resulting in a supply curve or a demand curve. When water is an input to a production process (an "intermediate good"), such as in irrigated agriculture or in industrial use, water demand is derived from the demand for the final output and from water's role in producing this output; thus it is a derived demand function. In this case, water demand is a function of the price of water and the price of the final product produced. Estimating water's economic value is equivalent to isolating the marginal contribution of water to the total output value [9] [15]. In general, the most scientifically accepted methods are those based on actual market behavior and information [16] [17]. In the case of Jordan, since farmers in the Jordan Valley are paying for water—a neglected portion of production costs; it is difficult to establish a relationship between price and demand from actual behavior to generate demand functions. Moreover, the fact that water is provided by the government with heavy subsidies, strategic biases or simply the belief among farmers that water is a free gift from God [18], could probably lead to erroneous estimations of water values when using direct methods such as contingent valuation [12]. Therefore, following Lange [19] and Speelman [20], the Residual Imputation Method (RIM) was used in this study. Although this method clearly has its shortcomings, it was considered the most suitable technique to estimate water values for the studied irrigation schemes [21].

2.2. Objectives of the Study

This study aims to analyze the famers' ability to pay for irrigation water in Jordan Valley from farmers' perspective that takes into account the full economic costs of production using RIM approach. The previous water valuation study conducted in Jordan provide a short-run disaggregated water value for fruits, vegetable, and field crops at the regional and country level. They take into consideration the water value added—according to different agro-climatological zones and the type of water used in the production process such as groundwater versus surface or blended waste water with surface water [22]. The major distinction in this study is in the estimation of water value through farmers' ability to pay for water in the long run where all fixed costs are variable costs in the long runs. The growing water scarcity causes increasing pressure on farmers to allocate water more efficiently. The JVA does not cover the O&M cost. The average tariff billed per cubic meter of water billed to irrigation in 2010 is JD 0.012 per m^3. Based on billed water volume, the average operation and maintenances cost per cubic meter billed are about are JD 0.12 per m^3 in the last three years. JVA is not able to cover its basic operating costs; its revenues fall far short. The operating cost coverage ratio is less than 30 percent for all purpose of water use and only 10 percent of irrigation water [3]. Therefore, to formulate a new water policy, water subsidies currently received by farmers will gradually decrease and become negative, $i.e.$ in the near future; farmers will have to pay for the water they use. This paper will try to estimate the economic value of water in agriculture by producing well-differentiated estimates of Farmers' Ability to Pay (FAP) for irrigation water in Jordan Valley.

2.3. Background to Economy and Agriculture in Jordan Valley

The agriculture sector is a major consumer of water, and the returns to water from crop production tend to be low in comparison to other sectors. The horticulture is becoming the main source of agriculture GDP. The horticulture sector can be divided into highland and Jordan Valley. Around 61 percent of Jordanian agriculture GDP is generated from Jordan Valley. About 19 percent and 17 percent of agricultural GDP is generated from MJV and NJV, respectively [23]. The irrigated agriculture is mainly based on surface and marginal water resources in the Jordan valley, where the highlands the irrigation is based on groundwater resources. The importance of the agricultural sector in MJV stems from the fact that it is the major source fruits and vegetables during winter season (35% of value added in winter vegetables), and also one of the sources of hard currencies originated from exports. One can use a rough estimate of value added of irrigation water by dividing the total value added (agriculture GDP) in each geographical location in JV by water consumed in each location. The NJV yielded the highest gross water value added with JD 1.4 per cubic meter [24]. The Jordan Rift Valley (JRV) is a low-lying strip that extends along Jordan's west border from northern Jordan near Lake Taiberia at an elevation of about 212 m b.s.l. to southern Jordan near Aqaba. The part of the JRV covered by this paper is that extends from northern Jordan to near the Dead Sea where elevation drops to about 420 m b.s.l., the lowest point on earth. The study area experiences a sharp gradient in rainfall from north to south. Average annual rainfall in the Northern Jordan Valley is about 377 mm and 77 mm in southern Shouneh [25]. The prevailing subtropical climate in the JRV and fertile soil allows for year around cultivation especially vegetables in winter. About 70% of Jordan's production of fruit and vegetables is from the Jordan valley which makes the valley Jordan's food basket. Total irrigable area in the Jordan Valley is about 363,000 dunum. The main water demand in the Jordan Valley is agricultural demand. Irrigation water demand in the Jordan Valley is distributed from north to south among five demand zones which are the Northern Jordan Valley (NJV), the Middle Jordan Valley (MJV), the Southern Jordan Valley (SJV), the extension project in the Southern Dead Sea (Safi) and recently Wadi Araba (**Table 1**). Irrigation water demand in the Jordan Valley is estimated at about 320 million cubic meter (mcm) per year based

Table 1. Distribution of farm units and irrigated areas by source of irrigation water in Jordan Valley.

Zone	Stage office	Development area (DA)	No. of farm units	Irrigated area (du)	Water source	Project name
Northern Jordan Valley (NJV), North Directorate	1	1, 2, 6, 7	365	12,892	KAC, Yarmouk	North Ghor Conversion Project
	2	11, 12	365	13,886	KAC, Yarmouk	North Ghor Conversion Project
	2	12, 13, 14, 15, 16	584	20,477	KAC, Yarmouk, KTR (ZCIII)	North Ghor Conversion Project
	7	3, 4, 5	379	12,525	KAC, Wadi Arab	Wadi Arab Project
	7	8, 9, 10, 33, 34, 35, 36, 37, 38, 39	756	28,791	KAC, Dam, Zeglab Dam, Al-Gorom Dam	North East Ghor Project
Sub Total			**2449**	**88,571**		
Middle Jordan Valley (MJV), Deir Alla Directorate	3	18, 19, 20, 21	498	20,339	KAC, KTR (ZCIII)	North Ghor Conversion Project
	4	23	683	25,623	KTR (ZCII), KAC	Middle Ghor Conversion Project
	5	24, 25, 30, 54	696	26,582	KTR (ZCII), KAC	Middle Ghor Conversion Project/ZorSheshaa
	8	22, 53	383	14,212	KTR (ZCII), KAC	Middle Ghor Conversion Project/Abo Zegan
Sub Total			**2260**	**86,756**		
Southern Jordan Valley (SJV), Karameh Directorate	8	29	337	11,454	KTR (ZCI)	Zarka Triangle Project
	6	26, 27, 28	1,026	39,794	KTR, KAC	18 km extension Project
	9	49, 50, 51, 52	1,382	56,600	KTR, KAC, Shueib	14.5 km Project
	10	31, 32	458	15,580	Kafrien Dam, Mujeb	Hisban-Kafrain Project
Sub-Total			**3203**	**123,429**		
Safi, Southern Ghor Directorate	11	40, 41, 42, 43, 44, 45, 46, 47, 48	1552	47,730	Ibn Hammad, & Al-Karak, Mujeb Dam	Southern Ghors Project /stage I
	12	55, 56, 57	326	10,217	WadiHisan, Fifa, Tanur dam	Southern Ghors Project /stage II
Sub-Total			**1,878**	**57,947**		
Wadi Araba	13	58	325	6,500	Ground Water Wells	New Wadi Araba Irrigation Project
Grand Total			**10,115**	**363,203**		

Source: MWI, 2013.

on crop water requirements (or roughly 360 mcm by the assumption of 1000 m^3 water requirements per dunums annually). Domestic demand in the Jordan Valley is minor compared to the total agricultural demand which is satisfied from groundwater sources. The JVA is a government organization in charge of the economic and social development of the valley and retains the responsibility for operation and maintenance of the irrigation system including water distribution. The JVA has a directorate for operation and maintenance stationed in the valley. It employs over 1500 workers, 1362 employees were in the Water Division, of which about 250 take care for the operations [26]. There are three "regional" JVA field offices which calculate the water flow to each farm unit. A control central office monitors the calculated consumption for the entire valley, compares it to the water inflows from the various sources and calculates the un-accounted for water. The cultivation of citrus and banana is allowed only through license from JVA. The Jordan Valley irrigation scheme consists of 58 development areas that each depend on one common source of water, either surface or blended surface with treated wastewater or groundwater (**Table 1**). The water source is either one of the 28 pumping stations along the KAC or a direct inflow from a side wadi or a dam. Water is conveyed from the source (KAC or side wadis) to farmers' fields through a pressure pipe network and distributed to each farm unit (ca. 35 dunum) by an end user outlet, called a farm turnout assembly. All sources of water (surface water from Yarmouk river, side wadis and dams, Lake Tiberias, the main conveyor, KAC, its pump stations, the distribution networks and farm turnout assemblies and drinking water off-take from the KAC to Amman) are centrally monitored but are decentralized in operation [27]. These features make the Jordan Valley irrigation system almost unique, being a dual irrigation and municipal water supply system which is very difficult to compare with irrigation systems in other countries.

The total cultivated areas in JV in 2011 is about 363 thousand dunums, of which 311 thousand dunums is fully irrigated. The vast majority of irrigated agricultural production is in the form of fresh fruits and vegetables. As indicated in **Table 2**, about 60 percent of the irrigated areas in Jordan are allocated to vegetables. About 45 percent of total production and irrigated area of vegetables in Jordan are in Jordan Valley. Irrigated areas of fruit trees in JV occupy 23 percent of total irrigated areas of fruit trees and produce about 51 percent of total fruit trees production.

Water resources in the Jordan Valley consist of ground water, surface water, and treated wastewater from KTD. The safe yield of the Jordan Valley basin is estimated at 20 mcm and the safe yield of the Jordan Valley side wadis basin is estimated at 31 mcm [28]. Other sources in the Jordan Valley are Yarmouk River, Taiberia Lake, and Mukheba wells. The amount received from Yarmouk River varies significantly from year to year which depends on rainfall and on the upstream use by Syria. For example Yarmouk River flow at Adasiya near the inlet to KAC for the year 2004 was about 69 mcm which dropped to about 12.6 mcm for the year 2011 [26]. Flow from Taiberia Lake is governed by the Peace treaty which is about 50 mcm per year. The total water inflows into Jordan Valley fluctuated from year to year depends on rainfall precipitation. The total water inflows estimated in year 2011 with 211 mcm. The flow of Mukheiba wells in 2011 was about 25.5 mcm. Water from Yarmouk River, Taiberia Lake and Mukheiba wells flows at the upstream of King Abdulla Canal, which is the main transfer system in the Valley and extends from the northern Jordan Valley about 110 km to the south and ends a short distance before the Dead Sea. In addition, several side wadis distributed along the valley flow from east to west, the base flow of which is estimated at about 19.6 mcm [26]. These side wadis flow to the Jordan River, however, some water from these side wadis is used for private irrigation in the upstream area. In addition, four small dams exist on these wadis, the storage capacity of which is about 30 mcm. Al-Karameh dam with a storage capacity of about 52 mcm is the largest dam in the JRV; unfortunately its water is saline and can't be used for irrigation. A desalination plant was constructed to desalinate about 12 mcm per year to be used for domestic purposes but due to some technical problems, mainly high salinity, it's not in operation. Furthermore, brackish springs and tube well are existed in the Jordan Valley, some of which are desalinated and used for irrigation by private farmers. To compensate for the loss of Yarmouk river water, the Jordan Valley Authority extended the existing Zarqa Carrier II (which conveys water from the Zarqa river downstream of the King Talal Dam (KTD) to the King Abdullah Canal (KAC) to the North Ghor irrigation area. The new pipeline, called Zarqa Carrier III, enables another 4000 ha to be irrigated with water from KTD. In 2010 about 2 mcm were delivered from KTD to Northern areas and increased to about 10.3 mcm in 2011 from a total water use of 37 mcm

Table 2. Irrigated areas under tree crops, field crops, and vegetables in 2012.

Crops	Irrigated area in Jordan (dunums)	Irrigated area in JV (dunums)	Percentage of irrigated area in JV to total irrigated areas	Production of irrigated area in Jordan (ton)	Production of irrigated area in Jordan Valley (ton)	Percentage of JV production to total production
Tree crops	469,751	107,672	23%	337,992	171,356	51%
Field crops	87,549	20,283	23%	168,268	30,570	18%
Vegetables	407,195	183,627	45%	1,915,149	876,058	46%
Total	964,495	311,581	32%	2,421,409	1,077,983	45%

Source: DOS, 2012.

[26]. The carrier is intended originally to be used in dry periods only as an emergency supply if fresh surface water resources in the KAC run out, alternating fresh water supply and supply from KTD. The Disi fossil water carrier is completed in August 2013 and it is starting to provide additional between 75 - 80 mcm of fresh water for municipal purpose in Amman-Zarka basin. Consequently an additional 50 - 60 mcm of treated wastewater are available in the KTD. However, the total water use for irrigation purpose in Jordan valley estimated with an average of 133 mcm. An average of about 57 mcm of water are lost and unaccounted for, either for system inefficiency, lost, evaporations and illegal uses. This transition comes with considerable problems. The water taken from KTD has a salinity of around 1400 mg/l, which negatively affects yields, especially citrus, when compared with irrigation from fresh water resources. Also, because of increasing availability of treated wastewater resources, the use of treated wastewater in the northern part of the Jordan Valley will be considered in the near future to reduce the stress on freshwater resources for rural and urban areas in the Northern governorates. It is therefore expedient to prepare farmers for the permanent use of treated wastewater for irrigation of citrus and other crops. Experiences from the southern part of the Jordan Valley cannot be extended without further action research to the northern part, as the northern part receives higher rainfall, different soil type and different cropping pattern, a major factor to be taken into account when adapting to the different quality of irrigation water.It is important to note that fresh water upstream of the KAC at Deir Alla is pumped to Zai Water Treatment Plant which provides drinking water to west Amman, pumping from KAC to Zai WTP for the years between 2010 and 2011 about 53.5 mcm. Furthermore, groundwater resources in the Jordan Valley are also used to satisfy part of the domestic demand in Irbid governorate. Treated wastewater generated in Amman and Zarqa is a main water resource in the Jordan Valley. Wastewater generated in Amman and Zarqa is treated at As-Samra WWTP and discharged to Zarqa River which ends in KTD. As-Samra WWTP flow to Zarqa River grew from about 61 MCM for the year 2007 to about 84 MCM for the year 2010. Water from KTD is released to KAC where it gets mixed with fresh water there and used for unrestricted irrigation in the Middle and southern Jordan Valley. The facility is acknowledged for being the first project in Jordan to be built under a build, operate and transfer (BOT) basis and the first to receive a grant from USAID. With a peak flow of 840,000 cubic meters each day, the facility treats an average flow of 267,000 cubic meters of wastewater on a daily basis, serving a population of about 2.2 million living in the Greater Amman and Zarqa areas. An expansion of the facility began in 2012 and is expected to be completed by 2016. It will increase the plant's average treatment capacity to 133 million cubic meters each year. The estimated treated wastewater use in 2010 is estimated with 100 mcm. About 55 mcm was used in Jordan Valley [29]. Beyond all this, the fact that Jordan, provide irrigation water in JV to farmers at subsidized prices implies that the governments involved consider water in the hands of farmers to have a greater value than the farmer's own willingness to pay. Such a view may be because of social effects—the desirability of social order and to avoid social unrest, and because of political considerations, as well as to keep the stability of food prices. The success of structural reforms in the water sector depends on sustained, determined political commitment to implement them, on the support of supplementary reforms in regulatory regimes, realistic and efficient tariff structure and on a clear policy on subsidy and its mechanisms to provide quality service to the poor. It is to be noted that effective regulation is a necessary but not a sufficient condition.

2.4. Methodology

This study focuses on the use of water as an intermediate good, used as an input in the production of other goods and services. When used as an intermediate good, the value of water must be assessed from the producers' point of view. The conceptual valuation framework for the welfare benefits of increases or decreases in water use is provided by the producers' demand for inputs, including water. Therefore, the deductive techniques include residual imputation approach are commonly used to derive real prices of irrigation water. A residual method for valuing irrigation water is a special case of the well-known process of performing farm budget or cost and return analysis. This method subtracts the incremental value added by all production inputs except the irrigation water from the value of total output. The method identifies the incremental contribution of each input to the value of the total output and is the most widely used methodology for valuing irrigation water [9]. All costs of production except water are subtracted from the value of production and the remaining (or residual) value provides an estimate of the value of water in irrigation. The resulting value sometimes termed "quasi-rent" [30] and can be assumed to be the net value of irrigation water [31]. The residual imputation method is most suitable where the residual contributes the largest fraction of the value of output. This method requires the subtraction of the economic cost of all the other production inputs except water from the sales revenue. The difference becomes the value of water in the production of commodity. The application of the principle of ability to pay is based on profitability of irrigated farms as the basis for water pricing. In estimating the ability of irrigators to pay for water, typically we uses farm budget studies for the area in question to determine the net productivity of irrigation water for various crops and various type of land. Once the crop pattern and the size of the average farm are estimated, the total net productivity of water is computed after subtracting all other costs (operational and fixed costs), including normal profits under the guise of a "farm family living allowance". Therefore, the use of water in a production process can be determined using the residual imputation approach. The residual value represents the maximum amount the producer would be willing to pay for water and still cover input costs [32]. If only variable input costs are subtracted, then a short-run measure of the value of water is derived. If the costs of all non-water inputs are subtracted (including a normal rate of return on capital), then a long-run value is obtained. Three methods can be used derive water values in agricultural sector, these are:

1. Water values based on the Gross Value Added (GVA): The GVA represents the difference between the gross output of the farm minus the intermediate consumption. The resulting water productivity allows for determining the farmers supply curve of the agricultural products in the short run. The farmer is willing to pay that price of water to avoid losses in the short run and to recover the variable cost. All the fixed cost does not recover and lost.

2. Water values based on the Net Value added (NVA): NVA is the value of output less the values of both intermediate consumption and annualized fixed capital. NVA is obtained by deducting consumption of fixed capital (or depreciation charges plus opportunity cost of invested capital) from GVA, NVA. Therefore equals gross wages, pre-tax profits net of depreciation, and indirect taxes less subsidies

3. Water values based on the Net Profitability (NP): The net profitability is the measure of the surplus or profit accruing from production after deducting all costs (direct and indirect) and thus a proxy for total pre-tax profit income. The resulting water value is an indication about the economic efficiency of water consumption and a proxy for farmer's ability to pay for water. If the farmers changed this value for water they reach an equilibrium in the long run and normal profit. This implies that the sales revenue exactly equals the sum of all inputs used. In this case, there is no reward to risk and uncertainties in doing business. However, the farmers receives normal rate of return on invested capital.

However, any costs incurred by a firm may be classed into two groups: fixed costs and variable costs. Fixed costs, which occur only in the short run, are incurred by the business at any level of output, including zero output. These may include equipment maintenance, rent, wages of employees whose numbers cannot be increased or decreased in the short run, and general upkeep. Variable costs change with the level of output, increasing as more products is generated. Materials consumed during production often have the largest impact on this category, which also includes the wages of employees who can be hired and laid off in the span of time (long run or short run) under consideration. Fixed costs and variable costs combined together equal total costs. Therefore, in this attempt we takes into account the sunk costs in the form of asset fixity such as amortization of capital investments, opportunities costs of invested capital, amortization of capital land, opportunities costs of family labors [33]. The management will attempt to maximize profits by employing just the right amount of each factor of production subject to a predefined budget constraint. At a much more general level, profit maximization may be viewed as an unconstrained or constrained optimization problem where the decision variable is the firm's level of output. The marginal product of water (MP_w) is the change in total output given a unit change in the amount of water used. The marginal revenue product of water (MR_w) is the change in the firm's total revenue resulting from a unit change in the amount of water used. The marginal revenue product is the marginal product of water times the selling price of the product (P_y), i.e., $MR_w = P_y \cdot MP_w$. The total cost of water is the price rate times the total amount of water used. The marginal resource cost of water (MRC_w) is the change in total water cost resulting from a unit change in the number of units of water used. If the price rate (P_w) is constant, then the price rate is equal to the marginal cost of water. A profit-maximizing firm that operates in perfectly-competitive output and input markets will employ additional units of water up to the point where the marginal revenue product of water is equal to the marginal water cost, i.e., $P_y \cdot MP_w = P_w$. The problem confronting the decision-maker is to choose an output level that will maximize profit. Define profit as the difference between total revenue and total cost, both of which are functions of output, i.e., $\pi(Q) = TR(Q) - TC(Q)$. The objective is to maximize this unconstrained objective function with respect to output. The first-order and second-order conditions for a maximum are $dp/dQ = 0$ and $d^2p/dQ^2 < 0$, respectively. The profit maximizing condition is to produce at an output level at which *MR* = *MC*. Although profit maximization is the most commonly assumed organizational objective, firms that are not owner-operated, or firms that operate in an imperfectly competitive environment often adopt an organizational strategy of total revenue maximization. The first-order and second-order conditions are d*TR*/d*Q* = 0 and d^2*TR*/d*Q*^{2} < 0, respectively. Assuming that firms are price takers in resource markets (the price of water is fixed); because price and output are always positive, it can be easily demonstrated that the output level that maximizes total revenue will always be greater than the output level that maximizes total profit. This is because of the law of diminishing marginal product guarantees that the rate of increase in marginal cost is greater than the rate of increase in marginal revenue [34]. To obtain the profit maximizing output quantity (*y*), we start by recognizing that profit is equal to total revenue *TR*(*y*) minus total cost *TC*(*y*). The profit-maximizing output is the one at which this difference reaches its maximum. Therefore, a firm's profit is its revenue minus its cost. If the price p_y at which the firm can sell its output is not significantly affected by the size of its output, it is reasonable to model the firm as taking the price as given. In this case, for a single product, its total revenue is *TR*(*y*) = p_y·*y*, where *y* is its output. Thus the firm's profit function is

$$\pi(y) = TR(y) - TC(y) = P_y \cdot y - TC(y) \tag{1}$$

where *TC* is either the firm's long run cost function, the firm chooses its output y to maximize its profit $\pi(y)$, taking price as given. The net profit $\pi(y_j)$ in term of (JD/dunum) equals the gross revenue less all annualized capital cost (*CC$_j$*) and other fixed costs (*FC$_j$*), which form total fixed costs (*TFC$_j$*), all variable costs (*VC$_j$*) including water charges ($Q_w \cdot P_w$). Assume the producer objective is to maximize profits of single output as a function on multiple inputs *Y* = *f*(*X$_i$*), leaving one single input, water in our case water (Q_w) separately. Then the profit equation is:

$$\pi = P_y \cdot Y - \sum_{i=1}^{n} Px_i \cdot X_i + TFC + P_w \cdot Q_w \tag{2}$$

To find the conditions for optimal profits, take the first derivative of π with respect to water and set that equal to zero:

$$\frac{d\pi}{dQ_w} = Py \cdot \frac{df(Y)}{dQ_w} - P_w = 0 \tag{3}$$

The $df(Y)/dQ_w$ is the marginal products of water, $Py \cdot df(Y)/dQ_w$ is the is the value of the marginal product (VMP) of water, where the value of marginal product is defined as output price multiplied by the marginal physical productivity of the input [34]. If all the inputs, including water, are exchanged in a competitive market and employed in the production process, the value of water of the last cubic meter used should equal its price. Rearrange equation [17] then

$$P_w \cdot Q_w = P_y \cdot Y - \left(\sum_{i=1}^{n} Px_i \cdot X_i + TFC + \pi \right) \tag{4}$$

When economic profit is equal to zero; this occurs when the difference between total revenue and total cost (explicit and implicit costs) equals zero. Normal profit is different than accounting profit because opportunity cost is taken into consideration. Normal profit is the minimum level of profit needed for a firm to remain competitive in the market. Profit in economics, is the return on conducting business and risk, also called earnings, minus the costs of maintaining land, labor, and capital. Thus, normal profit is the profit that could be earned in another activity elsewhere. It is the profit that could be earned in an alternative venture.

Residual valuation thus assumes that if all markets are competitive, except the one for water, the total value of production ($P_y \cdot Y$) equals exactly the opportunity costs of all the inputs. It is assumed that the opportunity costs of non-water inputs are given by their market prices (or their estimated shadow prices).The residual, obtained by subtracting the non-water input costs from total annual crop revenue equals the gross margin (water related contribution equal gross margin minus the water costs) and can be interpreted as the maximum amount the farmer who could pay for water and still cover costs of production. It represents the at-site value of water.

$$P_w = \left[P_y \cdot Y - \left(\sum_{i=1}^{n} P_i X_i + TFC \right) \right] / Q_w \tag{5}$$

In this study the water values were derived by three methods: these methods are:

1. Water values based on the gross value added,

$$P_w = \left[P_y \cdot Y - \left(\sum_{i=1}^{n} P_i X_i \right) \right] \Big/ Q_w . \tag{6}$$

2. Water values based on the net value added,

$$P_w = \left[P_y \cdot Y - \left(\sum_{i=1}^{n} P_i X_i + CC \right) \right] \Big/ Q_w \tag{7}$$

3. Water values based on the net profitability,

$$P_w = \left[P_y \cdot Y - \left(\sum_{i=1}^{n} P_i X_i + TFC \right) \right] \Big/ Q_w \tag{8}$$

The derived monetary amount derived from the last equation where net profit divided by the total quantity of water used on the crop production, determines the marginal value for water, corresponding to the irrigator's maximum ability to pay per unit of water for that crop [35]-[37]. However, the assumptions of the RIM are not overly restrictive, but care is required to assure that conditions of production under study are reasonable approximations of the conceptual model. The main issues can be divided into two types: 1) those relating to the specification of the production unction and 2) those relating to the market and policy environment (*i.e.*, the pricing of outputs and non-residual inputs), [9]-[19]. The residual value is assumed to equal the returns to water and represents the maximum amount the producer would be willing and able to pay for water and still cover input costs [32]. If only variable input costs are subtracted, then a short-run measure of the value of water is derived. If the costs of all non-water inputs are subtracted (including a normal rate of return on capital), then a long-run value is obtained.

2.5. Data Collection

Questionnaires were used to collect data for the period 2011-2012. These data encompass production (ton), cultivated area (du), yield (kg/du) by season in 4 districts in Jordan Valley. The estimation of the value of water for agriculture is performed on a per crop basis. The main field crops, vegetables and fruit trees in Jordan Valley are selected. In total 160 farmers were interviewed, spread over 4 irrigation schemes. The interviews gathered information on irrigation system, farm activities, quantities and costs of inputs used in production, quantities and value of output, quantity of water consumed and irrigation practices. Expert knowledge of the extension staff was used as a verification method to farmers' answers. This was particularly helpful for the estimation of water use and prices of inputs and outputs. After constructing an enterprise budget, a group of progressive producers is interviewed by the author in the targeted areas. The author and a farm management specialist work together to modify and develop a consensus estimate of enterprise costs and returns. It is fully realized by those involved in this process that the resulting enterprise budget does not represent any particular farm. The individual farmer must be modified it to fit his situation. However, the resulting budget is a reasonable estimate for each geographical location in the targeted area. Therefore, the ability to pay is expected to vary by location in JV, by time (winter vs. summer), and by water quality (surface, groundwater and blended water) and by individual based on his situation and endowments. The detailed collected data from farmers' field allows us to assess water values that are differentiated according to crop type, geographic area, seasons and water quality. Enterprise budgets are estimated for most of the fruits, field crops, and vegetable crops grown in Jordan Valley in different agro-ecological zones. The returns and costs were calculated on per dunum basis. The estimated enterprise budgets are based on the best and most accurate estimates on returns and costs available in 2012 for 226 observations. The net irrigation water applied by farmers is used to measure the value of irrigation water (which is subtracted later from calculation), fertilizers (trace elements, organic and compound or chemical fertilizer), pesticides and herbicides., containers and threads, plastic mulch used in vegetable production with drip irrigation, and under plastic houses, plastic sheet and cover used in plastic tunnels crop enterprises, fuel and electricity. The costs of hired machinery and seasonal hired labor expressed in hours, which includes planting, spraying, tillage, land preparation, rearing, and crop harvesting, have been calculated for all these operations. The gross margins were calculated without including irrigation water cost in the total variable cost. The fixed costs include expenses of all non-varying inputs required for the production process. The annual depreciation expenses for the crops are varying according to production systems. Total costs are the sum of both cost components, variable and fixed costs.

Net returns of the selected enterprises were obtained through deducting the total costs from the gross returns for each crop. In calculating labor costs for the enterprise budgets, operator and family labor are valued at their opportunity cost of being hired out to a neighboring farmer. The shadow price was calculated based on discussions with farmers and extension personnel and on the data on wage labor in the dataset. For land that is owned, the opportunity cost that is included in the budget is the net rental return that the producer would receive if the land was rented out rather than being used by the producer [38] [39].

3. Results and Discussion

The farmers' ability to pay will be presented separately with classified according to geographical location, crop type, water quality cropping season and production and irrigation technology. The residual water value is assumed to equal the returns to water and represents the maximum amount the producer would be willing to pay for water and still cover input costs. The approach is also extremely sensitive to small variations in assumptions concerning the nature of the production function or prices. The water values based on gross value added and net value added will be presented but will not be discussed, since it represents the water values in the short run. The farmers will be able to pay such value in order to avoid crop losses.

3.1. Farmers' Ability to Pay by Location

Farmer's ability to pay in term of water profitability vary from region to region depending on economic activity, climate zones, production season, soils and water qualities, in addition to many other factors. **Table 3** shows the average of water value and water profitability values in different regions. The highest water values are found in MJV with JD 1.59 m^{-3}. However, water profitability was found in MJV has the highest value of about JD 0.93 m^{-3}, Safi, and northern JV are similar with the value of about JD 0.79 m^{-3}. SJV is the lowest with about JD 0.62 m^{-3}. This might be because of the dominance of banana fruits in SJV, which require a high amount of water compared with MJV where vegetables are the dominant cropping pattern. The percent of water cost to total costs was the lowest value is found in MJV with 1.24%. The water costs ranges between 1.24% to 2.35% with an average of 1.88%. The results shows that water cost represent a neglected portion in the variable costs and total costs. This lowest portion does not encourage farmers to take a serious measure to conserve water and encourage a rational use of water. Farmers in highland areas, where water cost exceeds JD 0.7 m^{-3}, are taken serious measures to avoid any loss in the irrigation network and practicing deficit irrigation.

3.2. Farmers' Ability to Pay by Crop Type

The field crops include wheat, barley, alfalfa, maize and garlic. The water profitability in field crops is among the lowest in Jordan valley. However, the weighted average for ability to pay in field crop production is JD 0.134 m^{-3}. It is worth mentioning that field crop gown in JV for crop rotation purposes The cost of water represent a significant portion of total costs in field crop. The costs of water represent about 6% of the total variables cost and about 4.5% in the total costs as shown in **Table 4**. The average water profitability for the fruit trees is JD 0.53 m^{-3}. The average water value in Banana is JD 0.48 m^{-3} it ranged from JD 0.652 m^{-3} in the SJV to a lowest of JD 0.22 m^{-3} in NJV. Looking to the net profit to one cubic meter, it was found it is about JD 0.48 m^{-3} for banana crops. Therefore, it is economically rational for banana producers to install RO units to irrigate bananas, since the cost of desalination of one cubic meter is about the half of net profit from one cubic meter.

Table 3. Farmers' ability to pay and cost of water relative to total costs by location in JV.

Location	Water gross value added (JD m^{-3})	Water net value added (JD m^{-3})	Water net profitability (FAP) (JD m^{-3})	Percent cost of water to intermediate costs	Percent cost of water to total costs
NJV	1.327	1.157	0.791	2.59%	1.85%
MJV	1.591	1.333	0.931	1.81%	1.24%
SJV	1.341	1.143	0.622	2.95%	2.35%
Safi	1.279	1.156	0.794	2.18%	1.65%
Average	1.365	1.179	0.763	2.54%	1.88%

Table 4. Farmers' ability to pay and cost of water relative to total costs by crop type in JV.

Crop Type	Water gross value added (JD m^{-3})	Water net value added (JD m^{-3})	Water net profitability (FAP) (JD m^{-3})	Percent cost of water to intermediate costs	Percent cost of water to total costs
Winter Veg.	1.559	1.325	0.844	1.70%	1.24%
Summer Veg.	1.647	1.413	0.925	2.55%	1.92%
Field Crops	0.316	0.293	0.134	5.96%	4.46%
Fruit Tress	0.883	0.806	0.542	3.05%	2.26%
Average	1.365	1.179	0.763	2.54%	1.88%

Summer vegetables show the highest ability to pay (JD 0.95 m^{-3}). This could be due to water scarcity and banning summer cultivation in some years as a method of water rationing. The FAP for winter vegetables is JD 0.84 m^{-3}. The percentage of water cost to total cost is 1.24%. Therefore, the interviewed farmers does not consider water charges as a problem and they stress their willingness to pay higher water tariff for irrigation water in case of improvement in quality and quantity and most of farmers complain of water scarcity and not sufficient water delivery.

3.3. Farmers' Ability to Pay by Water Quality

The value of water quality can be looked at in several ways; poor water quality, for instance, can limit the crops a farmer is able to grow or reduces water use efficiency and yield [40]-[42]. Therefore, water quality is multi-dimensional, as it includes concentration of certain chemicals, level of salinity, concentration of bacteria and organic matter, as well as temperature. The Jordan Valley will divide in to four geographical locations by source of water, Northern areas irrigated with fresh water which is dominated by citrus crops, Middle areas irrigated with blended water dominated by winter vegetables, Southern Jordan Valley irrigated with blended water and side wadies in addition to brackish artisan wells. The dominant crops in these areas are winter vegetables and banana. The Southern Dead sea area (Safi area) irrigated with surface water and dominated by winter tomatoes. Surface fresh water is used in Northern Jordan Valley appears overall to be of acceptable quality and low salinity compared with other sources. The maximum farmer's ability to pay for surface water is the highest with JD 0.81 m^{-3} and fresh water in Safi area with JD 0.79 m^{-3}. Treated wastewater plays a major role in narrowing the gap between supply and demand in the agricultural sector in Jordan. The effluent of As Samra WWTP is discharged to Zarqa River where it is used for restricted irrigation upstream of KTD and for unrestricted irrigation downstream of KTD after mixing with its water. Poor water quality can limit the crops a farmer is able to grow. Low water quality also reduces water use efficiency and thus may reduce yield but increase water use. The result shows that the farmers' maximum ability to pay for blended water is JD 0.75 m^{-3}. This could be to lower yield of crops as a result of high water salinity. The farmers' maximum ability to pay for desalinated brackish water is JD 0.59 m^{-3}. The desalinated water is mainly used to irrigate banana and other cash crops such as strawberry. Looking to the net profit to one cubic meter, it was found by Al-Karablieh [22] that the net profit is about JD 0.51 m^{-3} for Banana crop. However, about 50 reverses osmosis plants are operated by cash crops farmers in Southern Jordan Valley desalinate about 7.6 mcm annually. The total brackish water abstraction was estimated with 11.8 mcm annually. The author estimated the average desalination costs for 50 Reverse Osmosis plant (operational and annualized capital costs) of about JD 0.28 per cubic meter with a standard deviation of JD 0.13 per cubic meter. Therefore, the current practice of banana producers is economically rational by installing RO unit to irrigate banana, since water value is twice the desalination costs of one cubic meter. The estimated values of brackish water desalination of JD 0.59 m^{-3} represent the maximum price that farmers might be willing to pay for water under the current market conditions. Regarding groundwater, the farmers' maximum ability to pay for tube well water is JD 0.56 m^{-3}. There is a trend of declining water tables and increasing salinity in most aquifers in JV, with resulting higher extraction costs (in terms of pumping as well as accelerated well replacement). Due to the increasing problem with water shortages experienced in Jordan Valley, the utilization of brackish water which was once not an attractive option has gained in prominence. The cost per unit of desalinated water has been dropping as advances have been made in desalination technology (**Table 5**).

Table 5. Farmers' ability to pay and cost of water relative to total costs by water qualities in JV.

Water quality	Water gross value added (JD m^{-3})	Water net value added (JD m^{-3})	Water net profitability (FAP) (JD m^{-3})	Percent cost of water to intermediate costs	Percent cost of water to total costs
Fresh KAC	1.355	1.182	0.812	2.46%	1.76%
BTWW	1.414	1.190	0.749	1.86%	1.35%
Tube well	1.324	1.150	0.562	5.53%	4.49%
Fresh Surface	1.279	1.156	0.794	2.18%	1.65%
Desalinated	1.069	0.982	0.591	26.2%	20.8%
Average	1.365	1.179	0.763	2.54%	1.88%

3.4. Farmers' Ability to Pay by Season

The result shows that the farmers' maximum ability to pay for water in four season. The water demanded to cultivate crops in spring season is the highest ability to pay (JD 1.2 m^{-3}), and the lowest was found for demanded in winter with about JD 0.72 m^{-3} as shown in (**Table 6**). The water demanded in summer has the highest percent of water cost to total production costs. The farmers' ability to pay for fruit trees and permanent crops is estimated with JD 0.53 m^{-3}.

3.5. Farmers' Ability to Pay by Production Technology

Plastic houses can provide protection from the weather, a major production challenge faced by vegetable growers. The serious potential loss of crops due to freezes and rain or wind is a major challenge and concern for all vegetable growers in climates such as Jordan Valley. Also, plastic structures can protect the crop from wind and rain, but also can protect from insects when fitted with insect exclusion screens. Therefore, plastic houses systems could reduce the use of pesticides. Protected vegetable production in plastic houses can afford several advantages to producers. They include the ability to moderate temperature during various seasons of the year, wind protection, insect protection, and rain protection. In 2012 about 67,000 plastic houses are installed in Jordan Valley. The most plastic houses are found in MJV for winter vegetables. The result shows that the farmers' maximum ability to pay for water according to production technology used by farmers. The water demanded to cultivate crops with plastic house technology has the highest ability to pay JD 1.34 m^{-3}, and the lowest was found for water used to cultivate in open field (JD 0.62 m^{-3}) as shown in (**Table 7**). Due to high production cost by using protected agriculture in plastic houses, the percent cost of water to other production cost is only about 1% of the total costs.

3.6. Farmers' Ability to Pay by Irrigation Technology

Irrigation technologies commonly used in Jordan include furrow, drip and sprinkler. Open space, greenhouse and plastic tunnels are the most technologies used for cultivation or production. About 75% of the Jordan Valley is now drip irrigated while 24% is surface irrigated and only less than 1% is sprinklers irrigated. Additional attention must now be paid to improve the management of on-farm systems and thereby increase their efficiency. In the JRV most farmers (75%) had a reservoir on their farms, 46% of farmers connected their drip irrigation systems directly to JVA pressure lines. About 90% of the drip irrigation systems use in-line emitters (G.R type) in 16 - 20 mm-diameter laterals; these can deliver 3 - 4 litres per hour. The result shows that the farmers' maximum ability to pay for water according to irrigation technology used by farmers. The water used in drip irrigation technology has the highest ability to pay (JD 0.84 m^{-3}), and the lowest was found for water used to cultivate open field with sprinkler irrigation (JD 0.07 m^{-3}) as shown in (**Table 8**). The open fields crops are irrigated with sprinkler irrigation are mainly wheat, barley and alfalfa. Those had the lowest value added and profitability.

3.7. Farmers' Ability to Pay by Individual Crop in Jordan Valley

To get insight about the maximum water ability to pay for each crop grown in Jordan Valley, the results were aggregate level for main crops for the purpose of policy recommendations. The list of the crops sorted from top

Table 6. Farmers' ability to pay and cost of water relative to total costs by production season in JV.

Production season	Water gross value added (JD·m⁻³)	Water net value added (JD·m⁻³)	Water Net profitability (FAP) (JD·m⁻³)	Percent cost of water to intermediate costs	Percent cost of water to total costs
Winter	1.435	1.213	0.724	1.92%	1.44%
Spring	1.866	1.615	1.193	1.69%	1.12%
Summer	1.486	1.302	0.873	3.55%	2.69%
Autumn	1.564	1.302	0.810	1.63%	1.16%
Permanent	0.874	0.798	0.534	3.12%	2.31%
Average	1.365	1.179	0.763	2.54%	1.88%

Table 7. Farmers' ability to pay and cost of water relative to total costs by production technologies in JV.

Production technologies	Water gross value added (JD·m⁻³)	Water net value added (JD·m⁻³)	Water net profitability (FAP) (JD·m⁻³)	Percent cost of water to intermediate costs	Percent cost of water to total costs
Open field	1.091	0.980	0.620	2.87%	2.15%
Plastic houses	2.413	1.948	1.343	1.46%	1.01%
Plastic tunnel	1.502	1.237	0.681	1.50%	0.99%
Average	1.365	1.179	0.763	2.54%	1.88%

Table 8. Farmers' ability to pay and cost of water relative to total costs by irrigation technologies in JV.

Irrigation technologies	Water gross value added (JD·m⁻³)	Water net value added (JD·m⁻³)	Water net profitability (FAP) (JD·m⁻³)	Percent cost of water to intermediate costs	Percent cost of water to total costs
Drip	1.500	1.291	0.843	2.36%	1.75%
Sprinkler	0.285	0.261	0.071	7.90%	5.69%
Surface	0.698	0.630	0.378	2.91%	2.20%
Average	1.365	1.179	0.763	2.54%	1.88%

to lowest in term of farmers' ability to pay for water as shown in (**Table 9**). Cucumber shows the highest ability to pay (JD 2.26 m⁻³). The percentage of water cost to total cost is 1.1%. Therefore, the interviewed farmers does not consider water charges as a problem and they stress their willingness to pay higher water tariff for irrigation water in case of improvement in quality and quantity and most of farmers complain of water scarcity and not sufficient water delivery. The weighted average for maximum farmers' ability to pay for irrigation water in Jordan valley estimated at for Okra (JD 1.46 m⁻³), for String beans at JD 1.37 m⁻³, for Green Beans at JD 1.35 m⁻³, for Tomatoes at JD 1.30 m⁻³, and for Dates at JD 1.18 m⁻³. The average FAP for water in Banana is about (JD 0.48 m⁻³) it ranged from JD 0.65 m⁻³ in the SJV to a lowest of JD 0.22 m⁻³ in NJV. The field crops are maize at JD 0.18 m⁻³, for olives at JD 0.13 m⁻³, for wheat JD 0.11 m⁻³, for alfalfa JD 0.09 m⁻³, and the last one is barley at JD 0.05 m⁻³. The water profitability in field crops is among the lowest in Jordan valley.

4. Conclusions and Policy Recommendations

The farmer's ability to pay for irrigation water in agriculture varies widely across crops, seasons, and production locations. Crop grown in plastic houses have the highest water profitability (JD 1.34 m⁻³), while field crops such as maize, barley, and wheat produce the lowest water profitability (JD 0.11 m⁻³). Among fruits, olives show consistently low water profitability (JD 0.13 m⁻³), while citrus is the highest (JD 0.46 m⁻³). The results showed that the weighted average of farmers' ability to pay for water used in field crops is JD 0.13 m⁻³ and JD 0.84 m⁻³ for winter vegetable crops, JD 0.92 m⁻³ for summer vegetables, and JD 0.54 m⁻³ for fruit trees. The overall weighted average water net profitability of irrigation was estimated at JD 0.78m⁻³. In general, the water costs

Table 9. Farmers' ability to pay and cost of water relative to total costs for main crop in Jv.

Crop	Water net profitability (FAP) (JD·m^{-3})	Percent cost of water to total costs	Crop	Water net profitability (FAP) (JD·m^{-3})	Percent cost of water to total costs
Cucumbers	2.266	1.1%	Lemons	0.464	1.73%
Okra	1.461	0.9%	Cabbages	0.456	2.80%
String beans	1.372	1.5%	Lettuce	0.447	1.26%
Green Beans	1.349	0.8%	Onion, dry	0.430	1.56%
Tomatoes	1.307	1.2%	Pomegranates	0.405	1.92%
Dates	1.181	2.1%	Clementine	0.390	2.18%
Sweet peppers	0.878	1.2%	Cauliflowers	0.379	2.58%
Grapes	0.871	2.0%	Pummels	0.377	1.90%
Potatoes	0.855	1.0%	Jew's mallow	0.364	1.85%
Onion green	0.799	1.9%	Prune	0.357	2.41%
Watermelons	0.780	2.49%	Peaches	0.326	2.17%
Green onion	0.712	2.76%	Eggplants	0.323	1.64%
Hot peppers	0.645	1.95%	Grapefruits	0.252	3.22%
Squash	0.644	1.84%	Maize	0.181	3.40%
Sweet melons	0.574	2.25%	Olives	0.130	4.07%
Mandarins	0.532	2.13%	Wheat	0.115	5.16%
Banana	0.483	3.28%	Alfalfa	0.093	4.76%
Oranges	0.465	1.91%	Barley	0.050	6.62%

represent less than 2% of the total production costs. Furthermore, farmers' ability to pay differs by production technology; therefore, water can be priced according to production technology to enhance water saving. Farmers cultivating crops under plastic houses are able to pay higher prices for water reaching up to JD 1.34 m^{-3}. Water profitability varies from region to region depending on the climate zones, production season, and water qualities. Water profitability in MJV has the highest profitability of about JD 0.93 m^{-3}. NJV and Safi are about JD 0.792 m^{-3} and the lowest in SJV with about JD 0.62 m^{-3}. This might be the dominance of citrus fruits in NJV, which requires higher amount of water compared with MJV where vegetables are the dominant cropping pattern. The average water profitability for surface water is the highest with JD 0.81 m^{-3} and groundwater is the lowest with JD 0.56 m^{-3}, whereas it reaches about JD 0.75 m^{-3} for blended water. The observed values of water were in the range of those found in other studies for irrigated vegetables in Jordan. In a study conducted by Al-Karablieh [22], they found that the average value of irrigation water is JD 0.51 m^{-3} at the country level. Haddadin [28] reports that the value of water is JD 0.48 m^{-3} for vegetables under plastic houses, JD 0.35 m^{-3} for citrus crops and JD 0.37 m^{-3} for fruit trees. The study revealed a high level of variability in irrigation water values. It was shown that the differences in water values and FAP can be mainly attributed to several factors that can be relevant for policy makers and extension services: 1) the irrigation technology system, 2) the type of crop grown and 3) the water quality. Current prices charged for water are substantially below both the average value of water for producing crops and JVA's cost of service provision. Higher water prices could (a) encourage more efficient water use by farmers, (b) encourage shifts to higher value crops, (c) encourage the adoption of protected agriculture and green houses, (d) encourage private development of desalinated brackish water sources and (e) provide sufficient funds for better irrigation system maintenance and more effective operation. Water pricing policy should be revised to provide incentives for water saving technology. Differential prices can be applied to account for irrigation water quality. Expected consequences of raising water tariff would include a loss in the variety of cultivated crops and an increase in requirement for investments by farmers to adopt new water saving techniques.

The increased requirement might also engender further negative secondary effects, since small-scale family enterprises, which constitute the majority of farms in the Jordan Valley, might not be able to cope with the increased financial demands. However, more efficient water use in response to higher prices would only occur if water is allocated and billed by volume rather than by area. Returning to a system of volumetric billing in the JVA service area would require retrofitting all connections with reliable meters and reestablishing a meter-reading and billing system, a process which should be closely integrated with the newly-established WUAs. If farmers have to pay the full cost of O&M, they need to pay at least JD 0.065 per cubic meter. The water value received by farmers is ten times higher than this suggested price of full cost recovery.

References

[1]   UNDP (2013) Water Governance in the Arab Region: Managing Scarcity and Securing the Future. United Nations Development Programme, Regional Bureau for Arab States (RBAS).

[2]   Karthikeyan, C. (2010) Economic and Social Value of Irrigation Water: Implications for Sustainability. *Fourteenth International Water Technology Conference*, Cairo, 21-23 March 2010, 823-835. http://www.iwtc.info/2010_pdf/13-02.pdf

[3]   World Bank (2013) Irrigation Water Pricing in the Jordan Valley. The World Bank, Washington DC.

[4]   Salman, A., Al-Karablieh, E., Regner, H.-J., Wolff, H.-P. and Haddadin, M. (2008) Participatory Irrigation Water Management in the Jordan Valley. *Water Policy*, **10**, 305-322. http://dx.doi.org/10.2166/wp.2007.051

[5]   IRG (International Resources Group) and Al-Karablieh, E. (2012) Disaggregated Economic Value of Water in Industry and Irrigated Agriculture in Jordan. United States Agency for International Development (USAID).

[6]   Chandrasekaran, K., Devarajulu, S. and Kuppannan, P. (2009) Farmers' Willingness to Pay for Irrigation Water: A Case of Tank Irrigation Systems in South India. *Water*, **1**, 5-18. http://dx.doi.org/10.3390/w1010005

[7]   Ashfaq, M., Jabeen, S. and Baig, I.A. (2005) Estimation of the Economic Value of Irrigation Water. *Journal of Agriculture and Social Sciences*, **1**, 270-272.

[8]   Gibbons, D. (1986) The Economic Value of Water. Resources for the Future, Washington DC.

[9]   Young, R. (2005) Determining the Economic Value of Water: Concepts and Methods. Resource for the Future, Washington DC.

[10]   Koss, P. and Khawaja, M.S. (2001) The Value of Water Supply Reliability in California: A Contingent Valuation Study. *Water Policy*, **3**, 165-174. http://dx.doi.org/10.1016/S1366-7017(01)00005-8

[11]   FAO (2000) Application of the Contingent Valuation Method in Developing Countries: A Survey. Economic and Social Development Paper 146, Food and Agriculture Organization of United Nation, Rome, 9.

[12]   Salman, A. and Al-Karablieh, E. (2004) Measuring the Willingness of Farmers to Pay for Groundwater in the Highland Area of Jordan. *Agricultural Water Management*, **68**, 61-76. http://dx.doi.org/10.1016/j.agwat.2004.02.009

[13]   Venkatachalam, L. (2004) The Contingent Valuation Method: A Review. *Environmental Impact Assessment Review*, **24**, 89-124. http://dx.doi.org/10.1016/S0195-9255(03)00138-0

[14]   Venkatachalam, L. (2006) Factors Influencing Household Willingness to Pay (WTP) for Drinking Water in Peri-Urban Areas: A Case Study in the Indian Context. *Water Policy*, **8**, 461-473. http://dx.doi.org/10.2166/wp.2006.055

[15]   Turner, K., Georgiou, S., Clark, R. and Brouwer, R. (2004) Economic Value of Water Resources in Agriculture. From the Sectoral to a Functional Perspective of Natural Resource Management. FAO Water Reports 27, Rome. http://www.fao.org/docrep/007/y5582e/y5582e00.htm#Contents

[16]   Birol, E., Karousakis, E.K. and Koundouri, P. (2006) Using Economic Valuation Techniques to Inform Water Resources Management: A Survey and Critical Appraisal of Available Techniques and an Application. *Science of the Total Environment*, **365**, 105-122. http://dx.doi.org/10.1016/j.scitotenv.2006.02.032

[17]   Hussain, I., Turral, H., Molden, D. and Ahmad, M. (2007) Measuring and Enhancing the Value of Agricultural Water in Irrigated River Basins. *Irrigation Science*, **25**, 263-282. http://dx.doi.org/10.1007/s00271-007-0061-4

[18]   Abu-Zeid, M. (2001) Water Pricing in Irrigated Agriculture. *International Journal of Water Resources Development*, **17**, 527-538. http://dx.doi.org/10.1080/07900620120094109

[19]   Lange, G.M. and Hassan, R., Eds. (2007) Case Studies of Water Valuation in Namibia's Commercial Farming Areas. The Economics of Water Management in Southern Africa: An Environmental Accounting Approach. Edward Elgar Publishing, Cheltenham.

[20]   Speelman, S., Farolfi, S., Perret, L., D'haese, L. and D'haese, M. (2008) Irrigation Water Value at Small-Scale Schemes: Evidence from the North West Province, South Africa. *International Journal of Water Resources Development*, **24**, 621-633. http://dx.doi.org/10.1080/07900620802224536

[21] Rigby, D., Alcon, F. and Burton, M. (2010) Supply Uncertainty and the Economic Value of Irrigation Water. *European Review of Agricultural Economics*, **37**, 97-117. http://dx.doi.org/10.1093/erae/jbq001

[22] Al-Karablieh, E., Salman, A., Al-Omari, A., Wolff, H.-P., Al-Assa'd, T., Hunaiti, D. and Subah, A. (2012) Estimation of the Economic Value of Irrigation Water in Jordan. *Journal of Agricultural Science and Technology*, **5**, 487-497.

[23] DOS (2012) Annual Agricultural Statistics 2012. Department of Statistics, Amman.

[24] Al-Karablieh, E., Jabarin, A.S. and Tabieh, M.A. (2011) Jordanian Horticultural Export Competitiveness from Water Perspective. *Journal of Agricultural Science and Technology*, **1**, 964-974.

[25] Al-Assaf, A., Salman, A., Fisher, F.M. and Al-Karablieh, E. (2007) A Trade-Off Analysis for the Use of Different Water Sources for Irrigation (The Case of Southern Shounah in the Jordan Valley). *Water International*, **32**, 244-253. http://dx.doi.org/10.1080/02508060708692204

[26] MWI (2012) Annual Report 2012. Ministry of Water and Irrigation, Amman.

[27] Salman, A., AL-Karablieh, E. and Haddadin, M. (2008) Limits of Pricing Policy in Curtailing Household Water Consumption. *Water Policy*, **10**, 295-307. http://dx.doi.org/10.2166/wp.2008.040

[28] Haddadin, M.J., Salman, A. and Al-Karablieh, E. (2006) The Role of Trade in Alleviating Water Shortage. In: Haddadin, M.J., Ed., *Water Resources in Jordan* (*Evolving Polices for Development, the Environment and Conflict Resolution*), Resources for the Future, Washington DC, 17.

[29] Wolff, H.-P., Al-Karablieh, E., Al-Assa'd, T., Subah, A. and Salman, A.Z. (2012) Jordan Water Demand Management Study: On Behalf of the Jordanian Ministry of Water and Irrigation in Cooperation with the French Development Agency (AFD). *Water Science & Technology: Water Supply*, **12**, 38-44. http://dx.doi.org/10.2166/ws.2011.114

[30] Hellegers, P. and Davidson, B. (2010) Determining the Disaggregated Economic Value of Irrigation Water in the Musi sub-Basin in India. *Agricultural*, **97**, 933-938. http://dx.doi.org/10.1016/j.agwat.2010.01.026

[31] Qureshi, M.E., Ranjan, R. and Qureshi, S.E. (2010) An Empirical Assessment of the Value of Irrigation Water: The Case Study of Murrumbidgee Catchment. *The Australian Journal of Agricultural and Resource Economics*, **54**, 99-118. http://dx.doi.org/10.1111/j.1467-8489.2009.00476.x

[32] Naeser, R. and Bennett, L.L. (1998) The Cost of Noncompliance: The Economic Value of Water in the Middle Arkansas River Valley. *Natural Research Journal*, **38**, 445-463.

[33] Heathfield, D.F. and Wibe, S. (1987) An Introduction to Cost and Production Function. MacMillan Education Ltd., London.

[34] Chambers, R.G. (1988) Applied Production Analysis—A Dual Approach. Cambridge University Press, Cambridge.

[35] Agudelo, J.I. and Hoekstra, A.Y. (2001) Valuing Water for Agriculture: Application to the Zamvezi Basin Countries. *Proceedings of the International Specialty Conference on Globalization and Water Resource Management: The Changing Value of Water*, Dundee, 6-8 August 2001, 9.

[36] Agudelo, J.I. (2001) The Economic Valuation of Water: Principles and Methods. Value of Water Research Report Series 5, IHE Delft, Delft.

[37] Moore, M.R. (1999) Estimating Irrigators' Ability to Pay for Reclamation Water. *Land Economics*, **75**, 562-578. http://dx.doi.org/10.2307/3147066

[38] Kletke, D. (1989) Enterprise Budgets. In: Tweeten, L., Ed., *Agricultural Policy Analysis Tools for Economic Development*, Ohio State University, Columbus, 196-206.

[39] Powers, L., Steve, I., Tim, W., Richard, T., John, S., Brent, R., Dave, S., Terry, J. and Winston, D. (1998) Horticulture Crop Enterprise Cost and Return Estimates for 1998. University of Kentucky, Lexington.

[40] Carr, G., Potter, R.B. and Nortcliff, S. (2001) Water Reuse for Irrigation in Jordan: Perceptions of Water Quality among Farmers. *Agricultural Water Management*, **98**, 847-854. http://dx.doi.org/10.1016/j.agwat.2010.12.011

[41] Majdalawi, M. (2003) Socioeconomic Impacts of Reuse of Water in Agriculture in Jordan Valley. *Farming & Rural Systems Economics*, **51**, 149-157.

[42] Bazza, M. (2003) Wastewater Recycling and Reuse in the Near East Region: Experience and Issues. *Water Science and Technology: Water Supply*, **3**, 33-50.

7

Evaluation of Groundwater Quality and Its Suitability for Domestic and Irrigation Use in Parts of the Chandauli-Varanasi Region, Uttar Pradesh, India

Shubhra Singh[1], N. Janardhana Raju[1*], Ch. Ramakrishna[2]

[1]School of Environmental Sciences, Jawaharlal Nehru University, New Delhi, India
[2]Department of Environmental Studies, GITAM University, Visakhapatnam, India
Email: [*]rajunj7@gmail.com

Abstract

The present study focused on the hydrochemistry of groundwater in parts of Chandauli-Varanasi region to assess the quality of groundwater for determining its suitability for drinking and agricultural purposes. Urbanization and agriculture activities have a lot of impacts on the groundwater quality of the study area. A total of 70 ground water samples were collected randomly from different sources *viz.* hand pump, dug wells and bore wells, and analyzed for major cations and anions. The domination of cations and anions was in the order of Na > Ca > Mg > K and HCO$_3$ > Cl > SO$_4$ > NO$_3$ > F. The Piper classification for hydrogeochemical facies indicates that alkaline earth exceeds alkalis and weak acids exceed strong acid. Water quality index rating was calculated to quantify overall water quality for human consumption. Out of 70 groundwater samples, 7% and 10% samples exhibit water unsuitable for drinking purposes in pre- and post-monsoon, respectively, due to effective leaching of ions, direct discharge of domestic effluents and agricultural activities. Residual sodium carbonate values revealed that 6% sample is not suitable for irrigation purposes in both the seasons due to low permeability of the soil. The calculated values of PI indicate that the water for irrigation uses is excellent to good quality in both seasons. As per Wilcox's diagram and US salinity laboratory classification, most of the groundwater samples are suitable for irrigation except one sample which is unsuitable for irrigation purposes. The overall quality of groundwater in post-monsoon season in all chemical constituents is on the higher side due to dissolution of surface pollutants during the infiltration and percolation of rainwater at few places due to agricultural and domestic activities.

---

[*]Corresponding author.

Keywords

Groundwater Quality, Hydrogeochemical Facies, Drinking Water, Irrigation Water

1. Introduction

Groundwater is a valuable natural resource; it occurs almost in all geological formations under the earth surface not in a single widespread aquifer but in thousands of local aquifer systems with similar characteristics. In tropical regions, groundwater plays an important role with context to fluctuating and increasing contamination of groundwater. The groundwater is used in study area for both domestic and agricultural purposes. Agriculture is the main source of livelihood of the population and ground water is the major source of irrigation. Groundwater gets contaminated with a variety of pollutants generated from diverse sources such as agriculture, industrial and domestic. The availability of this important natural resource has been taken for granted increasing ground water use and pollution generation has crossed the sustainable limits in many parts, due to fast changing land use pattern. There has been tremendous increase in demand for fresh water due to population growth and intense agriculture activities. With rapid increase in population and growth of industrialization, groundwater quality is being increasingly threatened by disposal of urban and industrial solid waste [1]. Open dumping is the most common way to dispose municipal and industrial wastes. Subsequent leaching of toxic contaminants through the dumping site also leads to extensive contamination of ground water at many places. It has been estimated that once pollution enters the subsurface environment, it may remain concealed for many years; rendering groundwater is unsuitable for consumption and other uses [2]. Rate of depletion of groundwater levels and deterioration of groundwater quality is of immediate concern in major cities and towns of the country. Variation of groundwater quality in an area is a function of physicochemical parameters that are greatly influenced by geological formations and anthropogenic activities.

The source of about 90% of drinking and irrigation water is from groundwater resources in study area. However, this water resource is facing problems including quality hazard in many areas where the exposure to pollution from agriculture and urbanization in shallow groundwater aquifers makes the water unfit for human consumption. Land use for urbanization and agricultural purpose in the Varanasi city has increased at an alarming rate during the last few decades. Borehole sediments revealed that multistoried sand bodies generated as a result of channel migration provide excellent aquifers confined by a thick zone of muddy sediments near the surface in Varanasi City [3]. In complex multilayered alluvial formations, the shallowest phreatic aquifer is often the most vulnerable to anthropogenic pollution and the most susceptible to saline intrusion. The aim of the study was to investigate the quality of ground water and to discuss the hydrochemical facies, water quality index and classify the groundwater for domestic and irrigation uses in the Chandauli-Varanasi, UP.

2. Study Area and Hydrogeology

The study area covers about 302 Km^2 and falls in Survey of India toposheets 63 O/3, 63 O/4, 63 K/15 and 63 K/16. Geographically this area lies between the latitude 25°20'N - 25°12.7'N and longitude 82°58'E - 83°10'E (**Figure 1**) is located in the middle Ganga plain of Indian sub-continent. The study area belongs to the middle Ganga plain with an average height of about 76.19 m above the mean sea level with even topography. The Ganga is the principal river flowing incised into its narrow valley from south to north direction. The study area falls in the subtropical climate region. The area experiences three distinct seasons namely summer, rainy and winter. The maximum temperature is recorded to be 48°C and 24°C and minimum 32°C and 8°C in summer and winter seasons, respectively. The study area receives greater of annual rainfall through south-west monsoon between June and September. The average annual rainfall of the study area is around 1020 mm.

The unconsolidated near surface Pleistocene to recent fluvial sediments underlying most of the Ganga plains are generally potential aquifers. The alternating sand and clay layers have created a multitier aquifer system in the study area [3]. In the near surface Holocene sandy aquifer, groundwater occurs under water table conditions while deeper aquifers occur in semi-confined to confined conditions and are mainly used for irrigation and domestic purpose in the study area. Both dug wells and hand tube wells are used for groundwater extraction for different purposes. The shallow tube wells puncturing unconfined aquifers at an about 20 to 60 m depth have

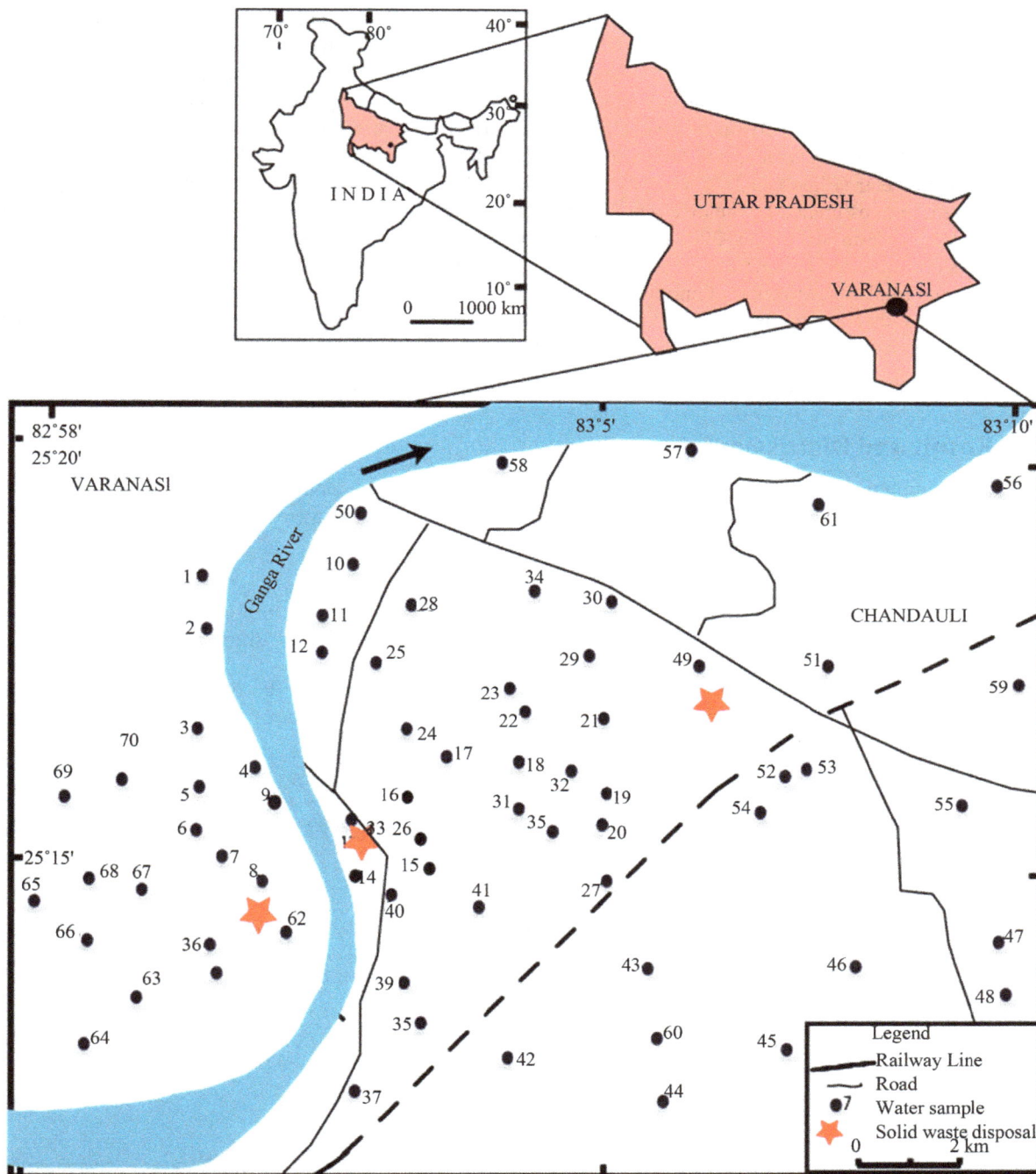

Figure 1. Physiographic and location map of study area.

water level fluctuation of 9 to 12 m. The general depth of deep tube wells ranges from 60 to 250 m below ground level. The intensive pumping of groundwater due to population increase and urbanization affect [4] led to the fall of groundwater levels about 1.43 m in the Varanasi area [5].

3. Methodology

Dugwells and borewells were selected for sampling, which are functional and continuously in use for drinking and domestic purposes. A total of 70 groundwater samples were collected from dug wells and bore wells (hand pumps) in the study area during November, 2011 (Post-monsoon) and April, 2012 (Pre-monsoon). Sample bottles were cleaned by rinsing them with distilled water followed by their treatment with 1M solution of the preservative acid. In the case of bore wells (hand pumps) the water samples were collected after pumping for 10 - 15 minutes in order to remove stagnant groundwater. Physico-chemical parameters such as pH, total dissolved solid (TDS) and electrical conductivity (EC) were measured by EC and pH meters in the field using the standard procedures. F was analyzed using Orion ion selective electrode 4 Star. Na and K were determined by using flame photometer (Elico CL-378). Ca, HCO_3, hardness, alkalinity and Cl were analyzed by titrimetric method and magnesium estimated by the difference in the hardness and calcium [6]. SO_4, PO_4, NO_3 and dissolved silica (SiO_2) were determined by UV-3200 double beam spectrophotometer model. The chemical data of groundwater samples are subjected to compute the ionic-balance-error between the total concentration of cations (Ca, Mg, Na and K) and total concentration of anions (HCO_3, Cl, SO_4, NO_3 and F) for testing accuracy of chemical analysis of each groundwater samples, before the interpretation of the chemical data is undertaken. The value of the ionic-balance-error is observed to be within the acceptable limit of ±5% [7].

$$\text{Electro neutrality}(\%) = [\text{total cations} + \text{total anions/total cations} - \text{total anions}]*100 \qquad (1)$$

4. Result and Discussion

Understanding the quality of groundwater is important because it is the main factor which decides its suitability for domestic, agricultural and industrial purposes. Range of chemical parameters in groundwater and their comparison with WHO standards are presented in **Table 1**.

4.1. Groundwater Classification and Hydrogeochemical Facies

As water flows through an aquifer it assumes a characteristic chemical composition as a result of interaction

Table 1. Ranges of chemical parameters and their comparison with WHO standards for drinking water.

Parameters	Range		Permissible limits	Sample number (% of sample exceeding permissible limit)		Undesirable effect
	Pre	Post	WHO (1997)	Pre-monsoon	Post-monsoon	
Ca^{2+}	30 - 260	34 - 267	200	29, 30 (3%)	29, 30, 58 (4%)	Scale formation
Mg^{2+}	1.1 - 109	1.9 - 110	150			-
Na^+	14 - 285	25 - 298	200	13, 16, 20, 24, 69 (7%)	1, 13, 16, 20, 24, 27, 69 (10%)	High blood pressure
K^+	1.3 - 109	1.5 - 91	12	1, 7, 14, 24, 28, 38, 45, 51, 58 (13%)	1, 7, 14, 24, 28, 38, 45, 51, 58 (13%)	Bitter taste
HCO_3^-	220 - 657	248 - 666	600	15, 20, 24, 34, 42, 58 (9%)	15, 20, 24, 34, 42, 47, 58 (10%)	-
Cl^-	54 - 494	76 - 513	600	-	-	Salty taste
PO_4^{3-}	0.2 - 7.3	0.2 - 14		-	-	-
F^-	0.2 - 5	0.2 - 6	1.5	7, 16, 17, 19, 20, 23, 32, 42, 46, 51 (14%)	7, 15, 16, 17, 19, 20, 21, 23, 25, 32, 42, 45, 46, 51, 52 (21%)	Mottling of tooth, deformation of bones
NO_3^-	1.3 - 90	2.2 - 106	50	7, 13, 19, 62, 66 (7%)	7, 13, 19, 62, 66 (7%)	Methaemoglobinaemia
SO_4^{2-}	2.2 - 302	3.9 - 316	600	-	-	Laxative effect
TH	140 - 720	144 - 776	500	26, 29, 30, 34, 46, 47, 58, 65 (11%)	15, 26, 29, 30, 34, 46, 47, 58, 65 (13%)	Scale formation
TA	180 - 539	203 - 546	-	-	-	Unpleasant
pH	6.9 - 7.9	7 - 7.9	9.2	-	-	Taste
TDS	258 - 1120	267 - 1281	1,500	-	-	Gastrointestinal irritation

with the lithologic framework. The term hydrochemical facies is used to describe the bodies of groundwater in an aquifer, that differ in their chemical composition. The facies are a function of the lithology, solution kinetics and flow patterns of the aquifer. Hydochemical facies can be classified on the basis of dominant ion using the piper's trilinear diagram. The concentrations of major ionic constituents of groundwater samples were plotted in the piper trilinear diagram [8] to determine the water type (**Figure 2** and **Figure 3**). Piper's trilinear diagram method is used to classify the groundwater, based on basic geochemical characters of the constituent ionic concentrations. The diagram consists of two triangular fields and a central diamond shaped field. The diamond shaped field between the two triangles is used to represent the composition of water with respect to both cations and anions. In the two tri-angular fields, percentage epm values of major cations and anions are plotted separately

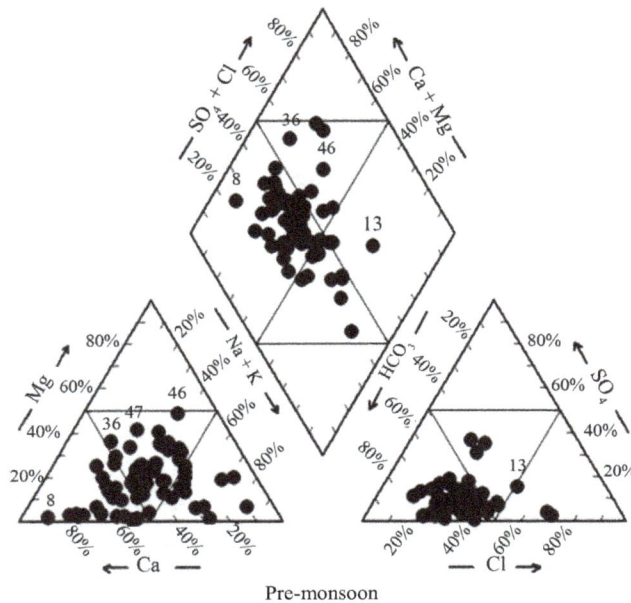

Figure 2. Piper diagram showing the relative cation and anion composition of groundwater samples.

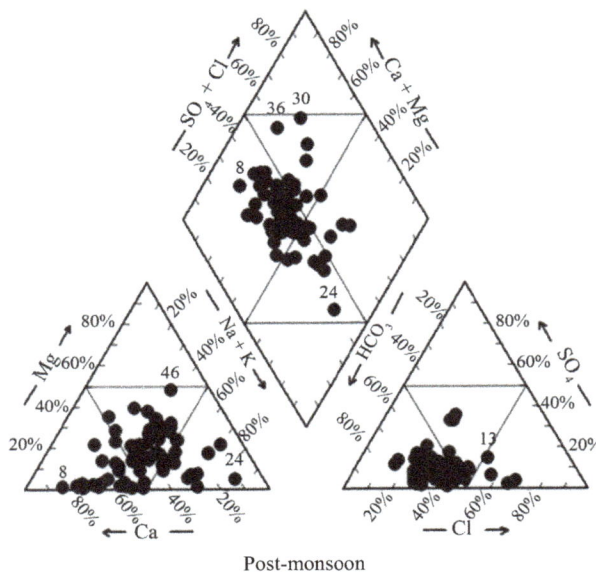

Figure 3. Piper diagram showing the relative cation and anion composition of groundwater samples.

and then projected on to the central field for the representation of overall characteristic of water. The classification for cation and anion facies, in terms of major ion percentages and water type, is according to the domain in which they occur on the diagram segments [9]. From the cationic and anionic triangular files of piper diagram, it is observed that 52%, 27% and 21% of groundwater samples fall into the no dominant, Ca and Na fields in cation facies of pre-monsoon respectively, whereas 50%, 27% and 23% of groundwater samples fall into the no dominant, Ca and Na fields in cation facies of post-monsoon, respectively. Conversely 87%, 4% and 9% of groundwater samples fall into the HCO_3, Cl and no dominant fields in anion facies of pre-monsoon, respectively, whereas 81%, 6% and 13% of groundwater samples fall into the HCO_3, Cl and no dominant fields in anion facies of post-monsoon, respectively.

The plot of the chemical data on diamond shaped trilinear diagram (**Figure 2** and **Figure 3**) reveals that majority of the groundwater samples fall in the field of 1, 3, 5 suggesting that alkaline earth exceeds alkalies, weak acids exceeds strong acids and the ions representing carbonate hardness (secondary alkalinity) exceeds 50% respectively (**Table 2**) in both season. From the data plots, it is apparent that the total hydrochemistry is dominated by alkaline earths and weak acids. However, some of the groundwater samples having high sulfate and chloride concentration fall in 2 and 4 fields indicating alkalies exceeds alkaline earths and strong acids exceeds weak acids. Some samples also fall in the field 9 indicating mixed water having no one cation-anion pair exceeds 50%. 2 samples fall in the field of 7 indicating that non-carbonate alkali (primary salinity) exceeds 50%. The diagram indicates dominance of the major ions Ca, Na and HCO_3 while other ions, such as Mg, SO_4 and Cl are comparatively less represented.

4.2. Groundwater Suitability for Drinking and Domestic Purposes

The water used for drinking should be free from colour, turbidity and microorganism. To understand the groundwater suitability for drinking and public health use, hydrochemical parameters of the study area are compared (**Table 1**) with the guidelines prescribed by World Health Organization [10]. From the **Table 1**, it is evident that 3% and 4% samples in pre- and post-monsoon for calcium ion, 7% and 10% samples for sodium ion in pre- and post-monsoon, 13% samples for potassium ion in pre- and post-monsoon, 14% and 21% samples for fluoride ion in pre- and post-monsoon, 7% sample for nitrate ion in pre- and post-monsoon and 9% and 10% samples for bicarbonate ion in pre- and post-monsoon are exceeding the permissible limit set by WHO.

The pH values of groundwater samples range between 6.9 and 7.9 in pre-monsoon and 7 to 7.9 in post-monsoon season, which indicates that the groundwaters are slightly alkaline in nature. In general, groundwater pH is slightly alkaline due to the influx of HCO_3 ions in the groundwater aquifer which is due to percolation of rain water through soil [11] [12]. Total dissolved solids (TDS) mainly consist of inorganic salts such as carbonates, bicarbonates, chlorides, sulfates, phosphates and nitrates of calcium, magnesium, sodium, potassium, iron etc and small amount of organic matter and dissolved gases. The concentration of (TDS) is ranging from 258 to 1120 mg/l in pre-monsoon and 267 to 1281 mg/l in post-monsoon (**Table 1**). To ascertain the suitability of ground-

Table 2. Distribution of groundwater samples (%) in the subdivisions of Piper diagram (Piper 1953).

Area	Sub-division	Sample fall (%)	
		Pre	Post
1	Alkaline earths exceed alkalies	80%	77%
2	Alkalies exceed alkaline earths	20%	23%
3	Weak acids exceed strong acids	87%	79%
4	Strong acids exceed weak acids	13%	21%
5	Carbonate hardness (secondary alkalinity) exceeds 50%	70%	64%
6	Non-carbonate hardness (secondary salinity) exceeds 50%	0%	0%
7	Non-carbonate alkali (primary salinity) exceeds 50%	3%	6%
8	Carbonate alkali (primary alkalinity) exceeds 50%	0%	0%
9	No one cation-anion pair exceeds 50%	27%	30%

water for any purposes, it is essential to classify the groundwater depending upon their hydro chemical properties based on their TDS values [13] [14] (**Table 3**). Based on total dissolved solids classification [13], 70% in pre-monsoon and 59% in post-monsoon are desirable for drinking purposes, 29% in pre- and 40% in post-monsoon are permissible for drinking and 1% in pre and post-monsoon are useful for agricultural purposes. According to Freeze and Cherry classification [14], 99% sample in pre and post monsoon belongs to fresh water category and 1% sample is brackish water. The classification of groundwater based on total hardness (TH) shows that 69% in pre- and 82% in post-monsoon samples (**Table 3**) fall in the very hard water category. Groundwater exceeding the limit of 300 mg/l is considered to be very hard water [15]. Hard water leads to incidence of urolithiosis, anencephaly, parental mortality, some types of cancer [16] and cardio-vascular disorders [17]. Such waters can also develop scales in water heaters, distribution pipes and well pumps, boilers and cooking utensils, and require more soap for washing clothes [18]-[20]. 30% and 17% groundwater samples in all the samples fall in moderately-hard category in pre- and post-monsoon, respectively and rest 1% groundwater sample indicates that slightly hard water in both the seasons. Among the cationic (Ca, Mg, Na and K) concentrations, sodium is the dominant ion (14 - 285 mg/l in pre and 25 - 298 mg/l in post) followed by calcium (30 - 260 mg/l in pre and 34 - 267 mg/l in post), magnesium (1.1 - 109 mg/l in pre and 1.1 - 110 mg/l in post) and potassium (1.3 - 109 mg/l in pre and 1.5 - 91 mg/l in post) in both the seasons. In general weathering, dissolution and base-exchange

Table 3. Classification of groundwater based on different parameters for different purposes.

Parameters	Range	Classification	% of sample	
			Pre-monsoon	Post-monsoon
TDS (David and De Wiest,1966)	<500	Desirable for drinking	70%	59%
	500 - 1000	Permissible for drinking	29%	40%
	1000 - 3000	Useful for agriculture	1.00%	1.00%
	>3000	Unfit for drinking and irrigation		
TDS (Freeze and Cherry, 1979)	<1000	Fresh water	99%	99%
	1000 - 10,000	Brackish water	1%	1%
	10,000 - 100,000	Saline water		
	>100,000	Brine water		
Hardness (Sawyer and Mc. Cartly, 1967)	<75	Soft		
	75 - 150	Slightly hard	1%	1%
	150 - 300	Moderately hard	30%	17%
	>300	Very hard	69%	82%
Chloride (Stuyfzand,1989)	<0.141	Extremely fresh		
	0.141 - 0.846	Very fresh		
	0.846 - 4.231	Fresh	64%	53%
	4.231 - 8.462	Fresh brackish	32%	40%
	8.462 - 28.206	Brackish	4%	7%
	28.206 - 282.064	Brackish salt		
	282.064 - 564.127	Salt		
	>564.127	Hyperhaline		
WQI	<50	Excellent water	0%	0%
	50 - 100	Good water	50%	54%
	100 - 200	Poor water	34%	24%
	200 - 300	Very poor water	9%	12%
	>300	Water unsuitable for drinking purposes	7%	10%

Where, TDS, hardness, and WQI in mg/l and chloride in meq/l.

processes control the levels of cationic concentrations in groundwater. High concentration of Na and Ca in the groundwater is attributed to cation exchange among minerals. Na was higher in both seasons indicating weathering from plagioclase bearing rocks. K was lesser in both the seasons indicating its lower geochemical mobility. Among the anionic (HCO_3, Cl, PO_4, F, NO_3, SO_4) concentrations bicarbonate is the dominant ion (220 - 657 mg/l in pre and 248 - 666 mg/l in post), followed by chloride (54 - 494 mg/l in pre and 76 - 513 mg/l in post), sulphate (2.2 - 302 mg/l in pre and 3.9 - 316 mg/l in post), nitrate (1.3 - 90 mg/l in pre and 2.2 - 106 mg/l in post) and fluoride (0.2 - 5 mg/l in pre and 0.2 - 6 mg/l in post) in both the seasons.

Chloride in all the samples is below the [10] limit in pre- and post-monsoon, respectively (**Table 1**). The chloride limits have been laid down primarily from taste view point. However, no adverse health effects on human being have been reported by the use of water having high chloride concentrations [21]. Excess concentration of Cl in drinking water gives a salty taste and has a laxative effect in people not accustomed to it. Based on Cl classification [22], none of the samples in both seasons are fall in extremely fresh and very fresh category. 64% sample in pre- and 53% in post-monsoon are fall in fresh water while 32% and 40% samples indicates fresh brackish category. 4% and 7% samples indicate that the brackish water type may be due to dumping of solid waste and intense agricultural practice (**Table 3**).

Five and seven samples, out of 70 collected groundwater samples are exceeding the permissible limits (>200 mg/l) of sodium content in the study area in pre-and post-monsoon season, respectively (**Table 1**). A sodium-restricted diet is recommended to patients suffering from hypertension or congenial heart diseases and also from kidney problems. For such people, extra intake of Na through drinking water may prove critical [23]. Na has different role in human body. It is related with the function of nervous system, membrane system and excretory system. Excess sodium causes high pressure, nervous disorder, etc. Ten and fifteen samples, out of 70 groundwater samples are exceeding the permissible limits (>1.5 mg/l) of fluoride content in the study area in pre-and post-monsoon season, respectively (**Table 1**). Fluoride is an essential element for maintaining normal development of healthy teeth and bones. Deficiency of F in drinking water below 0.6 mg/l contributes to tooth caries. An excess of over 1.2 mg/l causes fluorosis [24].

Four and five samples, out of 70 groundwater samples are exceeding the permissible limits (>50 mg/l) of nitrate content in the study area in pre- and post-monsoon season, respectively (**Table 1**). Concentrations of NO_3 are the result of different pollution processes involving municipal wastewaters, fertilizers (containing NPK) and the application of agricultural pesticides, among others. The highest NO_3 Concentrations were observed in areas where large amounts of N fertilizers (commonly urea, nitrate or ammonium compounds) are used due to intensive agricultural practices. Excessive NO_3 in drinking water can cause a number of health disorders, such as methemoglobinemia, gastric cancer, goitre, birth malformations and hypertension [25]. The high concentration of nitrate is due to the intensive urbanization and industrialization [9].

4.3. Water Quality Index (WQI)

Water quality index (WQI) is a simple and concise method useful for indicating impairment of water quality. WQI helps for the better management of water quality issues and improve the effectiveness of protective measures. It is an important parameter to classify water quality for suitability of drinking purposes [26]. The standards for drinking purposes as recommended by BIS 10,500 (2003) have been considered for the calculation of WQI (**Table 4**). WQI calculation has been done by assigning weights (w_i) according to relative importance of each chemical parameter for drinking purposes (**Table 4**). The parameters like chloride, nitrate, total dissolved solids, fluoride, iron and sulfate has been assigned maximum weight 5 because of the major importance in water quality assessment [27]. Bicarbonate and phosphate is given the minimum weight of 1 as it plays an insignificant role in the water quality. Other parameters like calcium, magnesium, sodium, total hardness (TH), manganese, silicate and potassium were assigned weight between 1 and 5 depending on their importance in water quality determination. The relative weight (W_i) is computed (**Table 4**) from the following equation:

$$W_i = \frac{w_i}{\sum_{i=1}^{n} w_i} \tag{2}$$

where, W_i is the relative weight, w_i is the weight of each parameter and n is the number of parameters.

A quality rating scale (q_i) for each parameter is assigned by dividing its concentration in each water sample by

Table 4. Relative weight of chemical parameters.

Chemical parameters	BIS (mg/l)	Weight (w_i)	Relative weight (W_i)
TH	300	2	0.047
Ca	75	3	0.07
Mg	30	3	0.07
Alkalinity	200	1	0.023
Cl	250	5	0.116
TDS	500	5	0.116
F	1	5	0.116
Mn	0.1	4	0.093
NO$_3$	45	5	0.116
Fe	0.3	5	0.116
SO$_4$	200	1	0.023
PO$_4$	0	4	0.093
Na	0	2	0.047
K	0	2	0.047
Silicate	0	2	0.047
		$\sum w_i = 49$	$\sum W_i = 1.14$

its respective standard according to the guidelines laid down in the [28] and the result is multiplied by 100:

$$q_i = (C_i/S_i) \times 100 \tag{3}$$

where, q_i is the quality rating, C_i is the concentration of each chemical parameter in each water sample in milligrams per liter, S_i is the Indian drinking water standard for each chemical parameter in milligrams per liter according to the guidelines of the [28]. For computing the WQI, the SI is first determined for each chemical parameter, which is then used to determine the WQI as per the following equation:

$$SI_i = W_i \times q_i \tag{4}$$

$$WQI = \sum SI_i \tag{5}$$

where SI_i is the sub-index of ith parameter, q_i is the rating based on concentration of ith parameter, n is the number of parameters. The WQI range and type of water can be classified as excellent water (<50); good water (50 - 100); poor water (100 - 200); very poor water (200 - 300); water unsuitable for drinking purposes (>300) (**Table 3**). The calculated WQI values of the study area ranges from 51.7 to 405 mg/l in pre- and 53 to 484 mg/l in post-monsoon (**Figure 4** and **Figure 5**).

Out of 70 groundwater sample, 50% in pre- and 54% in post-monsoon represents good water, 34% in pre- and 24% in post-monsoon indicate poor water, and 9% in pre- and 12% in post-monsoon shows very poor water and 7% in pre- and 10% in post-monsoon indicate water unsuitable for drinking purposes. This may be due to effective leaching of ions, overexploitation of groundwater, direct discharge of effluents, and agricultural impact. The high value of WQI at some locations has been found to be mainly from the higher values of iron, nitrate, total dissolved solids, hardness, fluorides, bicarbonate and manganese in the groundwater.

4.4. Water Quality for Irrigation Purposes

The concentration and composition of dissolved constituents in groundwater determine its quality for irrigation use. The suitability of groundwater for irrigation is liable on the effects of the mineral constituents in the water on both the plants and soil [29]. Higher salt content in irrigation water causes an increase in soil solution osmotic pressure [30]. Effect of salts on soil causing changes in soil structure, permeability and ae-

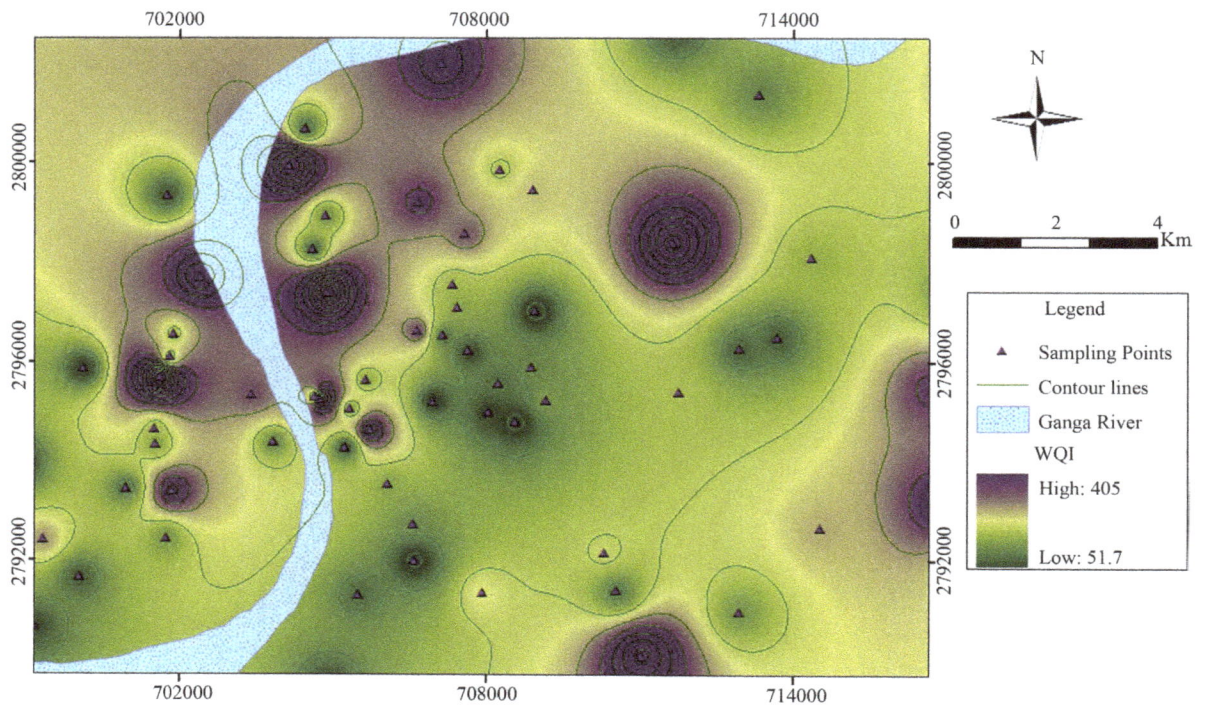

Figure 4. Spatial distribution of water quality index (mg/l) during pre-monsoon season.

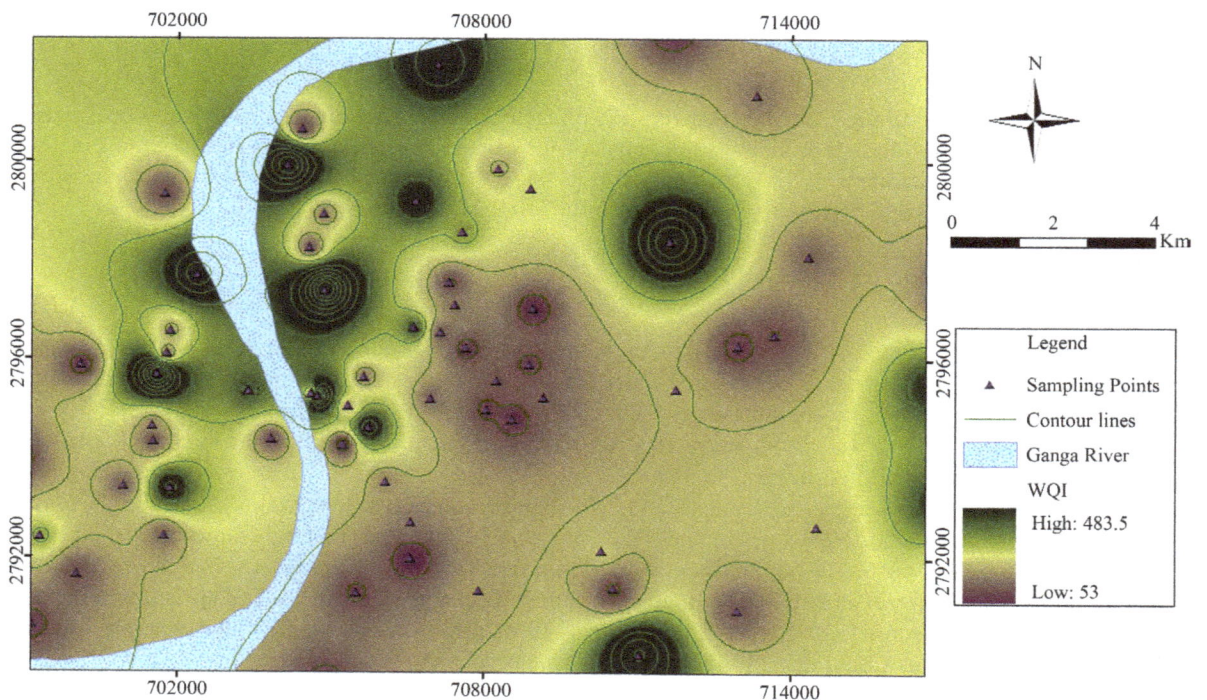

Figure 5. Spatial distribution of water quality index (mg/l) during post-monsoon season.

ration in directly affect plant growth. Since plant roots extract water osmosis, the water uptake of plants decreases. The osmotic pressure is proportional to the salt content or salinity hazard. The salts, besides affecting the growth of plants directly, also affect the soil structure, permeability and aeration, which indirectly affect the plant growth. An important factor allied to the relation of crop growth to water quality is drainage.

If a soil is open and well drained, crops may be grown on it with the application of generous amounts of saline water; on the other hand, a poorly drained area combined with application of good quality water may fail to produce as satisfactory a crop [18].

The important chemical parameter for judging the degree of suitability of water for irrigation are sodium (%Na), sodium adsorption ratio (SAR), residual sodium carbonate (RSC) and permeability index (PI).

4.5. Sodium Percentage (% Na)

Percent sodium (% Na) is also widely utilized for evaluating the suitability of water quality for irrigation [31]. Excess sodium concentration in groundwater produces the undesirable effects because Na reacts with soil to reduce its permeability and support little or no plant growth [9] [26]. The % Na is computed with respect to relative proportions of cations present in water, where the concentrations of ions are expressed in meq/l, using the following formula:

$$Na\% = (Na + K) \times 100 / \{(Ca + Mg + Na + K)\} \, (meq/l) \tag{6}$$

The calculated values of Na% ranges 10.14 to 83.77 in pre- and 15.43 to 83.61 in post-monsoon. Generally, % Na should not exceed 60% in irrigation waters. 8% sample in pre- and 11% sample in post-monsoon are higher than 60% of Na% are unsafe for irrigation. The Na% was higher may be dissolution of minerals from lithological composition, and the addition of chemical fertilizers by the irrigation waters [32].

4.6. Sodium Absorption Ratio (SAR)

The important chemical parameter for estimating the degree of suitability of water for irrigation as sodium content or alkali hazard for crops, which is expressed in sodium adsorption ratio (SAR). SAR is calculated from the ratio of sodium to calcium and magnesium. Calcium and magnesium ions are important since they are tending to counter the effect of sodium. Higher concentration of SAR leads to breakdown in the physical structure of the soil. Sodium is adsorbed and become attached to soil particles. The soil then become hard and compact when dry and impervious to water penetration. Sodium replacing adsorb calcium and magnesium is a hazard as it causes damage to the soil structure. The degree to which irrigation water tends to enter into cation exchange reaction in soil can be indicated by the SAR. The SAR recommended by the salinity laboratory of the US Department of Agriculture [31] is calculated using the formula:

$$SAR = Na^+ / \{(Ca + Mg)/2\} 0.5 \, (meq/l) \tag{7}$$

There is a close relationship between SAR values in irrigation water and the extent to which Na is absorbed by soils. If water used for irrigation is high in Na and low in Ca, the ion-exchange complex may become saturated with Na, which destroys soil structure, because of dispersion of clay particles. As a result, the soils tend to become deflocculated and relatively impermeable. Such soils can be very difficult to cultivate. The sodium hazard is expressed in terms of classification of irrigation water as low (S1: <10), medium (S2: 10 to 18), high (S3: 18 to 26) and very high (S4: >26). The SAR value in ground water sample range from 0.34 to 10.37 in pre-monsoon and 0.60 to 10.49 in post-monsoon indicating that all the groundwater samples are suitable for irrigation purposes (**Table 5**).

4.7. Residual Sodium Carbonate (RSC)

The excess sum of carbonate and bicarbonate in groundwater over the sum of calcium and magnesium also influences the suitability of groundwater for irrigation. When the excess carbonate concentration becomes too high, the carbonate combines with calcium and magnesium to form solid materials which settles out of the water. The relative abundance of sodium with respect to alkaline earths and the quantity of bicarbonates and carbonate in excess of alkaline earths also influence the suitability of water for irrigation. RSC is an important parameter to evaluate the suitability of irrigation water [9] [33], calculated using the formula.

$$RSC = [(HCO_3 + CO_3) - (Ca + Mg)] \, (meq/l) \tag{8}$$

Generally, >2.5 meq/l of RSC is unsuitable for irrigation purposes. The RSC value in ground water sample range from −9.15 to 7.13 in pre-monsoon and −8.70 to 7.15 in post-monsoon was observed (**Table 5**).

Table 5. Classification of groundwater for agricultural purposes.

Parameter	Sample range		Range	Classification	% of Sample	
	Pre	Post			Pre	Post
Na% (meq/l)	10 - 83.7	15.4 - 83.6	0 - 20	Excellent	6%	4%
			20 - 40	Good	54%	47%
			40 - 60	Permissible	32%	37%
			60 - 80	Doubtful	7%	10%
			>80	Unsuitable	1%	1%
SAR (meq/l)	0.34 - 10.37	0.60 - 10.49	0 - 10	Excellent (suitable for all types of crops and soil except for those crops sensitive to Na)	100%	100%
			10 - 18	Good (suitable for coarse textured or organic soil with permeability	-	-
			18 - 26	Fair (harmfully for almost all soils)	-	-
			>26	Poor (unsuitable for irrigation)	-	-
RSC (meq/l)	−9.15 - 7.13	−8.70 - 7.15	<1.25	Good	88%	87%
			1.25 - 2.5	Medium	6%	7%
			>2.5	Bad	6%	6%
EC (µS/cm)	525 - 2240	535 - 2562	<250	Low salinity hazard (good)	-	-
			250 - 750	Medium salinity hazard (moderate)	36%	33%
			750 - 2250	High salinity hazard (poor)	64%	66%
			>2250	Very high salinity hazard (very poor)	-	1%
PI (meq/l)	35.8 - 102	36 - 101	Class I	Max. permeability	11%	14%
			Class II	75% of Max. permeability	89%	86%
			Class III	25% of Max. permeability	-	-

The classifications of ground water for irrigation purpose according to the RSC values indicate that about 6% water sample in pre- and post-monsoon were beyond the permissible limit. 88% and 87% samples in pre- and post- monsoon have RSC value much less than 1.25 meq/l indicates that safe quality categories for irrigation while 6% and 7% in pre- and post-monsoon indicates that marginal quality for irrigation. Most of the sampling sites show the negative RSC value indicates that there is no complete precipitation of calcium and magnesium.

4.8. Permeability Index (PI)

Soil permeability is affected by long-term use of irrigation water with high salt content as influenced by Na^+, Ca^{2+}, Mg^{2+}, and HCO_3 contents of the soil. PI is defined by the following equation [34].

$$PI = Na + (HCO_3)^2 / Ca + Mg + Na * 100 \, (meq/l) \tag{9}$$

The PI values >75% comes under class I and indicates that the excellent quality of water for irrigation. The PI value between 25% - 75% comes under class II indicates that the good quality of water for irrigation and the PI value less than 25% comes under class III indicates that the unsuitable nature of water for irrigation. The calculated PI value ranges from 35.8 to 102 meq/l in pre-monsoon and 36 to 101 in post-monsoon respectively. According to the permeability index values, 11% and 14% of the groundwater in pre- and post-monsoon comes under class I (PI > 75%) category while 89% and 86% in pre- and post-monsoon comes under class II (PI ranges from 25% to 75%) category.

4.9. US Salinity Laboratory's Diagram

The US Salinity Laboratory's diagram [35] is widely used for rating the irrigation waters, where SAR is plotted against EC. The plots of chemical data of the groundwater samples in the US Salinity Laboratory's diagram are illustrated in **Figure 6**. The total concentrations of soluble salts in irrigation water can be classified into low (C1), medium (C2), high (C3) and very high (C4) salinity zones. The zones (C1 - C4) have the value of EC less than 250, 250 - 750, 750 - 2250 µS/cm and more than 2250 µS/cm, respectively. Higher EC in water creates a saline soil. The groundwater sample points, as shown as a cluster, fall in C2S1, C3S1, C3S2 and C4S1 zones. **Figure 6** shows that the 37% in pre- and 33% in post-monsoon of the groundwater samples fall in the category of C2S1, indicating medium salinity and low alkali water, which can be used for irrigating most of the soils and crops with little danger of exchangeable sodium. However, 60% in pre- and 63% in post-monsoon of the water samples fall in C3S1 class, which shows a high salinity hazard and low alkali hazards [29]. However, two samples (3% in pre and post-monsoon) fall in the C3S2 waters, indicating high salinity to a medium sodium type. This type of water may be used on coarse-textured or organic soils with good permeability [36]. 1% sample in pre- and post-monsoon fall is C4S1 category indicating very high salinity and low alkalinity hazards. This water will be suitable for plants having good salt tolerance and it restricts the suitability for irrigation, especially in soils with restricted drainage. Moderate and bad water quality types are increased in post-monsoon due to enrichment of Na and EC concentrations. The good waters can be used for irrigation with little danger of harmful levels of exchangeable Na. The moderate waters can be used to irrigate salt-tolerant and semi-tolerant crops under favourable drainage conditions. The bad waters are generally undesirable for irrigation and should not be used on clayey soils of low permeability. Bad waters, however, can be used to irrigate plants of high salt tolerance, when grown on previously salty soils to protect against further decline of fertile lands.

4.10. Wilcox's Diagram

Wilcox's diagram [31] is adopted for the classification of groundwaters for irrigation, wherein the EC is plotted against % Na. Data of pre- and post-monsoon groundwater samples of the area are plotted in the Wilcox's diagram (**Figure 7**). Out of the 70 groundwater samples, 37% in pre and 33% in post-monsoon of the groundwater

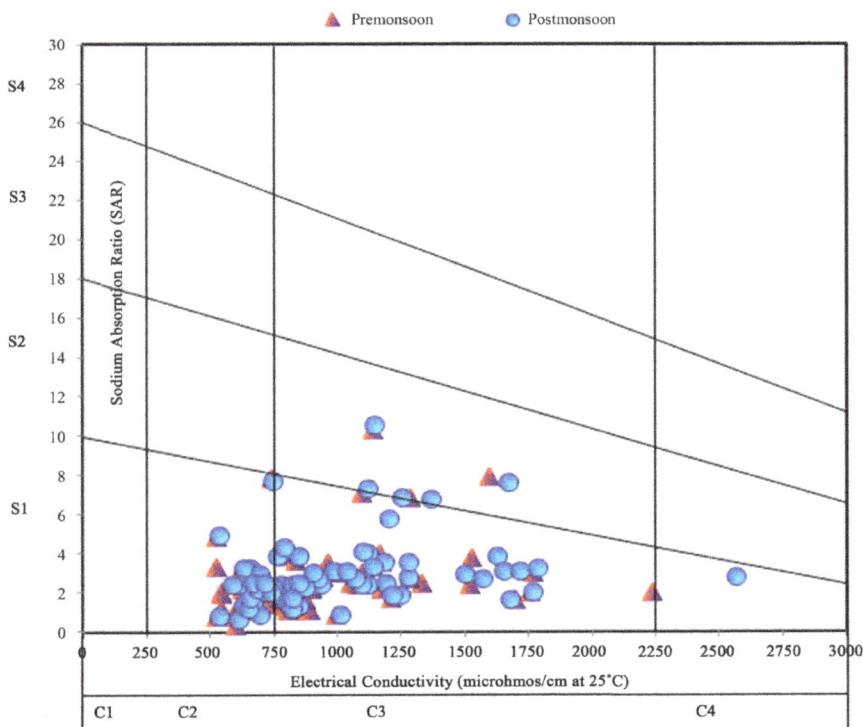

Figure 6. Classification of irrigation waters (after US Salinity Laboratory Staff 1954).

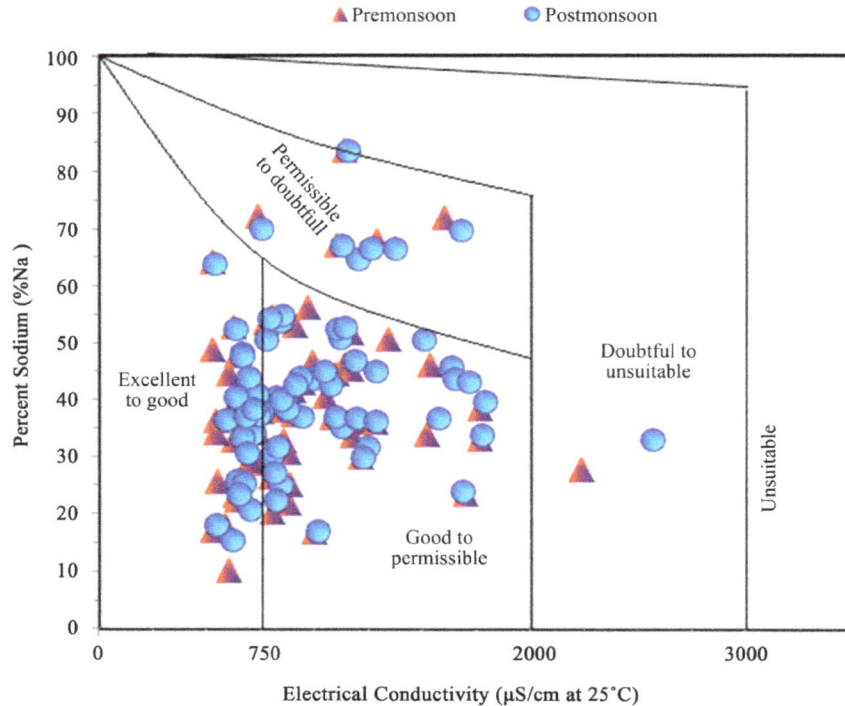

Figure 7. Classification of irrigation waters (after Wilcox 1948).

samples belong to excellent to good category followed by 52% in pre- and 56% in post-monsoon samples belonging to good to permissible category, 10% in pre- and post-monsoon belonging to permissible to doubtful, and 1% in pre- and post-monsoon sample belonging to doubtful to unsuitable category for irrigation use. The agricultural yields are generally low in lands irrigated with waters belonging to doubtful to unsuitable category. This is probably due to the presence of excess sodium salts, which cause osmotic effects on soil-plant system. When the concentration of sodium is high in irrigation water, sodium ions tend to be adsorbed by clay particles, displacing Mg and Ca ions by base-exchange process. This exchange process of Na in water for Ca and Mg in soil reduces permeability and eventually results in soil with poor internal drainage. Hence, air and water circulation is restricted during wet conditions, and such soils are usually hard when dry [37].

5. Conclusion

The study provides significant information on the development of ground water quality in parts of Chandauli-Varanasi region. Ground water is immensely important for water supply in both urban and rural areas of developing nations. The major ion chemistry data revealed that the ground water in the study area is slightly hard to very hard and fresh to brackish in nature. According to the WHO (1997) classification of water based on TDS, all the samples are within the permissible limit. The sequence of the abundance of the major ions is in the following order of $Na > Ca > Mg > K$ for cations and $HCO_3 > Cl > SO_4 > NO_3 > F$ in anions. The alkali earth (Ca + Mg) significantly exceeds alkalis (Na + K) and HCO_3 and Cl exceeds the other anions in both the seasons. The WQI calculated exhibits poor quality in less percentage indicating the effective ion leaching, overexploitation and anthropogenic activities from discharge of effluents from agricultural and domestic uses in both the seasons. Based on the classification of irrigation water according to SAR and PI values, all the sample locations are suitable for irrigation purposes. Irrigation water quality based on % Na indicates that 54% and 47% of the water samples belong to good category in pre- and post-monsoon, respectively. According to RSC values, 88% and 87% ground water samples are suitable for irrigation purposes in pre- and post-monsoon, respectively. Based on the Wilcox classification, 1% of the water samples belong to doubtful to unsuitable category for irrigation use due to the presence of excess sodium salts that cause deflocculating and reduce the permeability of soil. The analytical data plotted on the US Salinity diagram illustrate that 1% of the ground water samples fall in the field of C4S1, indicating high salinity, low sodium and 3% samples fall in the field of C3S2 indicating high salinity and medium sodium hazard rendering it usable only on coarse-textured or organic soils with good permeability.

References

[1] Raju, N.J., Shukla, U.K. and Ram, P. (2011) Hydrogeochemistry for the Assessment of Groundwater Quality in Varanasi: A Fast-Urbanizing Center in Uttar Pradesh, India. *Environmental Monitoring and Assessment*, **173**, 279-300. http://dx.doi.org/10.1007/s10661-010-1387-6

[2] Raju, N.J. (2012) Evaluation of Hydrogeochemical Processes in the Pleistocene Aquifers of Middle Ganga Plain, Uttar Pradesh, India. *Environmental Earth Sciences*, **65**, 1291-1308. http://dx.doi.org/10.1007/s12665-011-1377-1

[3] Shukla, U.K. and Raju, N.J. (2008) Migration of the Ganga River and Its Implication on Hydro-Geological Potential of Varanasi Area, UP, India. *Journal of Earth System Science*, **117**, 489-498. http://dx.doi.org/10.1007/s12040-008-0048-4

[4] Raju, N.J. and Reddy, T.V.K. (2007) Environmental and Urbanization Affect on Groundwater Resources in a Pilgrim Town of Tirupati, Andhra Pradesh, South India. *Applied Geochemistry*, **9**, 212-223.

[5] Sinha, T.K. (2003) Groundwater Conditions and Its Quality in Varanasi City. *Indian Journal of Geomorphology*, **8**, 153-154.

[6] APHA (2005) Standard Methods for the Examination of Water and Wastewater. 25th Edition, American Public Health Association, Washington DC.

[7] Domenico, P.A. and Schwartz, F.W. (1990) Physical and Chemical Hydrogeology. John Wiley & Sons, New York, 824.

[8] Piper, A.M. (1953) A Graphic Procedure in the Chemical Interpretation of Water Analysis. US Geological Survey Groundwater Note, 12.

[9] Raju, N.J., Ram, P. and Dey, S. (2009) Groundwater Quality in the Lower Varuna River Basin, Varanasi District, Uttar Pradesh, India. *Journal of the Geological Society of India*, **7**, 178-192. http://dx.doi.org/10.1007/s12040-008-0048-4

[10] World Health Organization (1997) Guideline for Drinking Water Quality. 2nd Edition, Vol. 2, WHO, Geneva, Health criteria and Other Supporting Information, 940-949.

[11] Mor, S., Ravindra, K., De Visscher, A., Dahiya, R.P. and Chandra, A. (2006) Municipal Solid Waste Characterization and Its Assessment for Potential Methane Generation: A Case Study. *Science of the Total Environment*, **371**, 1-10. http://dx.doi.org/10.1016/j.scitotenv.2006.04.014

[12] Alam, M., Rais, S. and Aslam, M. (2012) Hydrochemical Investigation and Quality Assessment of Ground Water in Rural Areas of Delhi, India. *Environmental Earth Sciences*, **66**, 97-110. http://dx.doi.org/10.1007/s12665-011-1210-x

[13] Davis, S.N. and De Wiest, R.J.M. (1966) Hydrogeology, Vol. 463. Wiley, New York.

[14] Freeze, R.A. and Cherry, J.A. (1979) Groundwater. Prentice Hall, Engle Wood Cliffs, 604.

[15] Sawyer, G.N. and McCartly, D.L. (1967) Chemistry of Sanitary Engineers. 2nd Edition, McGraw Hill, New York, 518.

[16] Agrawal, V. and Jagetia, M. (1997) Hydrogeochemical Assessment of Groundwater Quality in Udaipur City, Rajasthan, India. *Proceedings of National Conference on Dimensions of Environmental Stress in India*, Baroda, 151-154.

[17] Durvey, V.S., Sharma, L.L., Saini, V.P. and Sharma, B.K. (1991) Handbook on the Methodology of Water Quality Assessment. Rajasthan Agriculture University, India.

[18] Todd, D.K. (1980) Ground Water Hydrology. Wiley, New York, 535.

[19] Hem, J.D. (1991) Study and Interpretation of the Chemical Characteristics of Natural Water. 3rd Edition, Scientific Publishers, Jodhpur, 2254.

[20] Karanth, K.R. (1997) Groundwater Assessment, Development and Management. Tata McGraw-Hill Publishing Company Limited, New Delhi.

[21] Jain, C.K., Bandyopadhyay, A. and Bhadra, A. (2010) Assessment of Ground Water Quality for Drinking Purpose, District Nainital, Uttarakhand, India. *Environmental Monitoring and Assessment*, **166**, 663-676. http://dx.doi.org/10.1007/s10661-009-1031-5

[22] Stuyfzand, P.J. (1991) Non-Point Source of Trace Element in Potable Groundwater in Netherland. *Proceedings of the 18th International Water Supply Congress and Exhibition* (*IWSA*), Copenhagen, 25-31 May 1991, Water Supply 9.

[23] Holden, W.S. (1971) Water Treatment and Examination. John & Churchill Publishers, London.

[24] ISI (1983) Indian Standard Specification for Drinking Water. IS: 10500. Indian Standard Institute, India.

[25] Majumdar, D. and Gupta, N. (2000) Nitrate Pollution of Groundwater and Associated Human Health Disorders. *Indian Journal of Environmental Health*, **42**, 28-39.

[26] Vasanthavigar, M., Srinivasamoorthy, K., Vijayaragavan, K., Rajiv Ganthi, R., Chidambaram, S., Anandhan, P., Mani-

vannan, R. and Vasudevan, S. (2010) Application of Water Quality Index for Groundwater Quality Assessment: Thiru-manimuttar Sub-Basin, Tamilnadu, India. *Environmental Monitoring and Assessment*, **171**, 595-609.
http://dx.doi.org/10.1007/s10661-009-1302-1

[27] Srinivasamoorthy, K., Chidambaram, M., Prasanna, M.V., Vasanthavigar, M., Peter, J. and Anandhan, P. (2008) Iden-tification of Major Sources Controlling Groundwater Chemistry from a Hard Rock Terrain—A Case Study from Met-tur Taluk, Salem District, Tamilnadu, India. *Journal of Earth System Sciences*, **117**, 49-58.
http://dx.doi.org/10.1007/s12040-008-0012-3

[28] Bureau of Indian Standards (2003) Drinking Water-Specification IS: 10500. BIS, New Delhi.

[29] Richards, L.A. (1954) Diagnosis and Improvement of Saline Alkali Soils, Agriculture, 160, Handbook 60. US Depart-ment of Agriculture, Washington DC.

[30] Thorne, D.W. and Peterson, H.B. (1954) Irrigated Soils. Constable and Company Limited, London, 113.
http://dx.doi.org/10.1097/00010694-195411000-00021

[31] Wilcox, L.V. (1948) Classification and Use of Irrigation Waters. U.S. Department of Agriculture, Washington DC, 962.

[32] Subba Rao, N., Prakasa Rao, J., John Devadas, D., Srinivasa Rao, K.V., Krishna, C. and Nagamalleswara Rao, B. (2002) Hydrogeochemistry and Groundwater Quality in a Developing Urban Environment of a Semi-Arid Region, Guntur, Andhra Pradesh, India. *Journal of the Geological Society of India*, **59**, 159-166.

[33] Siddiqui, A., Naseem, S. and Jalil, T. (2005) Groundwater Quality Assessment in and around Kalu Khuhar, Super Highway, Sindh, Pakistan. *Journal of Applied Sciences*, **5**, 1260-1265. http://dx.doi.org/10.3923/jas.2005.1260.1265

[34] Raghunath, H.M. (1987) Groundwater. Wiley Eastern Ltd., Delhi.

[35] US Salinity Laboratory Staff (1954) Diagnosis and Improvement of Saline and Alkali Soils. Agricultural Handbook No. 60, USDA, USA, 160.

[36] Karanth, K.R. (1989) Hydrogeology. McGraw-Hill, New Delhi.

[37] Collins, R. and Jenkins, A. (1996) The Impact of Agricultural Land Use on Stream Chemistry in the Middle Hills of the Himalayas, Nepal. *Journal of Hydrology*, **185**, 71-86. http://dx.doi.org/10.1016/0022-1694(95)03008-5

8

Evaluating Groundwater Pollution Using Hydrochemical Data: Case Study (Al Wahat Area East of Libya)

Salam M. Rashrash[1], Bahia M. Ben Ghawar[1], Abdelrahim M. Hweesh[2]

[1]Faculty of Engineering, Geological Engineering Department, Tripoli, Libya
[2]General Water Authority, Tripoli, Libya
Email: srashrash@yahoo.com, gloriamuftah@yahoo.com, Ahweesh@yahoo.co.uk

Abstract

Water is one of the most challenging current and future natural resources, which will directly affect the environment and development by the changes in its quantity, quality and regional distribution. However, Water quality is the critical factor that influences human health and irrigation proposer. This work aims to investigate hydrochemical analysis and geochemical processes influencing the groundwater of Al Wahat area (Jalou, Awjla and Jukherra), which is located in central east Libya. Thirty four water samples collected from domestic and agricultural water wells were analyzed and used for conventional classification techniques which were Piper, Durov and Stiff diagrams to evaluate geochemical processes. Cluster analysis was used to identify the water type and ions concentration and distribution. Results show significant increase of dissolved salts, especially Nitrates. Elevated nitrates concentration can be attributed to either the disposal of untreated sewage water from disposal ponds and septic tanks or the infiltration of irrigation water saturated with fertilizing chemicals. Therefore, irrigation wells revealed that suffering from nitrate contamination caused an increase of the chance of nitrate pollution. In addition, contour maps present a sudden increase in the total dissolved salts (TDS) in the northeastern part coincident with the highest of secondary ions of NO_3 content, indicating the infiltration of irrigation water which is responsible partially for the groundwater degradation. Hydrogeochemical facie is NaCl type and enrichment of Na^+ and Cl^- can be attributed to urban untreated wastewaters and high rate of evapotranspiration. The concentrations of heavy elements such as Zn, Pb, Cu, Cd, Ni and Cr were low and within the WHO ranges.

Keywords

Al Wahat Area, Shallow Groundwater, Chemical Compound Analysis

1. Introduction

Fresh water in the Mediterranean regions represents 3% of the world's water resources though it gathers 7.3% of the world's population. 30 million Mediterranean citizens have no access to healthy water (PNUE [1]).

Much of the population in Libya is concentrated within a narrow strip along the Mediterranean coast; the bulk of the ground water potential is located to the south in the desert area such as the Murzuq and Al Kufra basin. Much of the ground water is used in irrigated agriculture, which represents 80% of total consumption (Alghariani [2]).

Groundwater can become contaminated from natural sources or numerous types of human activities. Waste from residential, commercial, industrial and agricultural activities can seriously affect groundwater quality. These contaminants may reach groundwater from activities on the land surface, such as industrial waste storage or spills, from sources below the land surface but above the water table, such as septic systems, from structures beneath the water table, such as wells, or from contaminated recharge from the aquifers.

The survey was performed cross Al Wahat area, about 6400 km^2 between 520,000 and 600,000 longitude E, and 3,174,000 and 3,254,000 latitude N (UTM WGS1984, zone 34). Data from ten water wells were not quite so extensive, but wells distributed throughout the region (**Figure 1**). Therefore, follow-up chemical variability of water, in particular, increased concentrations of dissolved salts of the Al Wahat area. Conducting chemical analyzes is to determine the extent of contamination and evaluate the quality and appropriateness of the use of urban and agricultural. In fact, understanding the origin and mechanisms of the salinization process is essential for preventing further deterioration of groundwater resources in the study area.

Consequently, the objective of this study is to understand the fluctuation and water quality of lower Middle Miocene aquifer. GWA [3] reported, for Jalo and Awjlah water situation study, that the main a semi-confined aquifer was post Middle Miocene. This aquifer, called the Marada Formation, is a series of fluviatile, medium- to coarse-grained sands with minor thicknesses of clay strata from the southwest, and grade finally into marine limestones, dolomites, shales, and clays with minor thicknesses of sandstones and sands beyond that area to the northeast (Wright *et al.* [4]). The water samples taken from wells have a depth ranging from 100 to 200 m.

2. Methodology

Thirty four water samples are taken from drinking, Piezometric and irrigation shallow aquifer. Two of these wells are used for domestic water supply, and the others are used to supply agricultural farms. Physical and chemical parameters of groundwater; pH, electrical conductivity (EC), total dissolved solids (TDS), total hardness (TH), Ca^{2+}, Mg^{2+}, Na^+, K^+, HCD_3^-, Cl^-, SO_4^{2-}, NO_3^- were measured by using the standard methods.

Delineating of the hydrochemical processes and defining groundwater types of hydrochemical facies was derived by constructing scatter plots, Piper, Durov and Stiff diagrams. This allowed to represent all chemical parameters (major and trace elements) and to study their relationship in the aquifer system. Using of these methods facilitates the interpretation of the evolutionary trends and the hydrochemical processes occurring in the groundwater system. In additional, Multivariate statistical analysis (cluster analysis or principal components analysis "PCA") is widely used to identify the sources of solutes in a groundwater system (Meng and Maynard [5]). It offers a better understanding of water quality and allows comparison of different samples of waters (Yidana *et al.* [6]). SURFER8 software used to elaborate the necessary maps.

3. Results and Discussion

The physical and chemical compositions of the Groundwater samples were statistically analyzed and the results (minimum, maximum, mean, and standard deviation of ions) obtained were summarized, and compliance with WHO [7] and EU [8] drinking water standards in **Table 1**.

Hydrochemical analysis presents the cations sodium (Na^+) is more abundant in the ground water (**Figure 2(a)**), while all the cations Ca^{2+}, Mg^{2+} and K^+ are decreases respectively. The second most abundant anion is Cl^- concentration increase (up to 2600 mg/L) to the central part of the study area, close to well 3 (**Figure 2(b)**), Considering that (Cl^-) is a major indicator that might be used to infer infiltration of waste water from cesspits into ground water (Foppen [9]). Sulfate (SO_4^{2-}) is the second anion present and has high concentration after the Cl^-. It increases on the north direction at wells 4 and 10. The presence of SO_4^{2-} frequently indicates a recharge in mixed water or a simple dissolution. Thus, the total anion charge of the samples decreases from the Cl^- to the

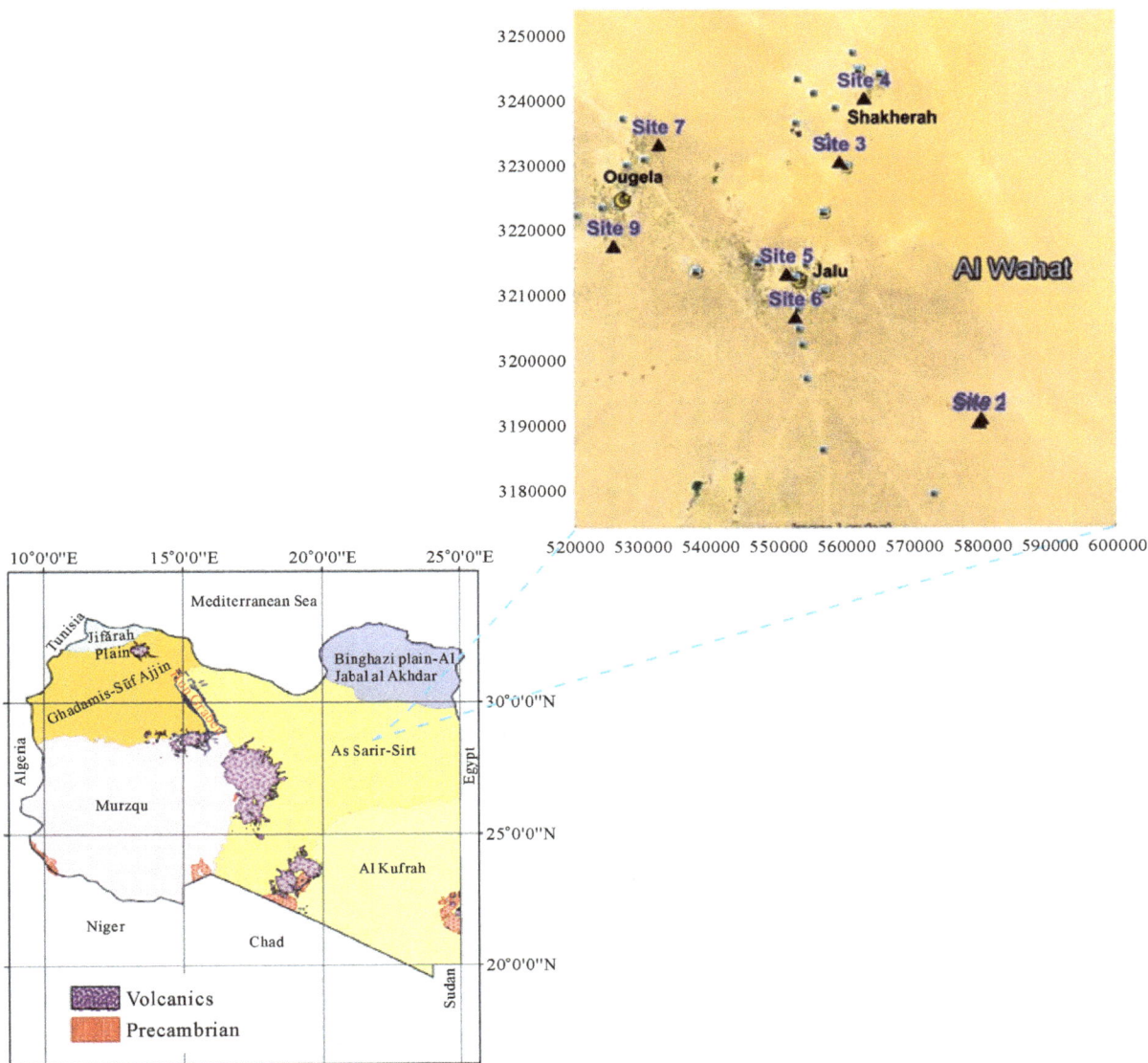

Figure 1. Location of the study area (after National Spatial Policy 2006).

Table 1. Evaluation of physical and chemical parameters of groundwater samples of the study area based on WHO (2006) and EU (1998) standards.

	TDS (mg/l)	PH	TH (mg/L)	Anions (mg/L)				Cations (mg/L)				
				Ca^{2+}	Mg^{2+}	Na^+	K^+	CO_3^{2-}	HCO_3^-	Cl^-	SO_4^{2-}	NO_3^-
Min	1045	6.77	360.07	64.016	38.416	320	9.8	0	3.2	6.768	5.081	0
Max	7216	7.71	2600.5	480.12	336.14	1900	35	300.0	341.71	2699	3338	199
Average	4563.8	7.31	1273.5	273.84	141.39	987.38	21.64	37.11	184.77	1444	991.9	86.3
SD	2498.7	0.27	749.51	142.12	93.118	620.29	8.372	98.93	111.93	982.2	1028	76.2
WHO (2006)	500	7 - 8.5	150	75	50	120	12	-	300	250	200	10
EU (1998)	NA	NA	100 - 500	NA	30 - 250	200	NA	-	NA	250	250	50 - 100

NA: Not available.

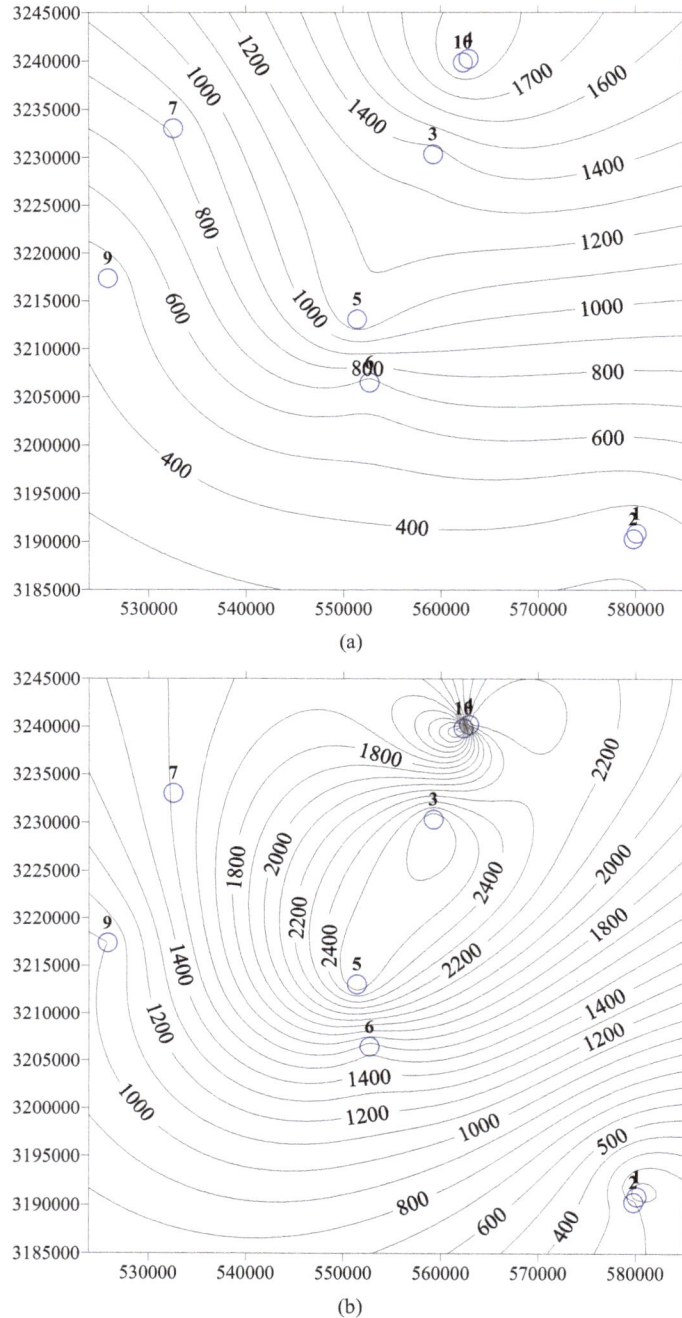

Figure 2. Cation sodium (Na^+) and anion chloride (Cl^-) ions concentration (mg/l) through the study area. (a) Na^+; (b) Cl^-.

SO_4^{2-}, HCD_3^-, NO_3^- to CO_3^-. This can be interpreted in terms of increasing weathering susceptibility from carbonates to halite. This trend also shows the presence of these diverse lithological components within aquifer systems. Generally, all cations and anions are increased toward centre of study area around well 5 and enrichment of Na^+ and Cl^- is also possible, related to urban wastewaters and high rate of evapotranspiration.

Cluster analysis is a data classification technique and one of the most powerful tools for analyzing water hydro-geochemical data (Reeve *et al.* [10]; Ochsenkuehn *et al.* [11]). This technique allows relationship investigation between the observations or the variables of a dataset, in order to recognize the existence of groups. The CA was used to split the standardized chemico-physical data into groups (clusters) based on similarities (or dissimilarities) so that each cluster represents a specific process in the aquifer system (Ragno *et al.*, [12]; Templ *et al.*, [13]. The result of the analyses is a graph, called dendrogram, which is a present Cl-Ca and SO_4-Na ions peers have close regression coefficient and the water type of samples were mainly HCO_3.

Nitrate levels for groundwater sources varied from 0 to 199 mg/l (**Table 1**). Contamination with waste water that might be infiltrating into it from surrounded cesspits. This inference might be supported [NO_3^-], elaborated within the same Table, where it is noted that range (highest in well 5). Nitrate concentration map (**Figure 3**) construct to illustrate major trend of increments and pollution trend.

The pH value is an important index of acidity or alkalinity and the concentration of hydrogen ion in GW Murugesan *et al.* [14]. The pH values of all water samples of different wards were found in permissible range of 7 - 8.5 according to WHO [7] recommended values. Electrical conductivity (EC) represents the total concentration of soluble salts in water. It is used to measure the salinity hazard to crops as it reflects the TDS in groundwater. Throughout the Al Wahat area, there is an increment from south to north.

TDS of groundwater salinity of post Middle Miocene values indicate a large range of variation from 1045 mg/L to about 7216 mg/L. It shows a strong mineralization of the water in the central and north direction of the study area. The water salinity in this area increases from south to north, which is illustrating existence of recharge processes at the southern part of the area. On other words, low TDS values, characterizing the southern and western border of the study area, reveal the dilution of the groundwater by the recharge coming from the southern border this region. Also, gradual increase of groundwater salinity is related to the abundance of evaporitic and marly deposits. Based on the total dissolved solids content in water after Todd [15] all the water in the study area was classified as brackish water because of TDS values between 1000 and 10,000 mg/L, as summarized in **Table 1**. According to Sawyer and MaCarty [16], total hardness classification scheme indicates the water samples were very hard (>300).

Piper-trilinear diagram permits the cation and anion compositions of samples to be represented on a single graph in which major groupings or trends in the data can be discerned visually (Freeze and Cherry [17]). Also, it is used to assess the hydrogeochemical facies. Based on the contents of major cations and anions, all samples fall within NaCl type, or type II (Na-K-Cl-SO_4) as shown in **Figure 4**. While **Figure 5** shows the comparison between the shapes of Stiff diagrams which reflects the concentration of water-quality constituents for the groundwater samples of the Al Wahat area. The samples 3, 4 and 5 show the highest concentration of chloride and (sodium + potassium).

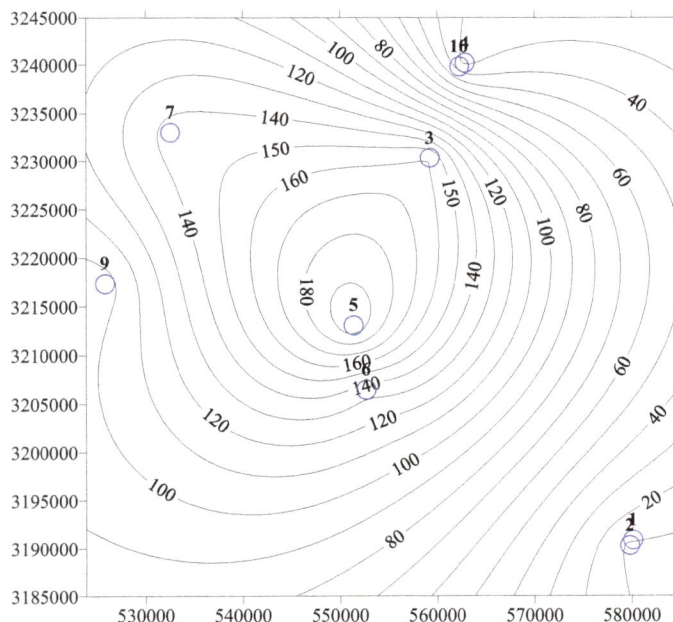

Figure 3. NO_3^- concentration (mg/l) map of the study area.

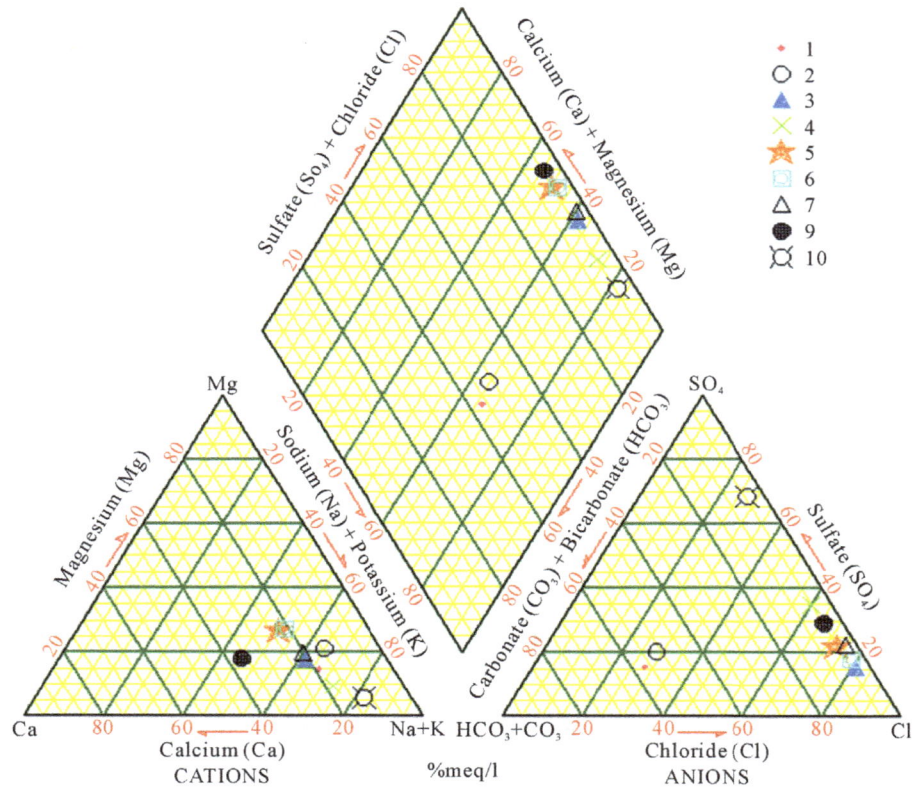

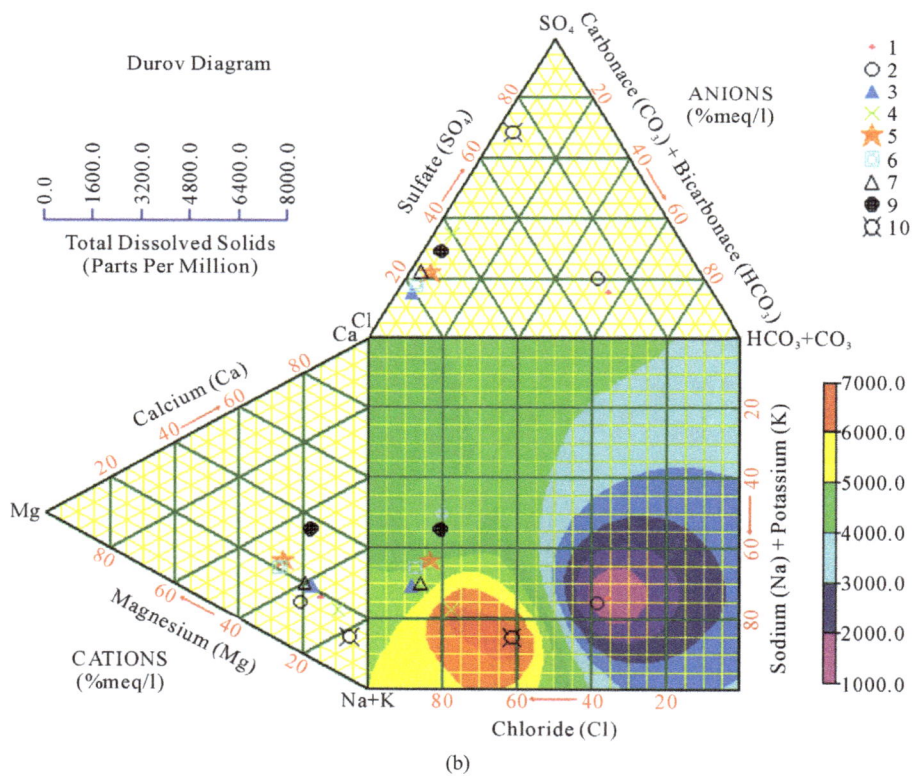

Figure 4. Piper-trilinear and Durov diagrams.

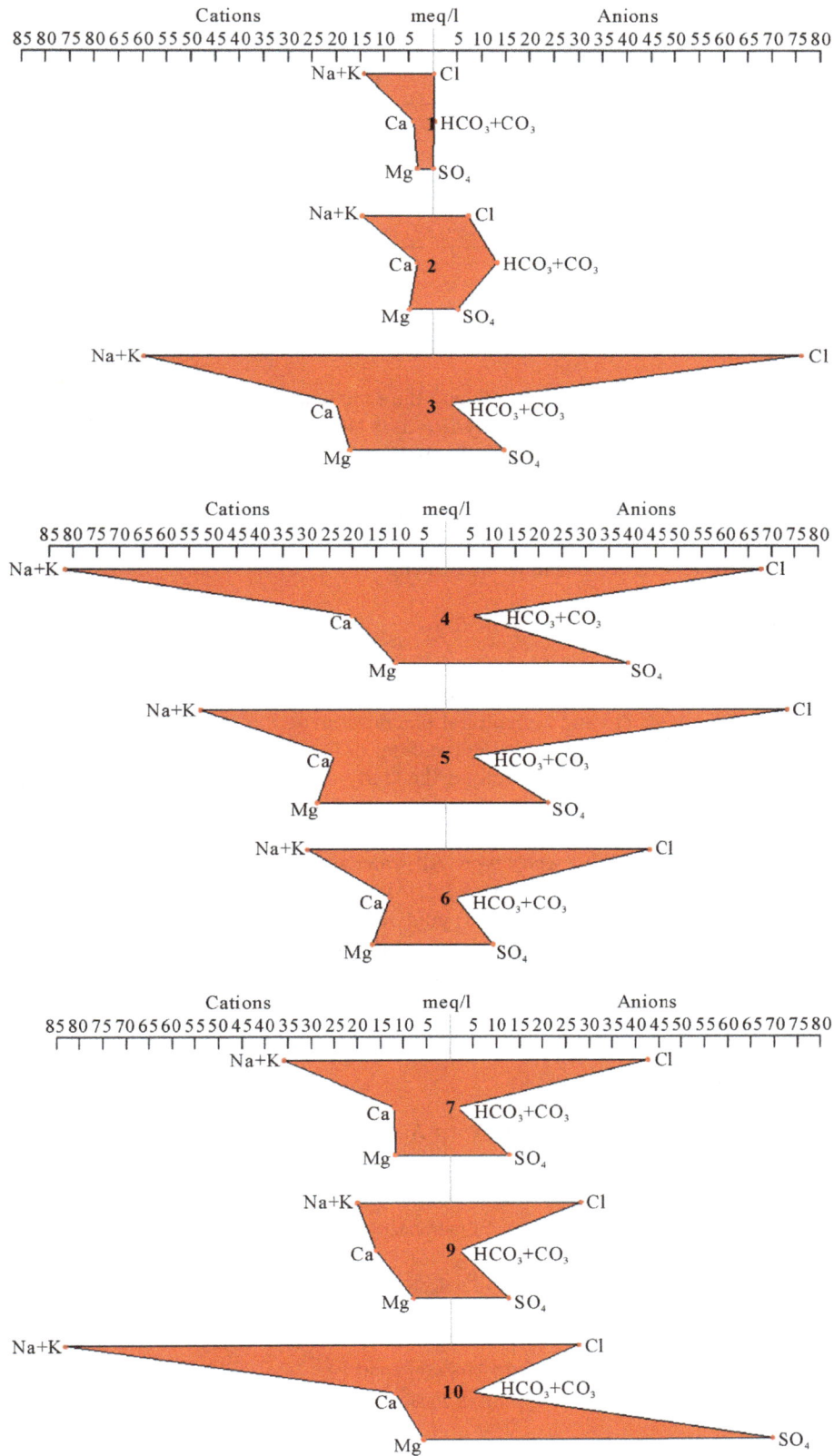

Figure 5. Stiff diagram of Al Wahat groundwater samples.

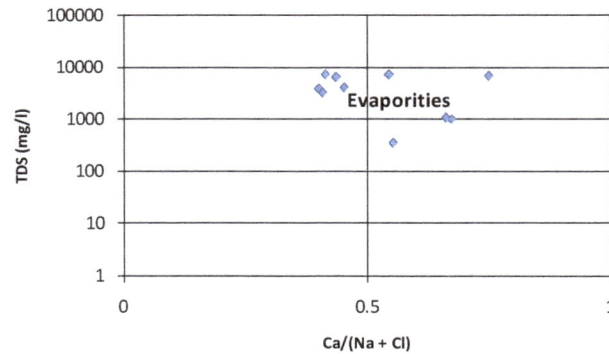

Figure 6. Gibbs diagram of Al Wahat groundwater samples.

The mechanism controlling water chemistry and the functional sources of dissolved ions can be assessed by plotting the ratios of $[Na^+/(Na^+ + Ca^{2+})]$ and $[Cl^-/(Cl^- + HCO^{3-})]$ as functions of TDS (Gibbs [18]). **Figure 6** presents Gibbs diagram of the water samples, clearly showing that the samples have become saline by evaporative enrichment. As noted, evaporation greatly increases the concentration of ions formed by chemical weathering, leading to higher salinity (Jalali [19]). The chemical composition of these water were mainly controlled by weathering reactions, as well as from dissolution of both carbonate and silicate minerals from them and by the interaction between the aquifer rocks and groundwater.

4. Conclusion

Hydrogeochemical studies are a useful tool which can help manage the quality of water resources. However, high concentration of NO_3^- is related to pollution, where NO_3^- has no known lithologic source, which is attributed to the urban wastewaters and agricultural practices involving chemical (nitrogenous) fertilizer applications. The cationic concentrations are ranged in the order of $Na^+ > Ca^{2+} > Mg^{2+} > K^+$, while it is $SO_4^{2-} > Cl^- > HCO_3^- > CO_3^- > NO_3^-$ for anions. Therefore, the study showed that all samples lay over normal chloride (<15 meq/l), and normal sulfate (<6 meq/l) except that water samples were taken from wells 1 and 2. However, they lay under normal bicarbonate water type. The Piper-trilinear diagram showed the predominance of fall within NaCl type, or type II (Na-K-Cl-SO$_4$) water type. All water samples fell within the recommended non permissible limit, except the water sample of well 1 and 2. All the physicochemical parameters of the water samples were not within the WHO (2006) and EU (1998) guidelines for drinking water and were non-suitable for human consumption in domestic uses except wells which were as mentioned above. According to the overall evaluation of the quality of the shallow ground water of the Al Wahat studied area, more hydrochemical investigations at central and north direction of the study area are required in general. Isotope analysis is recommended to determine the source of nitrates.

Acknowledgements

The authors like to acknowledge the General Water Authority for providing the laboratory data.

References

[1]   PNUE Programme des Nations Unies pour l'Environnement (2004) Plan d'Action pour la Méditerranée: MAP Technical Report Series No. 158.

[2]   Alghariani, S.A. (2002) Future Perspectives of Irrigation in Southern Mediterranean Region: Policies and Management Issues. In: Al-Rasheed, M., Singh, V.P. and Sheriff, M.M, Eds., *Proceedings of the International Conference on Water Resources Management in Arid Regions*, 313-320.

[3]   GWA (2002) General Water Authority. Unpublished Report.

[4]   Wright, E.P., *et al.* (1974) Jalu-Tazerbo Project: Phase I Final Report. Institute of Geological Sciences, London. (Unpublished)

[5]   Meng, S.X. and Maynard, J.B. (2001) Use of Statistical Analysis to Formulate Conceptual Models of Geochemical Behaviour: Water Chemical Data from the Botucatu Aquifer in Sa~o Paulotate, Brazil. *Journal of Hydrology*, **250**, 78-

97. http://dx.doi.org/10.1016/S0022-1694(01)00423-1

[6]   Yidana, S.M., Ophori, D. and Banoeng-Yakubob, B. (2008) A Multivariate Statistical Analysis of Surface Water Chemistry Data—The Ankobra Basin, Ghana. Journal of Environmental Management, **86**, 80-87. http://dx.doi.org/10.1016/j.jenvman.2006.11.023

[7]   WHO (2006) Guideline for Drinking Water Quality. 3rd Edition, World Health Organization.

[8]   EU (1998) Council Directive 98/83/EC of 3 November 1998 on the Quality of Water Intended for Human Consumption. *Official Journal of the European Communities*, **L330**, 32-54.

[9]   Foppen, J.W.A. (2002) Impact of High-Strength Was Tewater Infiltration on Groundwater Quality and Drinking Water Supply: The Case of Sana'a, Yemen.

[10]   Reeve, A.S., Siegel, D.I. and Glaser, P.H. (1996) Geochemical Controls on Peatland Pore Water from the Hudson Bay Lowland: A Multivariate Statistical Approach. *Journal of Hydrology*, **181**, 285-304. http://dx.doi.org/10.1016/0022-1694(95)02900-1

[11]   Ochsenkuehn, K.M., Kontoyannakos, J. and Ochsenkuehn, P.M. (1997) A New Approach to a Hydrochemical Study of Ground Water Flow. *Journal of Hydrology*, **194**, 64-75. http://dx.doi.org/10.1016/S0022-1694(96)03218-0

[12]   Ragno, G., De Luca, M. and Ioele, G. (2007) An Application of Cluster Analysis and Multivariate Classification Methods to Spring Water Monitoring Data. *Microchemical Journal*, **87**, 119-127. http://dx.doi.org/10.1016/j.microc.2007.06.003

[13]   Templ, M., Filzmoser, P. and Reimann, C. (2008) Cluster Analysis Applied to Regional Geochemical Data: Problems and Possibilities. *Applied Geochemistry*, **23**, 2198-2213. http://dx.doi.org/10.1016/j.apgeochem.2008.03.004

[14]   Murugesan, A., Ramu, A. and Kannan, N. (2006) Water Quality Assessment for Uttamapalayan Municipality in Theni District, Tamil Nadu, India. *Pollution Research*, **25**, 163-166.

[15]   Todd, D.K. (1980) Groundwater Hydrology. 2nd Edition, John Wiley and Sons Inc., Hoboken, 315.

[16]   Sawyer, G.N. and McCarthy, D.L. (1967) Chemistry of Sanitary Engineers. 2nd Edition, McGraw Hill, New York.

[17]   Freeze, R.A. and Cherry, J.A. (1979) Ground Water. Prentice Hall, Inc., Englewood Cliffs, 604 p.

[18]   Gibbs, R.J. (1970) Mechanisms Controlling World Water Chemistry. *Science*, **17**, 1088-1090. http://dx.doi.org/10.1126/science.170.3962.1088

[19]   Jalali, M. (2007) Salinization of Groundwater in Arid and Semi-Arid Zones: An Example from Tajarak, Western Iran. *Environmental Geology*, **52**, 1133-1149. http://dx.doi.org/10.1007/s00254-006-0551-3

9

Modeling of Infiltration Characteristics by Modified Kostiakov Method

Mahbub Hasan[1*], Tamara Chowdhury[1], Mebougna Drabo[2], Aschalew Kassu[1], Chance Glenn[3]

[1]Department of Engineering, Construction Management and Industrial Technology, Alabama A&M University, Normal, USA
[2]Department of Mechanical and Civil Engineering, Alabama A&M University, Normal, USA
[3]College of Engineering, Technology and Physical Sciences, Alabama A&M University, Normal, USA
Email: [*]mahbub.hasan@aamu.edu

Abstract

An infiltration characteristic model was developed by using the modified Kostiakov method for the Agricultural Engineering demonstration field of Bangladesh Agricultural Research Institute (BARI). The constant values a, α, and b of the equation for accumulated infiltration $y = at^\alpha + b$ were 9.12, 0.683, and 0.145, respectively. The average value of percentage of error between the actual and calculated values by the model was only 0.134 and showed very good agreement between the model and the field values of accumulated infiltration. This model will be very helpful for making a good irrigation scheduling and best water management.

Keywords

Infiltration, Cylindrical Infiltrometer, Accumulated Infiltration, Actual Infiltration, Calculated Infiltration

1. Introduction

Infiltration may be defined as the intake of water into the soil profile. The rate and cumulative infiltration amount are necessary to calculate the total water requirement for efficient irrigation system [1]. Infiltration is one of the most important components of hydrologic cycle. As the duration of rainfall continues to increase, the soil becomes increasingly saturated resulting in a decrease in infiltration capacity [2]. Consequentially, the excess rainfall from the infiltration process starts surface ponding in surface depressions which leads to surface runoff [3]. Irrigation scheduling involves two main considerations. They are 1) when to apply water, and 2) how

---

[*]Corresponding author.

much water to apply. Ensuring the answers of these two issues is very important for best crop and water management practices [1]. In different developing or underdeveloped countries, measurement of infiltration is not practiced or even shows any interest of measuring it. There are several reasons for this:
- lack of awareness of measuring actual volume and rate of infiltration;
- lack of skill and knowledge of infiltration measurement and its role in water management practices.

Considering the importance and economic benefit of irrigation and water management practices, a mathematical model may be prepared and available for serving the assessment and quantifying the amount of water needed for actual water requirement of the crops [1].

2. Factors Involving in Infiltration Rate

Infiltration capacity is dependent on soil texture, soil structure, and soil cover. Also, infiltration is dependent on existing soil moisture content, soil hydraulic conductivity, soil porosity, existing soil swelling colloids and organic matters, irrigation or rainfall duration, and viscosity of water [4]. For irrigating the crops, how much water needs to be applied to reach the water requirement by the crop needs, the information of infiltration and accordingly apply that amount of water for maximum water application and use efficiencies. Infiltration has the unit of velocity, like cm/hour. From this definition, it is well understood that the distance or height of entrance of water per unit time is dependent of soil characteristics. Compactness of soil, porosity, and soil type play a vital role to allow the water to enter and flow downward. Moreover, if the antecedent moisture content is already enough in the soil profile, there will be a less opportunity of the incoming water to get into the soil profile. Water will infiltrate until the water has enough space in the soil profile or if the soil becomes saturated, the excess water will start filling up the soil depressions and finally runoff will take place following the surface gradient into the nearest reservoir.

3. Methods of Measuring Infiltration

There are three methods for determining the infiltration characteristics for any irrigation system design and water management practices. They are: 1) cylindrical infiltrometer method, 2) Accumulation infiltration estimation from waterfront advance data, and 3) Depletion of free water surface measurement in a large basin.

Out of the above mentioned methods, cylindrical infiltrometer method is most commonly used. Cylindrical infiltrometer method offers the advantages over the other two avoiding the cumbersome procedure in collecting correct data from the field while estimating from waterfront advance data and there is necessity of considering the evaporation loss due to atmospheric influences on the large basin while measurement of water depletion is considered [3]. Hence, the cylindrical infiltrometer is comparatively reliable for measuring the infiltration rate and accumulated infiltration. Infiltration characteristics can be measured by using a metal cylindrical round shaped hollow drum driven to a certain length into the soil surface and then ponding this cylinder with water and simultaneously record the time required to deplete water and enter into the soil surface. In early days, only one cylinder was used to measure the height of water lowered in the cylinder. That procedure yielded several drawbacks and a higher degree of variability due to the uncontrollable movement of lateral seepage and movement of water to and from the cylinder. This lateral movement of water has been well controlled by another concentric cylinder similarly with ponded water as the inner cylinder.

Figure 1 shows the dimensions of a common Infiltrometer. There are two cylinders of diameters 30 cm and the 60 cm are driven concentrically into the soil surface about 10 cm and the total height of these two ring cylinders is 25 cm.

The material used for making these cylinders is of 2 mm rolled steel. The inner and the outer cylinders are both ponded with water. The outer cylinder is used as a buffer pond to avoid the lateral movement of water from and to the inner cylinder. Care should be taken against beveling of the cylinder bottoms. The cylinders are driven into the soil by a falling weight hammer striking on top of a wooden plank placing on top of the cylinders to avoid the damage at the edge of the cylinders.

The main objective of this study was to develop a model for infiltration characteristic by modified Kostiakov method and to calculate the accumulated infiltration and infiltration rate with specific focus on: 1) deriving the constant values of modified Kostiakov method for the soil under consideration, 2) judging the applicability of the model using the field data, and 3) find the percentage of error between the actual and the values calculated by the model.

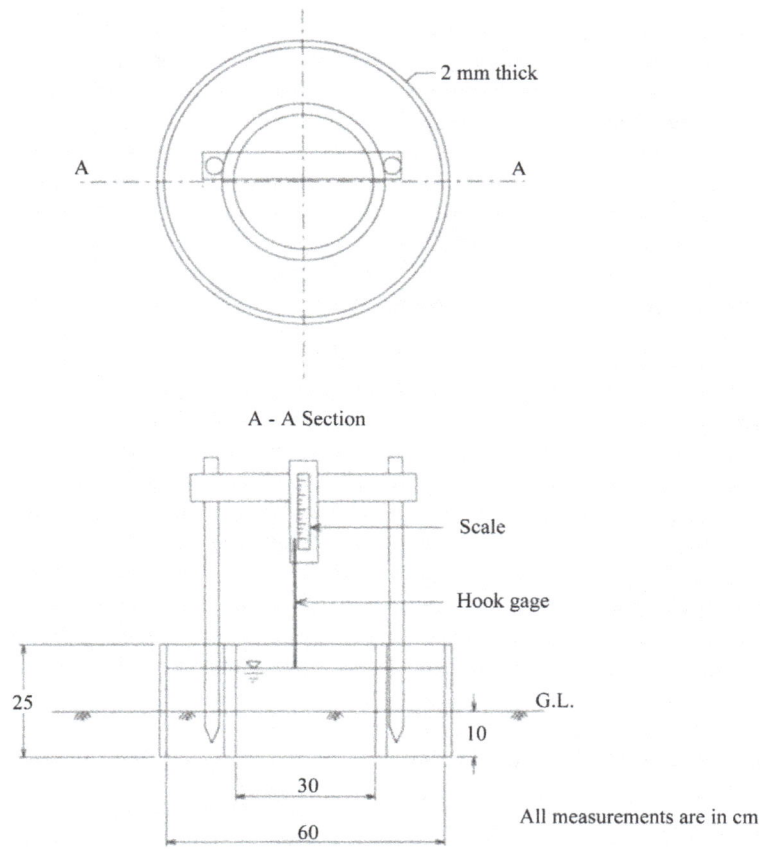

Figure 1. Plan and cross-sectional views of a cylindrical infiltrometer.

4. Methodology

This study was carried out in the Irrigation and Water Management demonstration field of BARI in 2012. This field is used for demonstration purposes and to exhibit different irrigation methods for training the farmers and agricultural extension officials. Sometimes, these plots are used for both exhibition and crop related research purposes too. The field was kept fallow at the time when the study was conducted. It was clay-loamy soil, without any tillage practices done and had common soil vegetation. Four infiltrometers were installed lengthwise with distances as shown in **Figure 2**.

The water levels of the inner cylinder were read by a needle type pointed hook gage whose sharp and pointed headend was just touching water level for initial water height reading and the tail end was set with a scale to read the difference after a predetermined time when a depletion of water height was there by adjusting the pointed head again touching the depleted water surface. The difference between initial and the final readings were the height of water that infiltrated during the predetermined time. This height divided by time is the one what is defined as the infiltration rate. Here times were recorded as minutes but they were converted into hours for calculation purposes and express the infiltration rate as cm/hr. After some period, there is no more depletion of water took place and the curve of accumulated infiltration (ordinate) vs. time (abscissa), is the constant infiltration rate the characteristic of this point and hereafter is called asymptote. At the initial stages, time vs. depletion of water recordings were taken frequently, refill of water were done as quickly as possible so that the pace of infiltration could be kept constant. Water levels in the inner and the outer cylinders were kept approximately same to keep up the water pressures same between the inner and outer cylinders and avoids the lateral water movement due to dissimilar height between the inner and the outer cylinders. The average values of accumulated infiltration y and average infiltration rate have been plotted against time t and shown in **Figure 3**.

The modified Kostiakov method [1] was tested if it suits the local soil condition and if it can be represented for the accumulated infiltration and infiltration rate. The relationship of accumulated infiltration y with respect

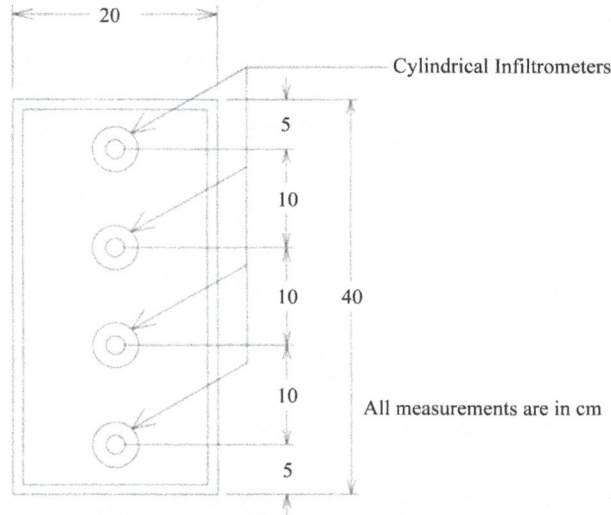

Figure 2. Field layout with the locations of the infiltrometers.

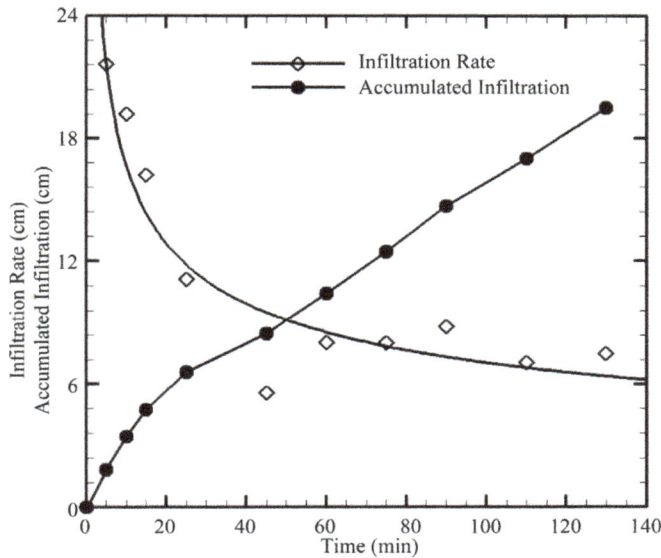

Figure 3. Average infiltration rate (cm/h) and accumulated infiltration (cm) vs. time (min).

to time t can be mathematically defined by the following equation, known as modified Kostiakov method:

$$y = at^\alpha + b \tag{1}$$

where
y = accumulated infiltration at time t, (cm),
t = elapsed time, (min), and
a, α, and b are three characteristic constants.

Values of a, α, and b can be calculated by the method suggested by Davis [5]. Values of a, α, and b are usually less than 1 [1]. The steps are described below:

1) plot the values of y and t;

2) select a pair of points of (t_1, y_1) and (t_2, y_2) from this plot, selection of the points on the plotted line should be near the extremities of the line to cover a wide range of interpolated values;

3) calculate a third value for time t_3 using the values of t_1 and t_2 from the procedure followed in the previous

step. The equation for calculating t_3 is as follows:

$$t_3 = \sqrt{t_1 \times t_2} = \sqrt{5.0 \times 130} = 25.495 \approx 25.5 \text{ min}$$

Values of t_1 and t_2 are available from the plotted line and that has been described in step 2. The corresponding value of accumulated infiltration $y_3 = 5.9$ cm when $t_3 = 25.5$ minutes. Value of b can be calculated by formula used in regression analysis. The formula can be shown as follows (refer to **Figure 4**):

$$b = \frac{y_1 \times y_2 - y_3^2}{y_1 + y_2 - 2y_3} = \frac{2 \times 18 - 5.9^2}{2 + 18 - 2 \times 5.9} = 0.145$$

Equation (1) can be rearranged and written as

$$y - b = at^\alpha \tag{2}$$

Taking log both sides

$$\log(y - b) = \log(a) + \alpha \log(t) \tag{3}$$

Table 1 shows the calculated values of the accumulated infiltration y in cm and their corresponding time t in minute for infiltrometers 1 and 2 and the same is represented in **Table 2** for Infiltrometer 3 and 4. **Table 3** shows the average values of accumulated infiltration, infiltration rate vs. time for Infiltrometer 1, 2, 3, and 4. These values are used to plot the accumulated infiltration.

Equation (3) yields the following equations when the values of b and t are substituted:

$$\log[1.80 - 0.145] = 0.2188 = \log a + \alpha \log(5) \text{ or } \log a + 0.699 \propto \tag{4}$$

$$\log[3.40 - 0.145] = 0.5126 = \log a + \alpha \log(10) \text{ or } \log a + 1.000 \propto \tag{5}$$

$$\log[4.75 - 0.145] = 0.6632 = \log a + \alpha \log(15) \text{ or } \log a + 1.176 \propto \tag{6}$$

$$\log[6.60 - 0.145] = 0.8099 = \log a + \alpha \log(25) \text{ or } \log a + 1.398 \propto \tag{7}$$

$$\log[8.45 - 0.145] = 0.9193 = \log a + \alpha \log(40) \text{ or } \log a + 1.602 \propto \tag{8}$$

$$\log[10.45 - 0.145] = 1.0130 = \log a + \alpha \log(60) \text{ or } \log a + 1.778 \propto \tag{9}$$

$$\log[12.45 - 0.145] = 1.0901 = \log a + \alpha \log(75) \text{ or } \log a + 1.875 \propto \tag{10}$$

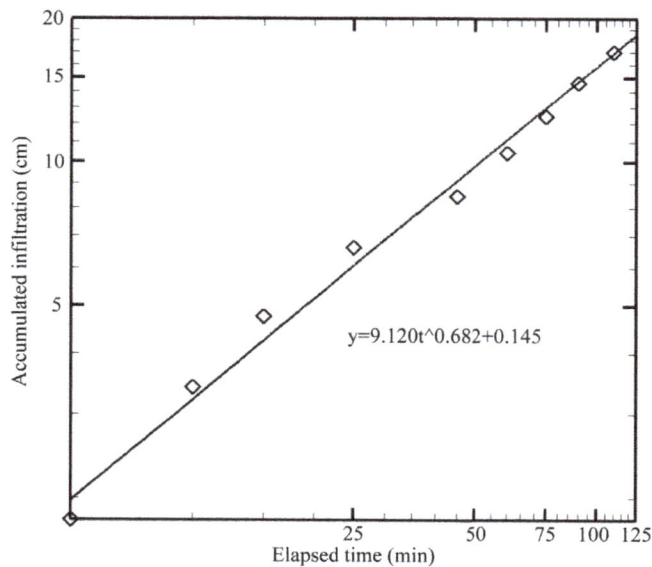

Figure 4. Accumulated infiltration versus time in log scale.

Table 1. Field trials of cylindrical infiltrometer, accumulated infiltration, infiltration rate vs. time for infiltrometer 1 and 2.

| | Infiltrometer No. 1 | | | | | Infiltrometer No. 2 | | | | |
| | Height of water surface from reference | | Infiltration during elapsed time | | | Height of water surface from reference | | Infiltration during elapsed time | | |
Time (min)	Before filling (cm)	After filling (cm)	Depth (cm)	Rate of infiltration (cm/hr)	Accumulated infiltration (cm)	Before filling (cm)	After filling (cm)	Depth (cm)	Rate of infiltration (cm/hr)	Accumulated infiltration (cm)
0	0	11.5	0	0	0	0	11.5	0	0	0
5	9.7	11.5	1.8	21.6	1.8	9.6	11.5	1.9	22.8	1.9
10	10.3	11.5	1.2	14.4	3.0	9.8	11.5	1.7	20.4	3.6
15	9.7	11.5	1.8	21.6	4.8	10.2	11.5	1.3	15.6	4.9
25	9.9	11.5	1.6	9.6	6.4	9.7	11.5	1.8	10.8	6.7
45	9.7	11.5	1.8	5.4	8.2	9.7	11.5	1.8	5.4	8.5
60	9.7	11.5	1.8	7.2	10.0	9.5	11.5	2.0	8.0	10.5
75	9.7	11.5	1.8	7.2	11.8	9.5	11.5	2.0	8.0	12.5
90	9.7	11.5	1.8	7.2	13.6	9.3	11.5	2.2	8.8	14.7
110	9.3	11.5	2.2	6.6	15.8	9.3	11.5	2.2	6.6	16.9
130	9.3	11.5	2.2	6.6	18.0	9.0	11.5	2.5	7.5	19.4

Table 2. Field trials of cylindrical infiltrometer, accumulated infiltration, infiltration rate vs. time for infiltrometer 3 and 4.

| | Infiltrometer No. 3 | | | | | Infiltrometer No. 4 | | | | |
| | Height of water surface from reference | | Infiltration during elapsed time | | | Height of water surface from reference | | Infiltration during elapsed time | | |
Time (min)	Before filling (cm)	After filling (cm)	Depth (cm)	Rate of infiltration (cm/hr)	Accumulated infiltration (cm)	Before filling (cm)	After filling (cm)	Depth (cm)	Rate of infiltration (cm/hr)	Accumulated infiltration (cm)
0	0	11.5	0	0	0	0	11.5	0	0	0
5	9.5	11.5	2	24	2.0	9.8	11.5	1.7	20.4	1.7
10	10.1	11.5	1.4	16.8	3.4	10.0	11.5	1.5	18.0	3.2
15	10.3	11.5	1.2	14.4	4.6	10.1	11.5	1.4	16.8	4.6
25	10.2	11.5	1.3	7.8	5.9	9.6	11.5	1.9	11.4	6.5
45	9.7	11.5	1.8	5.4	7.7	9.6	11.5	1.9	5.7	8.4
60	9.7	11.5	1.8	7.2	9.5	9.5	11.5	2.0	8.0	10.4
75	9.3	11.5	2.2	8.8	11.7	9.5	11.5	2.0	8.0	12.4
90	9.3	11.5	2.2	8.8	13.9	9.3	11.5	2.2	8.8	14.6
110	9.2	11.5	2.3	6.9	16.2	9.0	11.5	2.5	7.5	17.1
130	9.2	11.5	2.3	6.9	18.5	9.0	11.5	2.5	7.5	19.6

$$\log\left[14.65 - 0.145\right] = 1.1615 = \log a + \alpha \log\left(90\right) \text{ or } \log a + 1.954 \propto \qquad (11)$$

$$\log\left[17.00 - 0.145\right] = 1.2267 = \log a + \alpha \log\left(110\right) \text{ or } \log a + 2.041 \propto \qquad (12)$$

$$\log\left[19.50 - 0.145\right] = 1.2868 = \log a + \alpha \log\left(130\right) \text{ or } \log a + 2.114 \propto \qquad (13)$$

Adding Equations (4) to (8)

$$5\log(a) + 5.875\alpha = 3.1238 \tag{14}$$

Adding Equations (9) to (13)

$$5\log(a) + 9.762\alpha = 5.7781 \tag{15}$$

Solving Equations (14) and (15), the value of α becomes 0.683
The value of $\log(a) = 0.178$
Now substituting the values of a, b, and α in equation for individual elapsed times

$$\log(y - b) = \log a + \alpha \log(t)$$

At $t = 5$ min, $\log[y_{5\min} - 0.145] = -0.178 + 0.477 = 0.299$

$$y_{5\min} = 2.137 \text{ cm} \tag{16}$$

At $t = 10$ min, $\log[y_{10\min} - 0.145] = -0.178 + 0.682 = 0.505$

$$y_{10\min} = 3.344 \text{ cm} \tag{17}$$

At $t = 15$ min, $\log[y_{15\min} - 0.145] = -0.178 + 0.802 = 0.625$

$$y_{5\min} = 4.365 \text{ cm} \tag{18}$$

At $t = 25$ min, $\log[y_{25\min} - 0.145] = -0.178 + 0.953 = 0.777$

$$y_{25\min} = 6.126 \text{ cm} \tag{19}$$

At $t = 40$ min, $\log[y_{40\min} - 0.145] = -0.178 + 1.093 = 0.916$

$$y_{40\min} = 8.390 \text{ cm} \tag{20}$$

At $t = 60$ min, $\log[y_{60\min} - 0.145] = -0.178 + 1.213 = 1.036$

$$y_{60\min} = 11.021 \text{ cm} \tag{21}$$

At $t = 75$ min, $\log[y_{75\min} - 0.145] = -0.178 + 1.279 = 1.103$

$$y_{75\min} = 12.812 \text{ cm} \tag{22}$$

At $t = 90$ min, $\log[y_{90\min} - 0.145] = -0.178 + 1.333 = 1.157$

$$y_{90\min} = 14.492 \text{ cm} \tag{23}$$

At $t = 110$ min, $\log[y_{110\min} - 0.145] = -0.178 + 1.392 = 1.216$

$$y_{110\min} = 16.599 \text{ cm} \tag{24}$$

At $t = 130$ min, $\log[y_{130\min} - 0.145] = -0.178 + 1.442 = 1.266$

$$y_{130\min} = 18.588 \text{ cm} \tag{25}$$

5. Results and Discussions

The percentage of error was calculated by the following equation:

$$\text{Error} = \sum_{i=1}^{n} \frac{AI_a - CI_c}{AI_a} \times 100 \tag{26}$$

where, AI_a is the actual accumulated infiltration, AI_c calculated accumulated infiltration by the model, i is the number of data.

Table 4 shows the percentage of error between the actual and calculated values of accumulated infiltration with respect to time. The average value of percentage of error was 0.134 which is significantly lower range of acceptability. **Figure 5** also shows very good agreement between the actual and calculated values of accumulated infiltration. Individual error was calculated to see the deviation of the accumulated infiltration. The lowest value was only −0.71 and the highest value was 18.72 percent. Values of log (a), α, and b were 0.178, 0.683, and

Table 3. Average values of accumulated infiltration, infiltration rate vs. time for Infiltrometers 1, 2, 3, and 4.

Time (min)	Average	
	Rate of infiltration (cm/hr)	Accumulated infiltration (cm)
0	0	0
5	21.60	1.80
10	19.20	3.40
15	16.20	4.75
25	11.10	6.60
45	5.55	8.45
60	8.00	10.45
75	8.00	12.45
90	8.80	14.65
110	7.05	17.00
130	7.50	19.50

Table 4. Percentage of error between the actual and calculated values of accumulated infiltration vs. time.

Time (min)	Observed accumulated infiltration (cm)[*]	Calculated accumulated infiltration (cm)[**]	Percent of error (%)
	Observed	Calculated	
5	1.80	2.137	18.72
10	3.40	3.344	−1.65
15	4.75	4.365	−8.11
25	6.60	6.126	−7.18
40	8.45	8.39	−0.71
60	10.45	11.021	5.46
75	12.45	12.812	2.91
90	14.65	14.492	−1.08
110	17.00	16.599	−2.36
130	19.50	18.588	−4.68
		Average error	0.134

[*]Average of 1 to 4 infiltrometer reading; [**]Equations (16) to (25).

0.145, respectively, which are below 1 and follows and maintains the requirement of the modified Kostiakov method [1].

Figure 4 and **Figure 5** show very good agreements of the model values with the actual data from the field. Therefore, this model can be a very good tool to determine the infiltration rate and accumulated infiltration of the field. This will also be a good representative of the infiltration characteristic of the site. This information can be valued asset for irrigation scheduling for any crop cultivated in that field to ensure the best water management practices.

6. Conclusion

This model is developed to estimate the rate and accumulated infiltration of water in agricultural land is using

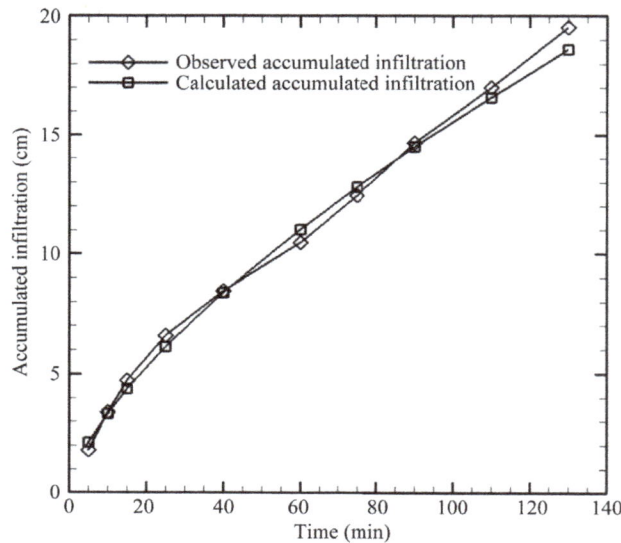

Figure 5. Observed and calculated accumulated infiltration vs. time.

modified Kostiakov method. Values for a, α, and b of modified Kostiakov equation are calculated to be 0.178, 0.683, and 0.145, respectively, which are below 1. This condition follows the principle of modified Kostiakov method [1]. It shows a good agreement between the model and the calculated values of the rate and accumulated infiltration. The percentage of error is 0.134 which is very promising. The characteristics constant values of modified Kostiakov equation are also determined and found to be within the acceptable ranges. One can easily identify the rate and the accumulated infiltration once the graphical results are produced. Also, this model will require no technical background for the users.

Acknowledgements

The authors would like acknowledge Prof. Rao Mentreddy and Prof. Michael Ayokanmbi of Alabama Agricultural and Mechanical University for reviewing this manuscript and making valuable suggestion.

References

[1]   Michael, A.M. (1997) Irrigation: Theory and Practice. Vikas Publishing House PVT Ltd., Delhi.

[2]   Dagadu, J.S. and Nimbalkar, P.T. (2012) Infiltration Studies of Different Soils under Different Soil Conditions and Comparison of Infiltration Models with Field Data. *IJAET*, **III**, 154-157.

[3]   Fetter, C.W. (2001) Applied Hydrogeology. 4th Edition, Prentice Hall, Upper Saddle River.

[4]   Charbeneau, R.J. (2000) Groundwater Hydraulics and Pollutant Transport. Prentice Hall, Upper Saddle River.

[5]   Davis, D.S. (1943) Empirical Equations and Monography. McGraw Hill Book Co., New York, 200.

10

Yield and Water Productivity of Chickpea (*Cicer arietinum* L.) as Influenced by Different Irrigation Regimes and Varieties under Semi Desert Climatic Conditions of Sudan

M. K. Alla Jabow[1]*, O. H. Ibrahim[2], H. S. Adam[3]

[1]Water Management Section, Agricultural Research Corporation (ARC), Hudeiba Research Station (HRS), Ed-Damer, Sudan
[2]Crop Agronomy Section, Agricultural Research Corporation (ARC), Hudeiba Research Station (HRS), Ed-Damer, Sudan
[3]Graduate Studies and Research Wad Medani Ahlia College, Wad Medani, Sudan
Email: *maie_kabbashi@yahoo.com

Abstract

A field experiment was conducted at Hudeiba Research Station Farm, located at Ed-Damer, Sudan during 2011/2012 and 2012/2013 winter seasons to investigate the effect of different irrigation regimes and varieties on chickpea (*Cicer arietinum* L.) yield, yield components and water productivity. The treatments include three irrigation regimes; irrigation every 10 days (I_1 = full irrigation), irrigation every 15 days (I_2 = moderate stress) and irrigation every 20 days (I_3 = severe stress) and two varieties (Borgieg and Wad Hamid). The treatments were arranged in factorial randomized complete block design (RCBD) with 3 replications. Irrigation water being applied, grain yield, yield components (number of pods per plant, number of seeds per pod and the 100 seeds weight) and crop water productivity (CWP) and irrigation water productivity (IWP) were recorded. Results showed that the number of pods per plant, number of seeds per pod, 100-seeds weight, grain yield and irrigation water applied were significantly ($p \leq 0.001$) affected by irrigation regimes. The highest values of these traits obtained with full irrigation, whereas the lowest values were recorded under severe water stress conditions. Results also indicated that, moderate and severe water stress regimes saved irrigation water by 24% and 32%, respectively compared with full irrigation. This study indicated that treatment I_1 which was irrigated every 10-days did

---

not produce the highest IWP, while treatment I_2 which irrigated every 15-days gave the highest IWP. The lowest IWP occurred at severe water stress regime (I_3). It could be concluded that moderate water stress might be adopted. Contrarily, the adoption of severe water stressed that produce high water savings would lead to yield losses that might be economically not acceptable. The late maturing chickpea variety of Borgieg significantly ($p \leq 0.05$) out-yielded the early maturing variety Wad Hamid by 11%. Borgieg displayed the highest values of CWP and IWP.

Keywords

Water Stress, *Cicer arietinum* L., Borgieg, Wad Hamid, Water Productivity

1. Introduction

The rapid increase of the world population and the corresponding demand for extra water by sectors such as industries and municipals, forces the agricultural sector to use its irrigation water more efficiently on the one hand and to produce more food on the other hand [1]. Although Sudan has sufficient potential water resources, it falls in water scarcity countries (economic water scarcity) because it is extremely difficult to find the financial resources to build enough water development projects [2]. Pump is the main source for irrigation water in Northern Sudan from River Nile (RN), the irrigation cost is considered as the most agricultural constraints and that may refer to the high cost pumping water from RN [3]. Such situation requires more efficient use of irrigation water as a pre-requisite for future agricultural expansion. One of the promising irrigation strategies to obtain "more crop per drop" is deficit irrigation [4]. Deficit irrigation is application of water below full crop water requirements (evapotranspiration) [5], and the crop is exposed to a certain level of water stress either during a particular period or through the whole growing season. The potential benefits of deficit irrigation arise from enhanced water productivity (WP) and lower production costs if one or more irrigation application can be eliminated. WP is useful for looking at potential increase in crop yield that may result from increased water availability [6] [7]. It provides a simple means of assessing whether yield is limited by water supply or other factors [8]. Quantitative information on WP is, therefore, necessary for effective planning of irrigation water management strategies in an area [9]. Crop water productivity (CWP) is generally defined as marketable yield (Y) to the volume of water consumed by the crop (ET) [10] [11], but economists and farmers are most concerned about the yield per unit of irrigation water applied [12].

Chickpea (*Cicer arietinum* L.) is an important source of protein, carbohydrates, vitamins, and certain minerals. While this pulse crop is an important source of dietary protein for human consumption, it is also important for the management of soil fertility due to its nitrogen-fixing ability [13]. Most chickpea producing areas are located in the arid and semi-arid zones, and approximately 90% of world's chickpea is grown under rain fed conditions [14] where terminal drought is one of the major constraints for its productivity. It is cultivated across the world in the Mediterranean Basin, the Near East, Central and South Asia, East Africa, South America, North America and Australia [15]. It has a total global production of 12 million tons from 13 million hectares [16]. In Sudan, chickpea faces competition with other winter legumes such as faba bean and common bean as well as other cash crops like spices. The major cultivated area is concentrated in the northern region of Sudan on basins and Islands along the River Nile and some small areas at Hawata and Jabel Marra. More recently, chickpea cultivation is extended to the central Sudan especially in the irrigated Gezira Scheme and New Halfa. In Sudan, it is either irrigated or utilizes the residual moisture stored in the soil after the River Nile flood recedes. Average area grown with this crop in the River Nile State for the period 2003-2012 was about 5500 ha with an average yield of 1.5 tha^{-1} [17]. In Sudan, many studies have been carried out to determine the response of chickpea to different irrigation levels. The results were based mainly on studies in three ways: 1) imposing different irrigation intervals throughout the crop cycle; 2) timing of the last irrigation; and 3) irrigation schedule during both vegetative and reproductive stages of the crop. Results indicated that frequent (7 - 10 day intervals) irrigation during the whole crop cycle always resulted in the highest grain yield [18]-[20]. It was also found that early termination of irrigation water drastically reduced grain yield [21]-[24]. Grain yield losses of 59% and 40% occurred when irrigation water was terminated after 50 and 70 days from sowing, respectively. Dealing with the crop life cycle as being composed of vegetative and reproductive phases, it was found, as expected, reproductive stage was the most sensitive stage to moisture stress developed through expanded irrigation intervals [22]-[25]. Consequently, the optimum irrigation schedule was established so as to irrigate the crop every 20 days during the vegetative stage and every 10 days during and the reproductive stages. Although data indicated that savings in irrigation water during less sensitive growth stages are possible, the information on quantity and cost of applied water for the different treatments is not adequate. Information pertaining to water productivity on chickpea in the northern Sudan is lacking

The objective of this research was to investigate the effect of different irrigation regimes and varieties on chickpea (*Cicer arietinum* L.) yield, yield components and water productivity yield, yield components and water productivity.

2. Materials and Methods

A field experiment was conducted under irrigation, for two consecutive seasons (2011/2012 and 2012/2013), at the Hudeiba Research Station Farm, Ed-Damer, Sudan, located at latitude (17.57°) N, Longitude (33.93°) E, and altitude 350 m above sea level. The local climate is semi-desert (26), very hot and dry in summer and relatively cool in winter. The average rainfall does not exceed 100 mm per year falling for only three months (July to September) with the rest of the year virtually dry. The prevailing thermal regime as daily mean temperature during the two growing seasons is displayed in **Figure 1**. According to soil profile (**Table 1**) the soil of the study site is clay in texture and is classified as Vertic Torrifluvent, fine Smectitic, calcareous, hyperthermic, Bergieg series (USA, Soil Taxonomy); with very low permeability, field capacity of 46% by volume and a permanent wilting point of 25% by volume. In general, the soil is non-saline and non-sodic, with alkaline reaction; and low in organic carbon and nitrogen content.

The experiment was a factorial design with three irrigation regimes (selected based on previous studies), namely, I_1 Irrigation every 10 days (full irrigation or normal), I_2 Irrigation every 15 days (moderate water stress), I_3 Irrigation every 20 days (severe water stress) and two varieties introduced from ICARDA, namely, Borgieg (erect, round seed shape, beige color seed, medium seed size, late maturing) and Wad Hamid (erect, round seed shape, beige color seed, large seed size, susceptible to stunt disease, early maturing). The treatments were arranged in randomized complete block design (RCBD) with 3 replications. Water was applied just below the surface of the top of the ridges. The gross plot size was 7 ridges × 0.6 m (ridge width) × 12 m (ridge length) = 50.4 m². The crop was sown manually in the third week of November in both seasons. All crops were planted in holes on top of 60 cm spaced ridges, with intra-row spacing of 0.1 m between holes and at the rates of 2 seeds

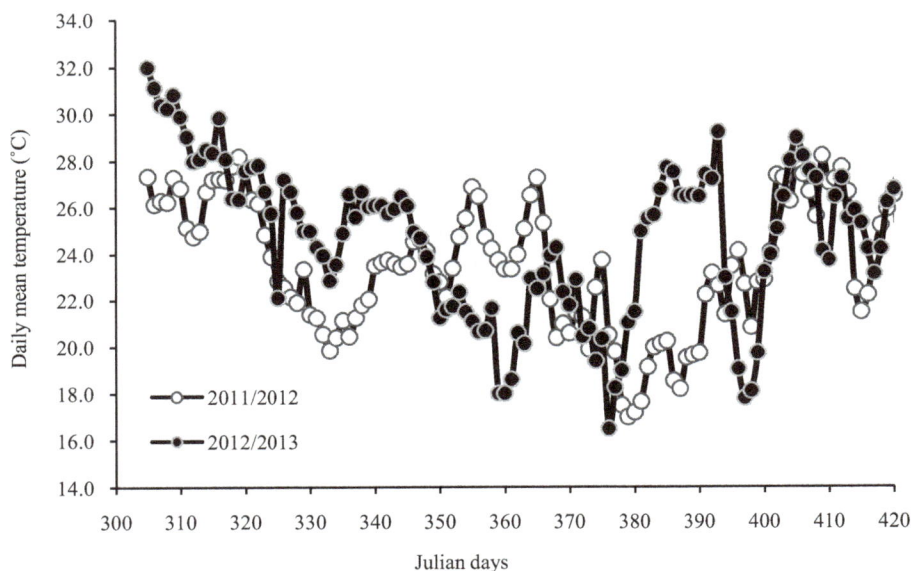

Figure 1. Prevailing thermal regime as daily mean temperature at Hudeiba Research Farm for the crop seasons 2011/2012 and 2012/2013.

Table 1. Selected physical and chemical properties of the soil at the experimental site in Northern Sudan.

Depth (cm)	0 - 23	23 - 44	44 - 87	87 - 120	120 - 157	157 - 203	Mean
Sand (%)	4	3	3	3	4	3	4
Silt (%)	47	42	39	37	40	37	40
Clay (%)	49	55	58	60	56	60	56
Hydraulic conductivity (cm/hr)	0.32	0.1	0.1	0.11	0.07	0.07	0.13
Moisture content at wilting point (m^3/m^3)	38	43	47	44	50	54	46
Moisture content at field capacity (m^3/m^3)	21	23	26	24	27	29	25
Soil bulk density (g/cm^3)	1.77	1.66	1.85	1.74	1.71	1.83	1.76
pH	7.8	8	7.9	7.7	8	7.9	7.9
Electrical conductivity (dS/m)	0.3	2.4	3.6	3.5	3.6	4.9	3.1
Calcium carbonate (%)	6	4.6	5.4	6	5.2	5.4	5.4
Total nitrogen (%)	0.045	0.04	0.045	0.03	0.035	0.035	0.038
Organic carbon (%)	0.499	0.312	0.203	0.265	0.187	0.218	0.281
Cation exchange capacity (meq/100g soil)	48	54	53	52	53	58	53
Sodium absorption ratio	1	7	10	12	7	7	7

per hole. Nitrogen at the rate of 43 kg N ha^{-1} in form of urea was applied uniformly, to all experimental plots before the second irrigation. Hand weeding of the experimental area was performed as required. The plots were irrigated by furrow irrigation method. The amount of irrigation water (m^3) for each plot in each irrigation event was measured directly in the field, using a current flow meter (type BFM001) connected to an irrigation pipe, using the following equation:

$$I = A \times T \times V \tag{1}$$

where, I = irrigation water (m^3), A = cross section area (m^2), T = total time (s) and V = velocity ($m \cdot s^{-1}$)

Evapotranspiration (ETc) was determined using a standard water balance Equation (2):

$$ETc = I + P + W - R - D \pm \Delta S \tag{2}$$

where, I = irrigation, P = rainfall, W = capillary rise, R = runoff, D = deep drainage, and S = soil moisture. For the period after irrigation and before the next irrigation, I = 0 as no irrigation water is added. During winter (November-February), the rainfall (P) is zero. The water table is deep so the capillary rise (W) is zero. The run-off (R) is negligible as the land is flat with a very gentle slope (1). The soil is impermeable so the deep drainage (D) is almost zero. Therefore, the evapotranspiration is equal to the change in soil moisture (ΔS). Soil moisture depletion (S) was calculated from soil water profile, measured in one replication for a depth of 60 cm with 20 cm intervals, 2 - 3 days after irrigation and immediately before each irrigation event. This was done from planting to harvesting, through gravimetric method. Soil samples were oven-dried at 105°C for 24 hours. Then, the calculated gravimetric moisture contents were converted into volumetric values, through multiplication with dry soil bulk density, viz:

$$\Delta S = \frac{\sum_{i=1}^{n} (\theta 1 - \theta 2) d}{\Delta t} \tag{3}$$

where, n = number of soil layers sampled in the effective root zone which is = 3 (0 - 20, 20 - 40, 40 - 60); $\theta 1$ volumetric moisture content within 2 - 3 days after irrigation; $\theta 2$ = volumetric moisture content before the next irrigation in the i-th layer; d = the thickness of i-th layer (mm), which is = 200 mm; and Δt = the time interval between two consecutive measurements (days).

Irrigation treatments were started from the third irrigation

At harvest in both seasons, grain yield was calculated from the central three ridges (8 m long) = 14.4 m^2 of each plot. A sub sample of ten plants was taken for determining the yield components (number of pods per plant, number of seeds per pod and the 100 seeds weight).

Crop water productivity is commonly expressed as the economic yield divided by the seasonal crop water use (seasonal evapotranspiration) [10] [11], while the Irrigation water productivity is the economic yield divided by the total irrigation water applied [12]-[27].

Crop water productivity (CWP) was calculated as

$$CWP = \frac{Y}{ET} \tag{4}$$

where, Y = yield (kg·ha^{-1}), ET = seasonal evapotranspiration (m^3·ha^{-1}). And Irrigation water productivity (IWP) was calculated as

$$IWP = \frac{Y}{I} \tag{5}$$

where, Y= yield (kg·ha^{-1}), I = irrigation water applied (m^3·ha^{-1}).

Analysis of variance (ANOVA) was carried out using MSTAT statistical package (1984). The data obtained were analyzed for each season separately, and then combined analysis was run for the two growing seasons because the homogeneity test was positive. As the soil moisture measurements were performed in one block, statistical analyses could not be performed for crop water productivity

3. Results and Discussion

3.1. Crop Growth Environment

The prevailing thermal regime as daily mean temperature during the two growing seasons is displayed in **Figure 1**. The second season experienced warm spells at the beginning and at the end of the season. However, it was comparatively cooler than the first season in the middle of the growing season.

3.2. Yield and Yield Components

Grain yield and yield components of chickpea as affected by irrigation regime and variety are presented in **Table 2**.

Analysis of variance showed that number of pods per plant and grain yield were significantly affected by irrigation regime and variety, but number of seeds per plant and 100 seeds weight were affected by irrigation regime. Statistic analysis indicated no significant interaction between irrigation regimes and varieties.

Table 2. Mean grain yield and yield components of chickpea as affected by irrigation regime and variety (averaged over seasons 2011/2012-2012/2013) at Hudeiba Research Farm.

	Grain yield (kg/ha)			No. of pods/plant			No of seeds/pod			100 seed weight (g)		
	Borgieg	Wad Hamid	Mean	Borgieg	Wad Hamid	Mean	Borgieg	Wad Hamid	Mean	Borgieg	Wad Hamid	Mean
I$_1$	1234	1096	1165	45	41	43	0.84	0.82	0.83	22.8	23.3	23.1
I$_2$	997	890	944	41	37	39	0.79	0.75	0.77	21.1	21.9	21.5
I$_3$	543	472	508	28	25	27	0.63	0.59	0.61	17.5	18.2	17.9
Mean	925	819	872	38	34	36	0.75	0.72	0.74	20.5	21.1	20.8
SE ± (I)	41.11***			0.82***			0.016***			0.404***		
SE ± (V)	33.57*			0.67**			0.013 ns			0.330 ns		
SE ± (I × V)	58.14 ns			1.16 ns			0.022 ns			0.571 ns		
C.V (%)	16.3			7.8			7.4			6.7		

Ns: Not significant. *, **, *** Significant at $p \leq 0.05$, 0.01 and 0.001 respectively.

Number of pods per plant decreased significantly ($p \leq 0.001$) with the increase in water deficit (**Table 2**). The highest number of pods per plant was observed in I₁ (full-irrigation). Similar results were reported by [28]. The variety Borgieg produced significantly ($p \leq 0.01$) more pods per plant than Wad Hamid (**Table 2**).

There were also significant ($p \leq 0.001$) reduction in number of seeds per pod and 100-seeds weight with water deficit and the trend was similar to the number of pods per plant trend (**Table 2**). The highest values of these traits obtained with full irrigation, whereas the lowest values recorded under severe water stress conditions. These results are in accordance with the finding of [29].

The two varieties were not significantly different in 100 seed weight. However, higher average weight (21.1 g) was recorded for Wad Hamid Grain yield was significantly decreased ($p \leq 0.001$) as water deficit increased (**Table 2**). The decrease in grain yield was more pronounced in severe water stress (irrigation every 20 days) than that in the moderate water stress (irrigation every 15 days). Application of moderate water stress (I₂) and severe water stress (I₃) caused 19% and 56% decrease in grain yield of water stressed plants, respectively when compared with the fully irrigated one (**Table 2**). Similar results were reported by [29] [30]. The variety Borgieg significantly ($p \leq 0.05$) out-yielded Wad Hamid by 11%. Wad Hamid was observed to be highly susceptible to stunt disease.

The results of this study indicated that the yield decrease due to water deficit was attributed to reduction in number of pods per plant, number of seeds per pod and 100-seeds weight. A positive and highly significant correlation was found between grain yield and these traits (**Figure 2**). Similar results were reported by [30]-[33].

No significant difference for variety X irrigation regime interaction indicates that the two varieties responded in a similar manner for water stress.

In this study, the unexpected low grain productivity of chickpea is attributed to the severe infestation of the crop by stunt disease.

3.3. The Amount of Irrigation Water Applied and Crop Water Use

Table 3 shows the number of irrigations, amount of irrigation water applied (including the first irrigation) and seasonal water used by the crop as an evapotranspiration (ET) in cubic meter per hectare. The total numbers of irrigations given in each irrigation regime in both seasons for I₁, I₂, and I₃ were 9, 6 and 5, respectively.

The mean seasonal ET varied between 3370 m³·ha⁻¹ and 2311 m³·ha⁻¹ (**Table 3**). The highest seasonal ET was recorded in treatment I₁, whereas the lowest seasonal ET recorded under I₃.

The analyses of variance (**Table 3**) revealed that irrigation water applied (I) was significantly ($p \leq 0.001$) affected by irrigation regime treatments. The highest amount of irrigation water was applied in the full irrigation and significantly ($p \leq 0.001$) reduced through the use of moderate and severe water-stress regimes with volume of water saved 1610 m³ and 2100 m³, respectively.

The amount of irrigation water applied to Borgieg was higher than that applied to Wad Hamid. This was due to less water requirement of short duration variety.

3.4. Yield-ET Relationship

The relationship between chickpea grain yield and seasonal ET is presented in **Figure 3** using all 12 data points

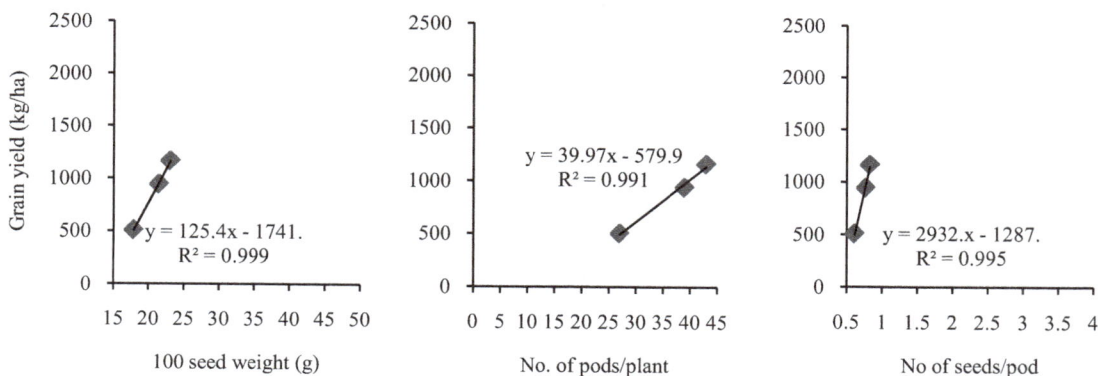

Figure 2. Relationship between grain yield and yield components of chickpea as affected by irrigation regime.

Table 3. Amount of irrigation water applied ($m^3 \cdot ha^{-1}$), number of irrigation events and crop evapotranspiration ($m^3 \cdot ha^{-1}$) of chickpea as affected by irrigation regime and variety (averaged over seasons 2011/2012-2012/2013) at Hudeiba Research Farm.

	Irrigation water applied ($m^3 \cdot ha^{-1}$) (number of irrigations)			Crop ET ($m^3 \cdot ha^{-1}$)		
	Borgieg	Wad Hamid	Mean	Borgieg	Wad Hamid	Mean
I_1	6843	6715	6779 (9)	3545	3195	3370
I_2	5217	5121	5169 (6)	2702	2437	2570
I_3	4750	4522	4636 (5)	2462	2160	2311
Mean	5603	5453	5528 (7)	2903	2597	2750
SE ± (I)	54***					
SE ± (V)	44*					
SE ± (I × V)	77 ns					
C.V (%)	3.4					

ns: Not significant. * and *** Significant at $p \leq 0.05$ and 0.001 respectively.

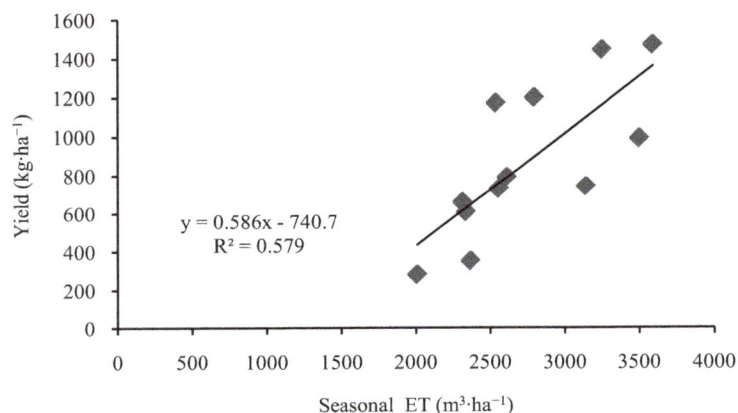

Figure 3. The relationship between grain yield of chickpea and seasonal evapotranspiration (ET).

obtained during the study period (6 treatments - 2 years). Grain yield varied from 238 to 1474 $kg \cdot ha^{-1}$ and ET values from 2009 to 3590 $m^3 \cdot ha^{-1}$. The linear regression between grain yield and ET showed that about 58% of the variation in grain yield could be attributed to variations in ET. Within the range of observed ET values, the regression slope predicts a yield increase of 58.6 $kg \cdot ha^{-1}$ for each 100 m^3 increase in ET. The negative value of the intercept indicates that a certain ET threshold value must be reached before any grain yield is obtained, which was 1264 $m^3 \cdot ha^{-1}$ in this study. Several previous studies have also shown a linear relationship between grain yield and ETc [34]-[37].

3.5. Water Productivity

Table 4 shows crop water productivity (CWP) and irrigation water productivity (IWP) of chickpea as affected by irrigation regime and variety.

CWP ranged from 0.220 $kg \cdot m^{-3}$ for treatment I_3 to 0.367 $kg \cdot m^{-3}$ for treatment I_2, while IWP for the same treatments ranged from 0.108 $kg \cdot m^{-3}$ for treatment I_3 to 0.182 $kg \cdot m^{-3}$ for treatment I_2 (**Table 4**). Treatment I_1 (full irrigation) did not produce the highest IWP, while treatment I_2 (moderate stress) gave the highest IWP. IWP for I_2 was 6% higher than that of I_1. Maximum CWP and IWP occur at crop water use less than the maximum. Moderate water-stress had improved IWP. However, reduction in grain yield occurred under this treatment.

Table 4. Mean irrigation water productivity (IWP) and crop water productivity (CWP) of chickpea as affected by irrigation regime and variety (averaged over seasons 2011/2012-2012/2013) at Hudeiba Research Farm.

	IWP (kg/m^3)			CWP (kg/m^3)		
	Borgieg	Wad Hamid	Mean	Borgieg	Wad Hamid	Mean
I_1	0.181	0.163	0.172	0.348	0.343	0.346
I_2	0.191	0.173	0.182	0.369	0.365	0.367
I_3	0.114	0.102	0.108	0.221	0.219	0.220
Mean	0.162	0.146	0.154	0.319	0.315	0.317
SE ± (I)	0.0078***					
SE ± (V)	0.0064 ns					
SE ± (I × V)	0.0110 ns					
C.V (%)	17.5					

ns: Not significant. * and *** Significant at $p \leq 0.05$ and 0.001 respectively.

Similar findings were reported by [38] who found that maximum wheat yields were obtained at full irrigation, though maximum water productivity was reached at two thirds of the seasonal irrigation water requirement. The lowest IWP occurred at severe water stress regime (I_3) (**Table 4**). This might be due to the fact that water savings at 20 = day intervals are not enough to overcome the concurrent yield losses. IWP for I_2 was 41% higher than that of I_3. Borgieg displayed the highest values of CWP and IWP.

4. Conclusion

Under the conditions of this study, grain yield and yield components were significantly ($p \leq 0.001$) affected by irrigation regimes. Exposing chickpea crop to water stress throughout the growing season significantly reduced grain yield. The low grain yield under water stress regimes was attributed to adverse effects of water stress on the yield components, mainly number of pods per plant, number of seeds per pod and 100 seeds weight. The highest seasonal ET was recorded in treatment I_1, which exceeded those of I_2 and I_3 by 24% and 31%, respectively. The highest amount of irrigation water was applied in the full irrigation regime and significantly ($p \leq 0.001$) reduced through the use of moderate and severe water-stress regimes. Treatment I_1 (full irrigation) did not produce the highest IWP, while treatment I_2 (moderate water stress) gave the highest IWP. Maximum CWP and IWP occurred at crop water use less than the maximum. The lowest IWP occurred at severe water stress regime (I_3). This might be due to the fact that water savings at 20 = day intervals are not enough to overcome the concurrent yield losses. In conclusion moderate water stress may be adopted. Contrarily, the adoption of severe water stress that produced high water savings would lead to yield losses that might be economically not acceptable. The late maturing chickpea variety of Borgieg significantly ($p \leq 0.05$) out-yielded the early maturing variety Wad Hamid by 11%. Borgieg displayed the highest values of CWP and IWP.

Acknowledgements

The authors thank the Land and Water Research Centre, Wad Medani, Sudan, the EU-IFAD PROJECT-ICARDA International Center for Agricultural Research in the Dry Areas for supporting and funding this work.

References

[1] Raes, D., Geerts, S., Kipkorir, E., Wellens, J. and Sahli, A. (2006) Simulation of Yield Decline as a Result of Water Stress with a Robust Soil Water Balance Model. *Agricultural Water Management*, **81**, 335-357. http://dx.doi.org/10.1016/j.agwat.2005.04.006

[2] Wilson, E. (1999) One Third of World's Population 2.7 Billion People Will Experience Water Scarcity by 2025, Says New Study Conducted by the International Water Management Institute, a Research Centre of the Consultative Group on International Agriculture Research (CGIAR). Washington DC.

[3]　Faki, H.H. (1999) Water Allocation and Its Effect on Faba Bean Technology Adoption in Shendi Area. Pages 72-75 in Nile Valley Regional Program on Cool-Season Food Legumes and Wheat, Annual Report 1990/91, Sudan ICARDA/ NVRP-DOC-017.

[4]　Pereira, L.S., Oweis, T. and Zairi, A. (2002) Irrigation Management under Water Scarcity. *Agricultural Water Management*, **57**, 175-206. http://dx.doi.org/10.1016/S0378-3774(02)00075-6

[5]　Fereres, E.M. and Soriano, A. (2007) Deficit Irrigation for Reducing Agricultural Water Use: Integrated Approaches to Sustain and Improve Plant Production under Drought Stress Special Issue. *Journal of Experimental Botany*, **58**, 147-159. http://dx.doi.org/10.1093/jxb/erl165

[6]　Burke, S., Mulligan, M. and Thornes, J.B. (1999) Optimal Irrigation Efficiency for Maximum Plant Productivity and Minimum Water Loss. *Agricultural Water Management*, **40**, 377-391. http://dx.doi.org/10.1016/S0378-3774(99)00011-6

[7]　Singh, R., van Dam, J.C. and Feddes, R.A. (2006) Water Productivity Analysis of Irrigated Crops in Sirsa District, India. *Agricultural Water Management*, **82**, 253-278. http://dx.doi.org/10.1016/j.agwat.2005.07.027

[8]　Augus, J.F. and van Herwaarden, A.F. (2001) Increasing Water Use and Water Use Efficiency in Dryland Wheat. *Agronomy Journal*, **93**, 290-298. http://dx.doi.org/10.2134/agronj2001.932290x

[9]　Igbadun, H.E., Mahoo, H.F., Tarimo, A.K.P.R. and Salim, B.A. (2006) Crop Water Productivity of an Irrigated Maize Crop in Mkoji Sub-Catchment of the Great Ruaha River Basin, Tanzania. *Agricultural Water Management*, **85**, 141-150. http://dx.doi.org/10.1016/j.agwat.2006.04.003

[10]　Zwart, S.J. and Bastiaanssen, W.G.M. (2004) Review of Measured Crop Water Productivity Values for Irrigated Wheat, Rice, Cotton and Maize. *Agricultural Water Management*, **69**, 115-133. http://dx.doi.org/10.1016/j.agwat.2004.04.007

[11]　Geerts, S. and Raes, D. (2009) Deficit Irrigation as an On-Farm Strategy to Maximize Crop Water Productivity in Dry Areas. *Agriultural Water Management*, **96**, 1275-1284. http://dx.doi.org/10.1016/j.agwat.2009.04.009

[12]　Vazifedoust, M., van Dam, J.C., Feddes, R.A. and Feizi, M. (2008) Increasing Water Productivity of Irrigated Crops under Limited Water Supply at Field Scale. *Agricultural Water Management*, **95**, 89-102. http://dx.doi.org/10.1016/j.agwat.2007.09.007

[13]　Maiti, R.K. (2001) The Chickpea Crop. In: Maiti, R. and Wesche-Ebeling, P., Eds., *Advances in Chickpea Science*, Science Publishers Inc., Enfield, 1-31.

[14]　Kumar, J. and Abbo, S. (2001) Genetics of Flowering Time in Chickpea and Its Bearing on Productivity in Semiarid Environments. In: Spaks, D.L., Ed., *Advances in Agronomy*, Vol. 2, Academic Press, New York, 122-124.

[15]　Soltani, A., Hammer, G.L., Torabi, B., Robertsonc, M.J. and Zeinali, E. (2006) Modeling Chickpea Growth and Development: Phenological Development. *Field Crops Research*, **99**, 1-13. http://dx.doi.org/10.1016/j.fcr.2006.02.004

[16]　FAO (2011) FAOSTAT. http://faostat3.fao.org/home/index.html

[17]　Ministry of Agriculture, River Nile State 2013.

[18]　Taha, M.B. (1987) Effect of Seed Rate, Phosphorous Fertilizer and Irrigation on Yield and Yield Components of Chickpea. Hudeiba Research Station, Annual Report, Ed-Damer.

[19]　Nourai, A.H. (1986) Effects of Frequency of Irrigation and Variety on Yield and Yield Components of Chickpea. Hudeiba Research Station, Annual Report, Ed-Damer.

[20]　Nourai, A.H. (1989) Effects of Variety, Frequency of Irrigation and Nitrogen Nutrition on Yield and Yield Components of Chickpea Grown at Borgieg. Hudeiba Research Station, Annual Report, Ed-Damer.

[21]　Taha, M.B. and Ali, M.E.K. (1991) Response of Chickpea to Methods of Planting and Date of Terminal Irrigation. Nile Valley Regional Program on Cool-Season Food Legumes. *Annual Coordination Meeting*, Cairo, 16-23 September 1991, 99.

[22]　Ibrahim, O.H. (1993) Effects of Irrigation Regime and Terminal Water Stoppage on Growth and Yield of Chickpea. Nile Valley Regional Program on Cool-Season Food Legumes and Wheat. Food Legumes Report, *Annual National Coordination Meeting*, Wad Medani, 29 August-2 September 1993, 202-203.

[23]　Ibrahim, O.H. (1994) Effects of Irrigation Regime and Terminal Water Stoppage on Growth and Yield of Chickpea. Nile Valley Regional Program on Cool-Season Food Legumes and Wheat. Food Legumes Report, *Annual National Coordination Meeting*, Wad Medani, 28 August-1 September 1994, 175.

[24]　Ibrahim, O.H. (1995) Effects of Irrigation Regime and Terminal Water Stoppage on Growth and Yield of Chickpea. Hudeiba Research Station, Annual Report, Ed-Damer.

[25]　Taha, M.B., Ali, M.E.K. and Ahmed, S.H. (1990) Effect of Sowing Method and Irrigation Frequency on Chickpea Yield and Yield Components. Hudeiba Research Station, Annual Report, Ed-Damer.

[26]　Adam, H.S. (2005) Agro-Climatology, Crop Water Requirement and Water Management. Water Management and Ir-

rigation Institute, University of Gezira, Wad Medani, 169 p.

[27] Pereira, L.S., Oweis, T. and Zairi, A. (2002) Irrigation Management under Water Scarcity. *Agricultural Water Management*, **57**, 175-206. http://dx.doi.org/10.1016/S0378-3774(02)00075-6

[28] Khurgami, A. and Rafiee, M. (2009) Drought Stress, Supplemental Irrigation and Plant Densities in Chickpea Cultivars. *African Crop Science Conference Proceedings*, **9**, 141-143.

[29] Ghassemi-Golezani, K., Mustafavi, S.H. and Shafagh-Kalvanagh, J. (2012) Field Performance of Chickpea Cultivars in Response to Irrigation Disruption at Reproductive Stages. *Research on Crops*, **13**, 107-112.

[30] Ghassemi-Golezani, K., Ghassemi, S. and Bandehhagh, A. (2013) Effects of Water Supply on Field Performance of Chickpea (*Cicer arietinum* L.) Cultivars. *International Journal of Agronomy and Plant Production*, **4**, 94-97.

[31] Al-Suhaibani, N.A. (2009) Influence of Early Water Deficit on Seed Yield and Quality of Faba Bean under Arid Environment of Saudi Arabia. *American-Eurasian Journal of Agricultural and Environmental Sciences*, **5**, 649-654.

[32] Emam, Y., Shekoofa, A., Salehi, F. and Jalali, A.H. (2010) Water Stress Effects on Two Common Bean Cultivars with Contrasting Growth Habits. *American-Eurasian Journal of Agricultural and Environmental Sciences*, **9**, 495-499.

[33] Szilagyi, L. (2003) Influence of Drought on Seed Yield Components in Common Bean. *Bulgarian Journal of Plant Physiology*, Special Issue, 320-330.

[34] Zhang, H. and Oweis, T. (1999) Water-Yield Relations and Optimal Irrigation Scheduling of Wheat in the Mediterranean Region. *Agricultural Water Management*, **38**, 195-211. http://dx.doi.org/10.1016/S0378-3774(98)00069-9

[35] Al-Jamal, M.S., Sammis, T.W. and Smeal, B.D. (2000) Computing the Crop Water Production Function for Onion. *Agricultural Water Management*, **46**, 29-41. http://dx.doi.org/10.1016/S0378-3774(00)00076-7

[36] Kipkorir, E.C., Raes, D. and Massawe, B. (2002) Seasonal Water Production Functions and Yield Response Factors for Maize and Onion in Perkerra, Kenya. *Agricultural Water Management*, **56**, 229-240. http://dx.doi.org/10.1016/S0378-3774(02)00034-3

[37] Igbadun, H.E., Ramalan, A.A. and Oiganji, E. (2012) Effects of Regulated Deficit Irrigation and Mulch on Yield, Water Use and Crop Water Productivity of Onion in Samaru, Nigeria. *Agricultural Water Management*, **109**, 162-169. http://dx.doi.org/10.1016/j.agwat.2012.03.006

[38] Oweis, T., Zhang, H. and Pala, M. (2000) Water Use Efficiency of Rainfed and Irrigated Bread Wheat in a Mediterranean Environment. *Agronomy Journal*, **92**, 231-238. http://dx.doi.org/10.2134/agronj2000.922231x

11

Wheat Yield Response to Water Deficit under Central Pivot Irrigation System Using Remote Sensing Techniques

M. A. El-Shirbeny*, A. M. Ali, A. Rashash, M. A. Badr

National Authority for Remote Sensing and Space Sciences (NARSS), Cairo, Egypt
Email: *mshirbeny@yahoo.com, m.elshirbeny@narss.sci.eg

Abstract

Scarcity of rainfall and limited irrigation water resources is the main challenge for agricultural expanding policies and strategies. At the same time, there is a high concern to increase the area of wheat cultivation in order to meet the increasing local consumption. The big challenge is to incerese wheat production using same or less amount of irrigation water. In this trend, the study was carried out to analyze the sensitivity of wheat yield to water deficit using remotely sensed data in El-Salhia agricultural project which located in the eastern part of Nile delta. Normalized Difference Vegetation Index (NDVI) and Land Surface Temperature (LST) were extracted from Landsat 7. Water Deficit Index (WDI) used both LST minus air temperature (Tair) and vegetation index to estimate the relative water status. Yield response factor (ky) was derived from relationship between relative yield decrease and relative evapotranspiration deficit. The relative Evapotranspiration deficit was replaced by WDI. Linear regression was found between predicted wheat yield and actual wheat yield with 0.2^{-6}, 0.025, 0.252 and 0.76 as correlation coefficient on 30th of Dec. 2012, 15th of Jan. 2013, 16th of Feb. 2013 and 20th of Mar. 2013 respectively. The main objective of this study is using a combination between FAO 33 paper approach and remote sensing techniques to estimate wheat yield response to water.

Keywords

Normalized Difference Vegetation Index (NDVI), Land Surface Temperature (LST), Water Deficit Index (WDI), Yield Response Factor (ky), Arid Region and Egypt

1. Introduction

About 21% of the world's food depends on the wheat crop, which grows on 200 million hectares of farmland worldwide. Most of developing countries including Egypt are wheat importers. About 81% of wheat in the developing world is produced and utilized within the same country, if not the same community [1]. With a rapidly growing world population, the pressure on limited fresh water resources increases. Egypt consumes 14 million

---

ton of wheat yearly, on the other hand, it produces about 8 million ton yearly. Although, wheat is low water consumption crop compared with other crops in Egypt, it is sensitive to water stress in some phonological stages.

Agriculture is the largest water consuming sector. It faces competing demands from other sectors, such as the industrial and the domestic sectors. With an increasing population and less water available for agricultural production, the food security for future generations is at stake. The great challenge of the agricultural sector is to produce more food from less water, which can be achieved by increasing Crop Water Productivity (CWP) [2]. Agronomical research and improving land and water management practices succeeded to increase CWP significantly during the years. Limited water is the principal factor responsible for reduced cereal yields globally and especially in Mediterranean environment [3]. Crop productivity is determined by the total amount of precipitation and also by its distribution during the growing season [4] [5].

Yield is defined as the marketable part of the total above ground biomass production; for wheat, maize and rice total grain yield is considered, and for cotton the total lint yield and total seed yield. Unfortunately, very few sources give the moisture content at which the yield is measured, which inevitably means an error that exists in the final results [2]. Grain yield depends on number of plants per unit land area, spikes per plant, spikelets per spike, grains per spikelet, and single grain weight (SGW). These yield determining components are interrelated and increase a crop's capacity to compensate for losses in any of the components caused by temporary, unfavorable conditions [6]. The response of yield to water supply is quantified by [7] through the yield response factor (ky) which relates relative yield decrease (1-Ya/Ym) to relative evapotranspiration deficit (1-ETa/PET). Water deficit of a given magnitude was expressed in the ratio actual evapotranspiration (ETa) and potential evapotranspiration (PET) may either occur continuously over the total growing period of the crop or it may occur during any of the individual growth periods (*i.e.* establishment (0), vegetative (1), flowering (2), yield formation (3), or ripening (4) period). Determination of PET that is substantially different from ETa has been made on the basis of a so-called two-step approach. *ETo* is first estimated and semi-empirical coefficient (crop coefficient, Kc) is then applied to take into account all other crop and environmental factors [8]-[11]. Similarities between Kc curve and a satellite-derived vegetation index showed potential for modeling a Kc as a function of the vegetation index [11]. Therefore, the possibility of directly estimating Kc from satellite data was investigated [9] [12]-[15].

Crop development and growth are subject to drought stress at different stages of the growth cycle, which results in differences in composition of yield components. This interrelationship was also evident when the role of grain number and weight was compared in yield determination in winter and spring wheat (*Triticum aestivum* L.) [16]. Generally, water deficit during the rapid spike-growth phase from booting to anthesis reduces floret set due to decreased shoot water status and increased accumulation of abscisic acid [17]. Number of grain bearing tillers and grain set may be also reduced [18] [19]. WDI is a function of actual evapotranspiration (ETa) to potential evapotranspiration (PET) ratio [20]:

$$WDI = 1 - ETa/PET \tag{1}$$

where: ET (mm/day) is the product of an uptake coefficient (α, mm/day) and available water ($\theta - \theta WP$) when ET is less than PET (mm/day) [21]: If ET < PET, ET = $\alpha(\theta - \theta WP)$ and If ET $\geq$ PET, ET = PET. PET occurs when the availability of soil water does not limit transpiration. It is estimated using the FAO 56 Penman-Monteith model [8]. [22] developed the WDI that used both surface minus air temperature and a vegetation index to estimate the relative water status of a field. The crop begins to experience some level of stress when the WDI falls to the right of a line formed between points 1 and 4 [23].

Productivity response to water stress is different for each crop and this response is expected to vary with the climate. Therefore, the critical values of WDI should be determined for a particular crop in different climates and soils to use it in yield prediction and irrigation scheduling. Many satellite data were used to calculate WDI. [24] used NOAA/AVHRR to calculate WDI in eastern part of Nile delta-Egypt. The main objective of this work is studying wheat yield response to water deficit under central pivot irrigation system using remote sensing techniques.

2. Materials and Methods

2.1. Study Area Location

El-Salhia project is located at the eastern part from Nile Delta. It is bounded by 30°22'35" and 30°31'19" latitudes and 31°55'24" and 32°02'38" longitudes as shown in (**Figure 1**). The whole area of the project is about 13,800 ha. Two irrigation systems are used in the project; the central pivots and the drip irrigation. The project has about 100 pivots irrigation units. Each pivot unit irrigates an area of about 63.6 ha when pivot length is 450 meter. The farms that are cultivated by orchards are irrigated by drip irrigation system. Climate in the study area is Dry Arid according to Köppen Climate Classification System, where precipitation is less than 50% of potential evapotranspiration. Annual average temperature is over 18°C. The average rainfall is approximately 20 mm/year. The maximum values of rainfalls are registered in January with average of 6.9 mm. The average maximum temperatures reach (34.6°C) in June and January represents the coldest month (19.0°C). The minimum temperatures range between 8.0°C in January to 21.5°C in August.

2.2. Remote Sensing Data

Remote sensing provides spatial coverage by measurement of reflected and emitted electromagnetic radiations, across a wide range of wavebands, from the earth's surface and surrounding atmosphere. Landsat ETM+ imageries, (path 176/row 039) around 10 a.m. local time with 30 meter ground resolution, on 30th of Dec. 2012, 15th of Jan. 2013, 16th of Feb. 2013 and 20th of Mar. 2013 were used in the current study to estimate LST, NDVI and WDI.

2.3. Land Surface Temperature (LST)

For landsat ETM+ data, the recorded digital numbers (DN) were converted to radiance units (Rad) using the calibration coefficients specific for each band.

$$\text{Radiance} = \text{Gain}*\text{DN} + \text{offset} \tag{2}$$

Surface emissivity (E_o) was estimated from the NDVI using the empirical equation developed from raw data on NDVI and thermal emissivity [25].

$$E_o = 0.9932 + 0.0194 \ln \text{NDVI} \tag{3}$$

The radiant temperature (T_o) can be calculated from band 6 radiance (Rad6) using calibration constants K1 = 666.09 and K2 = 1282.71 [26].

$$T_o = K2/\ln((K1/\text{Rad6}) + 1) \tag{4}$$

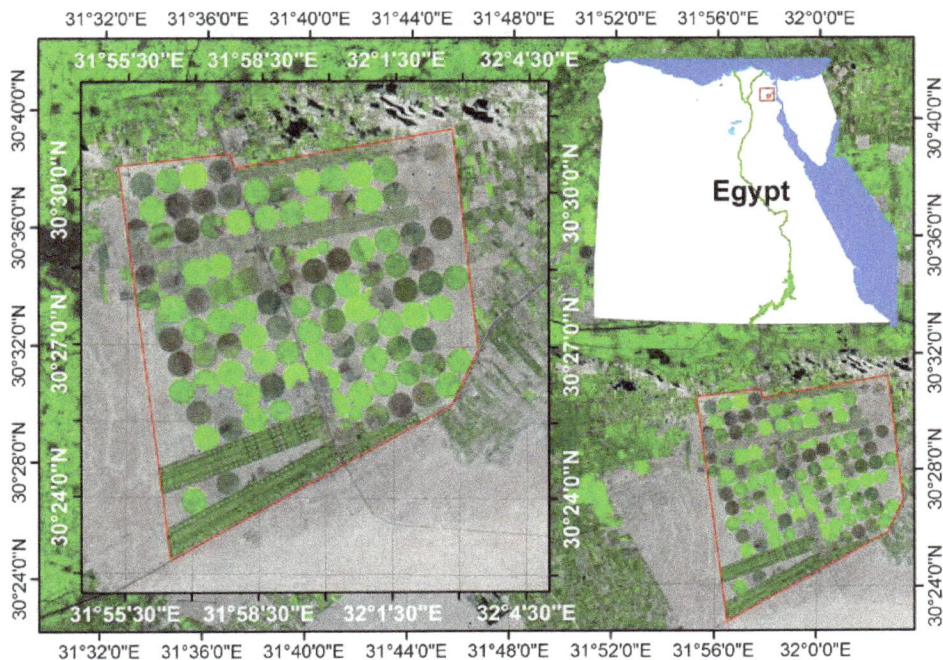

Figure 1. Location map of the study area.

The resulting temperature (Kelvin) is satellite radiant temperature of the viewed Earth atmosphere system, which is correlated with, but not the same as, the surface (kinetic) temperature. The atmospheric effects and surface thermal emissivity have to be considered in order to obtain the accurate estimate of surface temperature from satellite thermal data [27]. The surface temperature is calculated from the top of atmosphere radiant temperature (T_o) and estimated surface emissivity (E_o) as:

$$T = T_o/E_o \tag{5}$$

2.4. Water Deficit Index (WDI)

[22] developed the WDI (Equation (6)) that uses both surface minus air temperature and a vegetation index to estimate the relative water status of a field (**Figure 2**).

$$WDI = (dT - dTL_{13})/(dTL_{24} - dTL_{13}) \tag{6}$$

where: dT is the measure of surface subtracting air temperature at a particular percent cover, dTL_{13} is the surface minus air temperature determined by the line from points 1 to 3 for the percent cover of interest ("wet" line), and dTL_{24} is the temperature difference on the line formed between points 2 and 4 ("dry" line). Graphically, WDI can be viewed as the ratio of the distances AB to AC in the previous figure. As the WDI considers evaporation from a soil surface as well as the crop, it can be interpreted as a measure of the amount of ETa occurring relative to PET (Equation (1)). While WDI could be used to estimate ET, it does not provide a direct measure of crop water stress. As an index, it is vary according to soil-water evaporation as well as crop transpiration. The crop begins to experience levels of stress when the WDI falls to the right of a line formed between points 1 and 4 [23].

2.5. Yield Response to Water

The major importance in production planning is the yield response to water deficit. The response of yield to water supply is quantified through the yield response factor (ky) which relates relative yield decrease to relative evapotranspiration deficit. Water deficit of a given magnitude, is expressed in the ratio ETa and PET, may either occur continuously over the total growing period of the crop or it may occur during any one of the individual growth periods. The yield response to water deficit in different individual growth periods has a major importance in the scheduling of limited supply in order to obtain highest yield. Generally, crops are more sensitive to water deficit during emergence, flowering and early yield formation than early (vegetative, after establishment) and late growth periods (ripening) [7]. For Wheat, spring wheat is more sensitive than winter wheat and flowering period is more sensitive than yield formation and yield formation is more sensitive than vegetative period. In order to quantify the effect of water stress, it is necessary to derive the relationship between relative yield decrease and relative evapotranspiration deficit given by the empirically derived yield response factor (ky) [7].

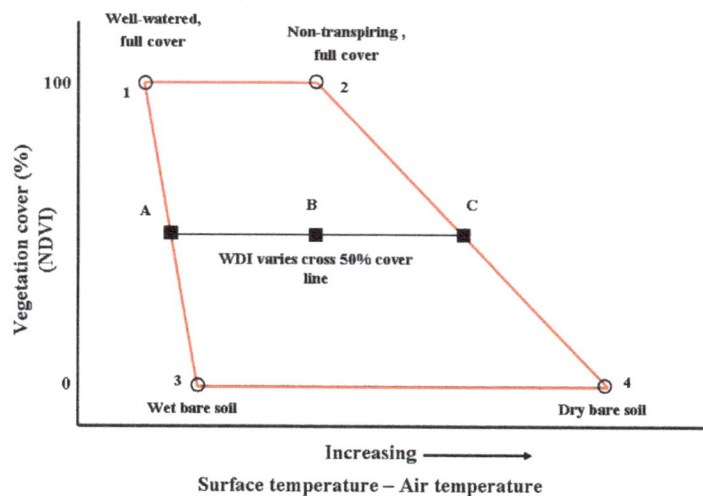

Figure 2. Water Deficit Index (WDI) trapezoid.

$$(1 - Ya/Ym) = ky \ (1 - ETa/PET) \tag{7}$$

where: Ya is actual harvested yield; Ym is maximum harvested yield; Ky is yield response factor; ETa is actual evapotranspiration; PET is potential evapotranspiration. Relative Evapotranspiration deficit could be replaced by WDI. From equation (1 and 7) remote sensing can take a place in FAO 33 equation as follows.

$$(1 - Y_a/Y_m) = Ky \ (WDI) \tag{8}$$

Figure 3 is a flowchart that shows the whole process to derive yield response from remote sensing data.

3. Results and Discussion

3.1. Water Deficit Index (WDI)

WDI is a function of ETa to PET ratio [28]. WDI of 0 indicates no water stress, and a value of 1 represents maximum water stress (**Figure 4**). Water stress causes stomatal closure and interruption in energy dissipation which results in rise of leaf temperature. The leaf or canopy temperature is used as an indicator of plant water stress [29].

WDI has been developed for the reference crop as a generic index for quantifying crop water stress for various crops. It explores how reliable of water stress estimations would be for various crops. WDI represents the suffering of crop from water shortage or/and thermal stress. WDI in the study area varied from stage to another and from year to year. It is affected by applied irrigation system, soil type and climatic conditions.

The minimum values of WDI for wheat in study area were 0.08, 0.03, 0.03, and 0.02 and maximum values were 0.51, 0.27, 0.2, and 0.14 on 30th of Dec. 2012, 15th of Jan. 2013, 16th of Feb. 2013 and 20th of Mar. 2013 respectively. The values of WDI were high in the first stage because the canopy was not 100% coverage and the temperature of soil was higher than temperature of canopy.

3.2. Yield Prediction

Actual wheat yield can only be determined by accurately measuring the area and determining the weight of grain harvested. Environmental stress always reduces Ym. In (**Figure 5**) the predicted Wheat yield was plotted against actual yield, the relations were varied during different phenological stages where R^2 were 0.2^{-6}, 0.03, 0.3 and 0.76. The yield response factor (ky) for most crops is derived through the assumption that the relationship between relative yield (Ya/Ym) and relative evapotranspiration (ETa/ETm) is linear and is valid for water deficits of up to about 50 percent or (1 − ETa/ETm = 0.5). The values of ky are based on an analysis of experimental field data covering a wide range of growing conditions. The experimental results used represent high-producing

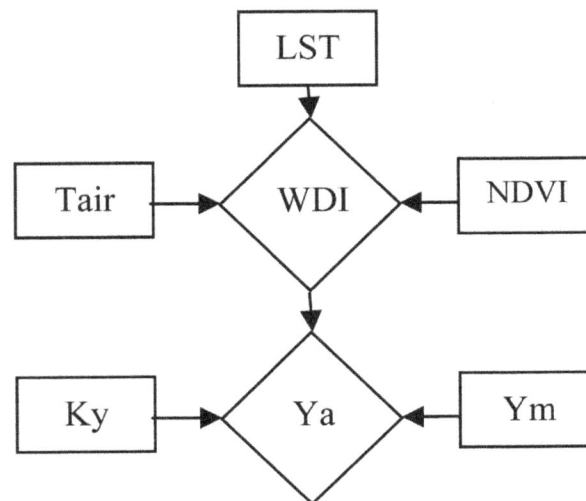

Figure 3. Illustrates the combination of FAO 33 paper approach and remote sensing techniques to estimate yield response to water.

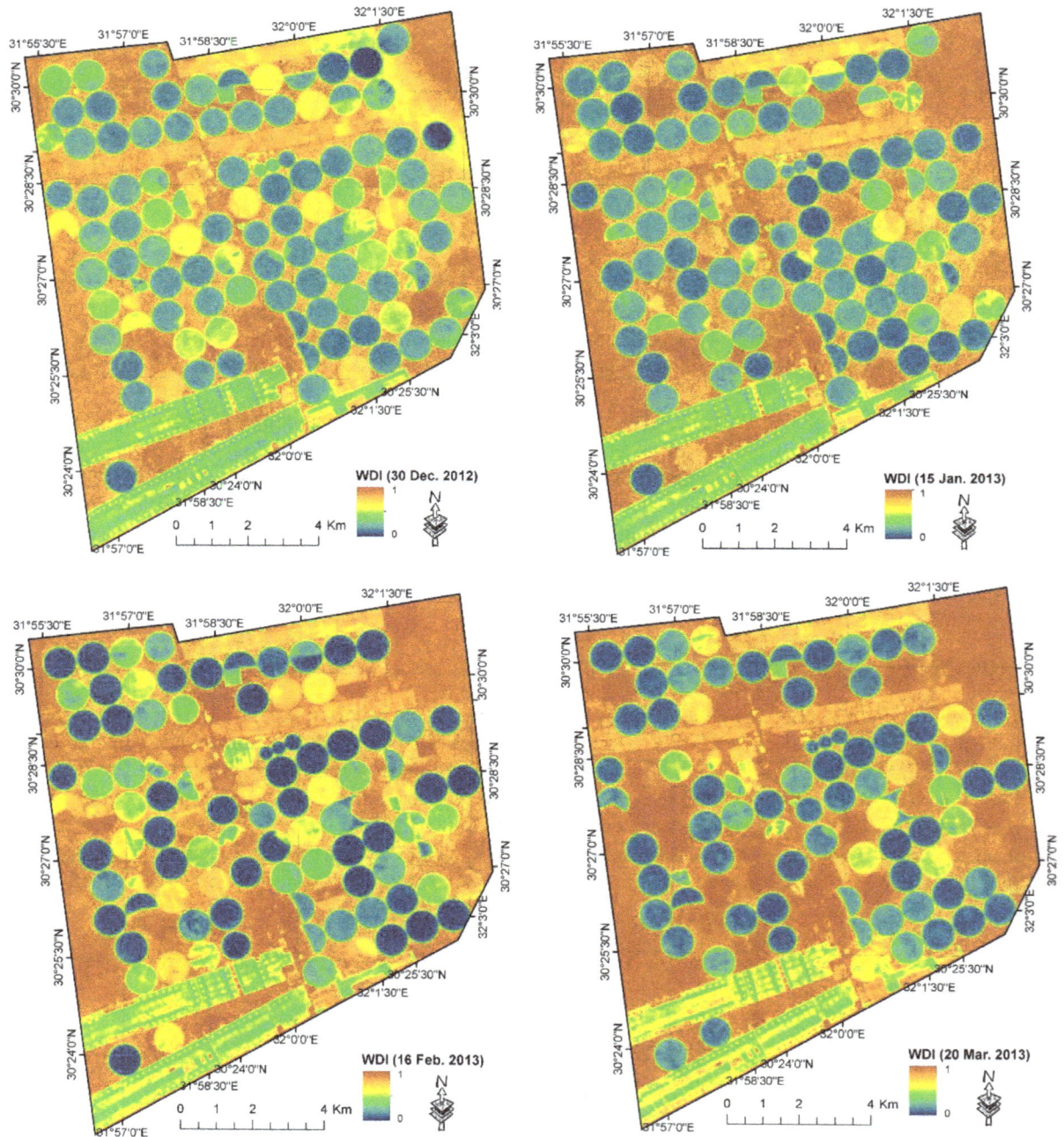

Figure 4. WDI calculated from remotely sensed data for study area.

crop varieties, well-adapted to the growing environment and grown under a high level of crop management [7]. In the current work, yield response factor (ky) values of FAO 33 paper for wheat were used to predict the wheat yield through remotely sensed data.

4. Conclusion

The values of WDI were higher in establishment and vegetative stages than flowering and yield formation because the soil was not fully covered with canopy and the temperature of soil was higher than temperature of canopy. This factor reduces the accuracy of this method in partial canopy coverage case. The predicted wheat

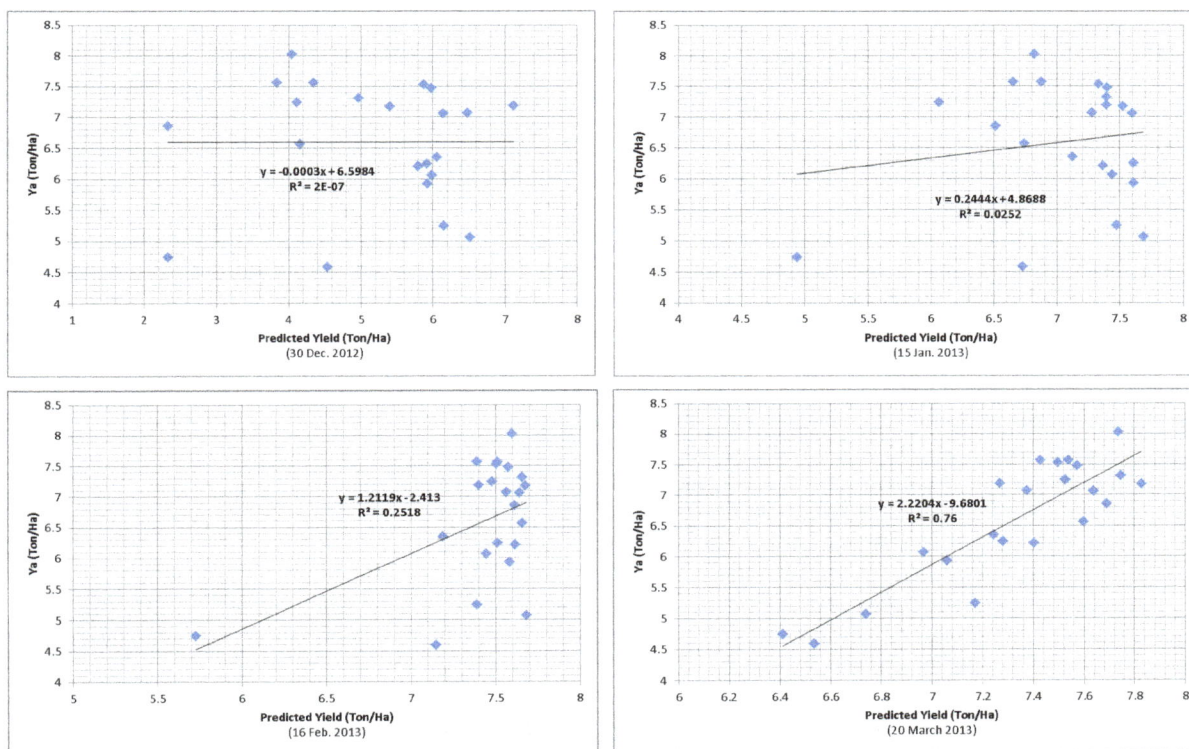

Figure 5. shows the relation between predicted wheat yield (calculated according to Equation (8)) and actual wheat yield.

yield was plotted against actual yield, the relations were varied during different phenological stages where $R^2 = 0.2^{-6}$, 0.03, 0.3 and 0.76. There was no correlation in establishment and vegetative stages but it improved in flowering stage and it was good in yield formation stage. The combination of FAO 33 paper approach and remote sensing techniques is a good idea to estimate yield response to water but it needs to improve.

References

[1]   http://www.fao.org

[2]   Zwart, S.J. and Bastiaanssen, W.G.M. (2004) Review of Measured Crop Water Productivity Values for Irrigated Wheat, Rice, Cotton and Maize. *Agricultural Water Management*, **69**, 115-133. http://dx.doi.org/10.1016/j.agwat.2004.04.007

[3]   Acevedo, E.H., Silva, P.C., Silva, H.R. and Solar, B.R. (1999) Wheat Production in Mediterranean Environments. In: Satorre, E.H. and Slafer, G.A., Eds., *Wheat Ecology and Physiology of Yield Determination*, Food Products Press, New York, 295-323.

[4]   Loss, S.P. and Siddique, K.H.M. (1994) Morphological and Physiological Traits Associated with Wheat Yield Increases in Mediterranean Environments. *Advances in Agronomy*, **52**, 229-276. http://dx.doi.org/10.1016/S0065-2113(08)60625-2

[5]   Slafer, G.A. (2003) Genetic Basis of Yield as Viewed from a Crop Physiologist's Perspective. *Annals of Applied Biology*, **142**, 117-128. http://dx.doi.org/10.1111/j.1744-7348.2003.tb00237.x

[6]   Royo, C., Villegas, D., Rharrabti, Y., Blanco, R., Martos, V. and Garcia del Moral, L.F. (2006) Grain Growth and Yield Formation of Durum Wheat Grown at Contrasting Latitudes and Water Regimes in a Mediterranean Environment. *Cereal Research Communications*, **34**, 1021-1028. http://dx.doi.org/10.1556/CRC.34.2006.2-3.233

[7]   Doorenbos, J. and Kassam, O.W. (1979) Yield Response to Water. FAO Irrigation And Drainage, Paper No. 33, FAO, Rome.

[8]   Allen, R.G., Perrier, L.S., Raes, D. and Smith, M. (1998) Crop Evapotranspiration: Guidelines for Computing Crop Requirements,. FAO Irrigation and Drainage, Paper No. 56, Rome.

[9]   Magliulo, V., d'Andria, R. and Rana, G. (2003) Use of the Modified Atmometer to Estimate Reference Evapotranspiration in Mediterranean Environments. *Agricultural Water Management*, **63**, 1-14.

http://dx.doi.org/10.1016/S0378-3774(03)00098-2

[10]   da Silva, V.de.P.R., Borges, C.J.R, Farias, C.H.A., Singh, V.P., Albuquerque, W.G. and da Silva, B.B. (2012) Water Requirements and Single and Dual Crop Coefficients of Sugarcane Grown in a Tropical Region, Brazil. *Agricultural Sciences*, **3**, 274-286. http://dx.doi.org/10.4236/as.2012.32032

[11]   Kamble, B., Kilic, A. and Hubbard, K. (2013) Estimating Crop Coefficients Using Remote Sensing-Based Vegetation Index. *Remote Sensing*, **5**, 1588-1602. http://dx.doi.org/10.3390/rs5041588

[12]   El-Shirbeny, M.A., Ali, A. and Saleh, N. (2014) Crop Water Requirements in Egypt Using Remote Sensing Techniques. *Journal of Agricultural Chemistry and Environment*, **3**, 57-65. http://dx.doi.org/10.4236/jacen.2014.32B010

[13]   El-Shirbeny, M.A., Ali, A.M., Badr, M.A. and Bauomy, E.M. (2014) Assessment of Wheat Crop Coefficient Using Remote Sensing Techniques. *World Research Journal of Agricultural Sciences*, **1**, 12-17.

[14]   El-Shirbeny, M.A., Saleh, N.H. and Ali, A.M. (2014) Estimation of Potential Crop Evapotranspiration Using Remote Sensing Techniques. *Proceedings of the 10th International Conference of AARSE*, 460-468.

[15]   El-Shirbeny, M.A., Alsersy, M.A.M., Saleh, N.H. and Abu-Taleb, K.A. (2015) Changes in Irrigation Water Consumption in the Nile Delta of Egypt Assessed by Remote Sensing. *Arabian Journal of Geosciences* (in Press). http://dx.doi.org/10.1007/s12517-015-2005-2

[16]   Peltonen-Sainio, P., Kangas, A., Salo, Y. and Jauhiainen, L. (2007) Grain Number Dominates Grain Weight in Temperate Cereal Yield Determination: Evidence Based on 30 Years of Multi-Location Trials. *Field Crops Research*, **100**, 179-188. http://dx.doi.org/10.1016/j.fcr.2006.07.002

[17]   Westgate, M.E., Passioura, J.B. and Munns, R. (1996) Water Status and ABA Content of Floral Organs in Drought-Stressed Wheat. *Australian Journal of Plant Physiology*, **23**, 763-772. http://dx.doi.org/10.1071/PP9960763

[18]   Foulkes, M.J., Sylvester-Bradley, R., Weightman, R. and Snape, J.W. (2007) Identifying Physiological Traits Associated with Improved Drought Resistance in Winter Wheat. *Field Crops Research*, **103**, 11-24. http://dx.doi.org/10.1016/j.fcr.2007.04.007

[19]   Rajala, A., Hakala, K., Makela, P., Muurinen, S. and Peltonen-Sainio, P. (2009) Spring Wheat Response to Timing of Water Deficit through Sink and Grain Filling Capacity. *Field Crops Research*, **114**, 263-271. http://dx.doi.org/10.1016/j.fcr.2009.08.007

[20]   Hiler, E.A. and Clark, R.N. (1971) Stress Day Index to Characterize Effects of Water Stress on Crop Yields. Transactions of Hydrology (210-VI-NEH).

[21]   Dardanelli, J.L., Ritchie, J.T., Calmon, M., Andrianiand, J.M. and Collino, D.J. (2004) An Empirical Model for Root Water Uptake. *Field Crops Research*, **87**, 59-71. http://dx.doi.org/10.1016/j.fcr.2003.09.008

[22]   Moran, M.S., Clarke, T.R., Inoue, Y. and Vidal, A. (1994) Estimating Crop Water Deficit Using the Relation between Surface Air Temperature and Spectral Vegetation Index. *Remote Sensing of Environment*, **49**, 246-263. http://dx.doi.org/10.1016/0034-4257(94)90020-5

[23]   Clarke, T.R. (1997) An Empirical Approach for Detecting Crop Water Stress Using Multi-Spectral Airborne Sensors. *Horticulture Technology*, **7**, 9-16.

[24]   El-Shirbeny, M.A., Aboelghar, M.A., Arafat, S.M. and El-Gindy, A.G.M. (2014) Assessment of the Mutual Impact between Climate and Vegetation Cover Using NOAA-AVHRR and Landsat Data in Egypt. *Arabian Journal of Geosciences*, **7**, 1287-1296. http://dx.doi.org/10.1007/s12517-012-0791-3

[25]   Valor, E. and Caselles, V. (1996) Mapping Land Surface Emissivity from NDVI: Application to European, African and South American Areas. *Remote Sensing of Environment*, **57**, 167-184. http://dx.doi.org/10.1016/0034-4257(96)00039-9

[26]   Goetez, S.J., Halthore, R.N., Hall, F.G. and Markham, B.L. (1995) Surface Temperature Retrieval in Temperate Grassland with Multi-Resolution Sensors. *Journal of Geophysical Research*, **100**, 397-410.

[27]   Norman, J.M., Divakarla, M. and Goel, N.S. (1995) Algorithms for Extracting Information from Remote Thermal-IR Observations of the Earth's Surface. *Remote Sensing of Environment*, **51**, 157-168. http://dx.doi.org/10.1016/0034-4257(94)00072-U

[28]   Hiler, E.A. and Clark, R.N. (1971) Stress Day Index to Characterize Effects of Water Stress on Crop Yields. Transactions of Hydrology (210-VI-NEH).

[29]   Jackson, R.D., Idso, S.B., Reginato, R.J. and Pinter, P.J. (1981) Canopy Temperature as a Crop Water Stress Indicator, *Water Resource Research*, **17**, 1133-1138. http://dx.doi.org/10.1029/WR017i004p01133

12

Planting Cotton in a Crop Residue in a Semiarid Climate: Water Balance and Lint Yield

Robert J. Lascano[1], Dan R. Krieg[2], Jeffrey T. Baker[3], Timothy S. Goebel[1], Dennis C. Gitz III[1]

[1]Wind Erosion and Water Conservation, USDA-ARS[*], Lubbock, TX, USA
[2]Department of Plant and Soil Science, Texas Tech University, Lubbock, TX, USA
[3]Wind Erosion and Water Conservation, USDA-ARS, Big Spring, TX, USA
Email: Robert.Lascano@ars.usda.gov

Abstract

Cotton (*Gossypium hirsutum* L.) is planted on more land area than any other crop on the Texas High Plains. Much of this area is considered highly erodible and requires a conservation compliance program to participate in government farm programs. Because this region is semiarid and because irrigation water is increasingly limited, water conservation and efficient use of water are necessary to maximize cotton lint yields. One popular conservation compliance practice used is to plant cotton into a chemically terminated small grain crop, *i.e.*, residue that provides wind protection to the cotton seedlings. Our hypothesis was that in a semiarid region the use of a small grain cover crop under irrigated conditions would use more water than it conserves compared to conventional tilled cotton, thus reducing cotton lint yields. To test the hypothesis separate field studies over two growing seasons and on two soil textures, a loamy fine sand and a clay loam, were conducted. The main treatments were tillage systems (conventional and conservation using terminated wheat residue). The two split plot treatments were water supply based on replacement of calculated grass reference evapotranspiration (ET_o). Tillage did not affect the amount of water used by the cotton crop at either location (< 7% difference, P > 0.05) except for an 80% ET_o irrigation treatment at a single location where the bare soil treatment used 10% more water than the residue treatments for both years. The residue treatment decreased (P < 0.05) cotton lint yields at both locations by 12% except for the 50% ET_o single irrigation treatment in which the residue treatment yielded 14% more lint than the bare soil treatment. The use of terminated wheat residue had no impact on soil water storage during any part of the year. During a 5-month period associated with wheat growth, the wheat evapotranspiration was 20 to 40 mm more water (P < 0.05) than that lost through soil water evaporation from the conventional treatments. The use of termi-

---

[*]The US Department of Agriculture (USDA) prohibits discrimination in all its programs and activities on the basis of race, color, national origin, age, disability, and where applicable, sex, marital status, familial status, parental status, religion, sexual orientation, genetic information, political beliefs, reprisal, or because all or part of an individual's income is derived from any public assistance program.

nated wheat residue did not benefit the water balance of the cotton crop, and was associated with decreased cotton lint yields. The results were consistent with our working hypothesis, and disproved the idea that planting cotton into wheat stubble cover increases water use efficiency.

Keywords

Soil Water, Evapotranspiration, Cropping Systems, Semi-Arid, Wheat Residue

---

1. Introduction

In 2013 and 2014, about 1.2 million ha were planted with upland cotton on the Texas South Plains [1]. The annual mean precipitation in this region is 467 ± 159 mm and about 50% of this rainfall occurs during the 4-month (May to August) growing season. About half of the planted area is irrigated with a water supply ranging from 2.5 to 7.5 $mm \cdot d^{-1}$ and the potential evaporative demand is on average 9.8 ± 0.8 $mm \cdot d^{-1}$ during the four-month growing season [2]. The source of the irrigation water is the Ogallala aquifer, where withdrawals exceed natural recharge, resulting in a declining water table [3]. Much of the water applied to the cotton is lost to bare soil water evaporation (E_{soil}) in conventional tillage systems [4] [5]. In this semiarid region, where cotton is planted in a wide row spacing (1-m) the crop seldom reaches a leaf area index of 3 $m^2 \cdot m^{-2}$ that is necessary to achieve canopy closure [6] [7]. Furthermore this exposed and bare soil is conducive to increase the amount of E_{soil} [8].

Given the decline of irrigation water from the Ogallala aquifer [3] [9] with a decreasing pumping capacity, particularly during periods of low rainfall which are frequent in this region, led to develop agronomic practices to maximize cotton's water use efficiency (WUE), i.e., production of lint yield per unit of water taken up by the plant [10]. An example of this conservation practice is to plant cotton into a winter wheat (*Triticum aestivum L.*) crop that is chemically terminated in the spring before the cotton crop is planted [11]. Residues may protect cotton seedlings by providing shelter to wind damage and sand abrasion [12]. Also, the residue may reduce E_{soil}; however, in some cases it may use more water than it conserves [5] [8] via wicking [13]. This is the process whereby the wheat residues may contribute to additional water loss via capillary action, i.e., wicking effect [14]. In this situation, the wheat residue may increase the ability to transport water from the seedbed to the atmosphere via evaporation and thus resulting in a reduction of soil water and negatively affecting cotton emergence when planted in the wheat residue.

In the cotton cropping systems of the Texas High Plains, the effect of crop residues on water conservation and lint yield, are a function of the water supply, from irrigation and rainfall, soil type, and tillage practices [15] [16]. Planting cotton into a wheat residue on a sandy loam soil only increased cotton lint yield when the crop was irrigated at the grass reference evapotranspiration demand (ET_o). However, this practice is not suited to the availability of irrigation water from the Ogallala aquifer. In field trialsand over a five-year period, in a loamy fine sandy soil the residue did not impact lint yield under irrigated conditions, but significantly and adversely affected dryland lint yields. In a clay loam soil, the lint yield of cotton planted in a wheat residue increased on average by 13% under irrigation and reduced dryland cotton lint yield by 8% [17] [18]. From these results, it can be concluded that the effects of wheat residues on cotton lint yield depend primarily on the water supply. The texture effect is related to the amount of water that can be stored in the soil, particularly from the time the wheat is terminated to when the cotton is planted. For example, in the surface 0.20 m of the soil profile, a clay soil can store about twice as much water as a sandy soil and for dryland conditions the additional stored water would be available during cotton planting and favor seed germination and emergence of the cotton plant.

In first analysis for the Texas High Plains, the amount of water required to establish the wheat residue is often ignored in the water balance of dryland and limited irrigated cotton cropping systems [4] [5] [8]. To establish a wheat residue that would not impact the germination and emergence of the subsequent cotton crop planted in the residue, and of sufficient growth to be beneficial to cotton seedlings requires a net gain of water. In the Texas High Plains, winter wheat is planted in months (November to February) where the average monthly rainfall is 16 ± 2 mm [2]. Clearly, in a dryland system the establishment of the residue comes with the risk of using stored water that otherwise would be available at the time needed when the cotton crop is planted [19]. Furthermore, the winter wheat crop requires water to grow and be of benefit to the cotton crop. The measured daily ET rate of irrigated winter wheat varied between 1 and 2 mm·d^{-1} with a maximum of 13 mm·d^{-1} during the late spring [20]. Therefore, it was concluded that for the Texas High Plains under dryland conditions the establishment of a wheat residue is perhaps not practical and will require irrigation to be of value [19].

The short- and long-term benefits of residues on soil physical and chemical properties, and crop production has been widely investigated, e.g., [15] [21]-[23]. Residues increase infiltration of water, from either rain or irrigation, by increasing the surface roughness and flow path tortuosity and thus increasing the time for more water to infiltrate by surface ponding [19] [24] [25]. Residues, also modify the kinetic energy from raindrops impacting the soil surface [26] [27], which leads to less erosion [26] and increasing rain interception [28] by as much as 10% of the annual rainfall [29]. A long-term benefit of residues is to increase water storage and thus providing more water for crop production [30] [31].

The impact of residues on E_{soil} is well documented [4] [5] [8] [32]-[34]. Residues can reduce the rate of E_{soil} in the so-called first stage [25] [30] [34]-[37]. The type of residue also affects the E_{soil} rate and wheat, for example, is more effective than cotton [35]. The presence of a residue reduces E_{soil}; however, the total evaporation equilibrates between residue covered soil and bare soil after a long period of time between water inputs [35]. For example, the rate of E_{soil} from a bare soil and a residue-covered soil will be different; however, the total amount of water lost from the bare and residue-covered soil will be very similar after a long period of time [5] [8]. The rate of E_{soil} is defined by exchanges of energy and water between the soil surface and the surrounding environment [32] and determined by the combined water and energy balance of the soil surface. The aerodynamic resistance to vapor and latent heat flux from the soil surface to the atmosphere is increased by the presence of a residue, which decreases the rate of E_{soil} [5] [8] [38] [39]. Another effect of a residue is to reduce the amount of irradiance that reaches the surface, decreasing soil temperature and E_{soil}.

The presence of a surface residue modifies the E_{soil} and affects the WUE [8]. The daily E_{soil} and crop transpiration (T) for a 100-day growing season of cotton planted into terminated wheat and conventionally tilled soil where compared by Lascano et al. [8]. In this experiment daily E_{soil} was measured for a seven-day period using microlysimeters [32] and calculated using the mechanistic ENWATBAL model [4] [40]. Both cotton cropping systems, residue and conventional, had the same seasonal ET. However, the conventional tilled cotton evaporated more E_{soil} and had a lower amount of plant T than the cotton planted into the terminated wheat. The wheat residue increased the seasonal crop T, i.e., more water used for plant growth, and reduced the seasonal E_{soil}, resulting in an increase of lint yield of 35% compared to the cotton planted in the conventional tillage with the same amount of water. The WUE of the cotton planted in the wheat residue was 2.6 g lint per kg of water used in ET, which was 27% more lint than the conventionally planted cotton [8]. In these field experiments the amount of water used to grow and establish the wheat residue were not measured and thus the calculated value of WUE is biased towards the cotton lint yield from the residue.

Our hypothesis is that despite the many benefits that residue offer, in a semiarid environment establishing the residue would use more water than conserved when used along a cotton crop. In our hypothesis the assumption was made that the cover crop that provides the residue and cotton were supplemented as needed with irrigation and each tillage treatment received the same amount of water annually. Further, it was postulated that the use of chemically terminated winter wheat residue would not significantly increase lint yields because less water would available for cotton growth and lint production. The main objective of this study was to examine and compare the water balance of two cotton cropping systems for a two-year period for conventionally tilled cotton, and for cotton planted into terminated wheat. Specifically, our objectives were to 1) compare the water used by wheat vs. soil water lost to evaporation from the conventionally tilled bare soil; 2) determine if wicking occurred after the wheat was terminated and before the cotton was planted; and 3) determine the effects that bare soils and residue-covered soils have on the water available for cotton growth and lint yield.

2. Materials and Methods

This study was done during two growing seasons, 1994 and 1995, at experimental fields within two research farms managed by Texas Tech University, Lubbock, TX. One farm was in central Terry County, TX near Brownfield (33°10'47"N - 102°16'15"W) where the site has an Amarillo loamy fine sand soil [fine-loamy, mixed, thermic, Torrertic Paleustalf]. The other farm was in northern Lubbock County, TX near New Deal (33°44'13.76"N, 101°43'58.04"W) with a Pullman clay loam soil [fine, mixed, thermic, Torrertic Paleustoll]. These two research farms were part of a long-term (1991-2001) study to evaluate dryland cropping systems, cotton and sorghum

(*Sorghum bicolor L.*) in the semiarid climate of the Texas High Plains. Results presented here are partially based on work by Vorheis [41] and Ralston [42].

2.1. Irrigation

Irrigation-water at each location was applied using sprinkler irrigation methods. At Brownfield, TX a center pivot, 400 m long, irrigation system provided 50, 75, and 100% of the weekly crop water use with Low Energy Precision Application (LEPA) spray nozzles [43]. The linear (300 m in length) irrigation system used at New Deal provided 40% or 80% of weekly crop water use with LEPA spray nozzles. Half the field at New Deal was irrigated at 40% ET_o and the other half irrigated at 80% ET_o.

At each location, the amount of water applied was based on calculated daily grass reference ET_o using as weather input net irradiance, air temperature and humidity, and wind speed measured at each site, and using the Penman-Monteith method as given by Allen *et al.* [44]. The daily water use was determined as ET_o multiplied by a locally developed crop coefficient (K_c) that depended on the ground cover [45]. The weekly crop water use was calculated using the appropriate ET_o and K_c for that site. Date and amount of water applications for Brownfield are given in **Table 1** and for New Deal in **Table 2**. Rainfall was measured at each site using a tipping bucket. Irrigation treatments at both sites did not begin until cotton emergence. The irrigation levels were different between the two locations because the two studies were independent of each other, with different soil textures and environmental conditions.

2.2. Experimental Design

The experimental design at both locations was a randomized block, split plot setup. The experimental field at each site was about 46 ha. The major plots were tillage effects and the irrigation treatments were the split plots. All treatments were replicated four times and the plot size of each replicate was 156 × 183 m. The irrigation treatments (*i.e.*, 50% and 100% ET_o at Brownfield, and 40% and 80% ET_o at New Deal) were started after the cotton emerged (late May through early June). Therefore, all plots for each location received identical irrigation amounts between cotton harvest from the previous season until cotton emergence the next season. One half of each irrigation treatment at each location was planted conventionally with cotton (left fallow in winter) and the other half was planted with cotton in terminated winter wheat. Locations were previously used for cotton production prior to this study with half of each field in terminated wheat residue and half of each field under conventional tillage. Treatments were replicated, n = 4, and mean separation was done using the PROC MIXED procedure and with t-type confidence interval of 0.05 using SAS on an Apple® computer (WMware and SAS version 9.2).

2.3. Cultural Practices

Winter wheat at both locations was drilled at a rate of 30 kg·ha^{-1} at 0.2-m row spacing into listed soil resulting in two rows of wheat in each furrow on 0.75-m centers. The tops of the beds were left bare in preparation for the planting of cotton. The wheat (TAM-200, Texas A&M University Foundation Seed, College Station, TX)# grew until it was tall, about 0.3 m, enough to provide wind protection, which was usually in mid- to late-April and was then terminated with the herbicide glysophate (Roundup®, Monsanto Company, St. Louis, MO) at a rate of 2 L·ha^{-1}. Cotton (HS-26, Paymaster Technology Corp., Scott, MS) was planted into the terminated wheat and on the fallow plots during early to mid-May. A complete fertilizer blend consisting of 50-50-0-10 was applied to the entire area after each cotton harvest and prior to planting the wheat in the fall each year. Supplemental N was provided through the irrigation water at a rate of 0.2 kg N mm^{-1}·ha^{-1} with an irrigation volume of 25 mm for the 100% and 80% irrigation treatments through the third week of flowering. Important dates for different cultural operations at both locations are given in **Table 3**.

2.4. Measurements

Soil water contents were measured gravimetrically at various times throughout the year, usually one to two

---

#Mention of this or other proprietary products is for the convenience of the readers only and does not constitute endorsement or preferential treatment of these products by the USDA-ARS.

Table 1. Irrigation applied and rain for two irrigation treatments and two growing seasons in Brownfield, TX.

Date	Irrigation Treatment [mm]		Rain [mm]	Date	Irrigation Treatment [mm]		Rain [mm]
	50% ET$_o$	100% ET$_o$			50% ET$_o$	100% ET$_o$	
28-Apr-94			7.4	04-Jan-95			4.6
29-Apr-94	25.4	25.4		20-Jan-95			20.1
29-Apr-94			8.4	27-Jan-95			3.3
30-Apr-94			18.0	18-Feb-95			13.0
3-May-94			1.0	01-Mar-95			1.0
10-May-94			15.5	10-Mar-95			5.6
11-May-94			29.2	07-Apr-95			8.1
12-May-94			54.1	09-Apr-95	25.4	25.4	
13-May-94			32.3	15-Apr-95	25.4	25.4	
18-May-94	6.4	6.4		02-May-95			1.3
25-May-94			3.3	05-May-95			10.9
26-May-94			1.5	06-May-95			4.1
27-May-94			5.6	11-May-95			11.4
31-May-94	25.4	25.4		15-May-95			1.5
3-Jun-94	13.2	13.2		24-May-95			1.1
10-Jun-94	25.4	25.4		30-May-95			41.0
12-Jun-94			1.5	04-Jun-95			6.4
21-Jun-94	25.4	25.4		10-Jun-95			13.5
12-Jul-94	10.8	21.6		23-Jun-95			18.0
13-Jul-94			1.8	24-Jun-95			7.0
14-Jul-94			3.8	27-Jun-95			21.0
22-Jul-94	25.4	50.8		05-Jul-95	14.0	28.0	
28-Jul-94			14.5	15-Jul-95			1.3
30-Jul-94			14.5	18-Jul-95	10.0	20.3	
3-Aug-94			3.0	20-Jul-95			5.3
8-Aug-94	25.4	50.8		21-Jul-95			2.8
27-Aug-94	12.5	25.4		25-Jul-95	10.0	20.3	
1-Sep-94			11.7	28-Jul-95	6.3	12.7	
9-Sep-94			2.0	31-Jul-95			1.3
15-Sep-94			3.8	01-Aug-95			5.1
16-Sep-94			7.9	14-Aug-95			58.0
7-Oct-94			10.7	21-Aug-95			6.0
8-Oct-94			5.3	09-Sep-95			19.3
15-Oct-94			3.6	10-Sep-95			13.7
18-Oct-94			2.3	13-Sep-95			3.0
5-Nov-94			5.8	16-Sep-95			134.0
24-Nov-94			4.3	22-Sep-95			11.0
				28-Sep-95			5.0
Total [mm]	**195.3**	**269.8**	**272.8**		**91.1**	**132.1**	**458.7**

Table 2. Irrigation applied and rain for two irrigation treatments and two growing seasons, in New Deal, TX.

Date	Irrigation Treatment [mm]		Rain [mm]	Date	Irrigation Treatment [mm]		Rain [mm]
	40% ET_o	80% ET_o			40% ET_o	80% ET_o	
26-May-94			14.0	22-Jan-95			6.9
13-Jun-94	25.4	25.4		29-Jan-95			3.0
16-Jun-94			3.0	10-Feb-95	38.1	38.1	
29-Jun-94			19.0	18-Feb-95			3.3
13-Jul-94			51.0	5-Mar-95			1.5
14-Jul-94			10.0	12-Mar-95			6.9
19-Jul-94	25.4	50.8		18-Mar-95			1.3
21-Jul-94			5.0	25-Mar-95			1.3
2-Aug-94			1.0	1-Apr-95			2.3
11-Aug-94	30.8	63.5		8-Apr-95			18.3
14-Aug-94			3.6	15-Apr-95	50.8	50.8	
8-Sep-94			7.4	16-Apr-95			8.1
14-Sep-94			11.4	18-Apr-95			1.0
15-Sep-94			8.4	5-May-95			18.5
7-Oct-94			4.3	6-May-95			5.1
14-Oct-94			1.5	15-May-95			14.5
15-Oct-94			4.6	23-May-95			2.3
18-Oct-94			1.3	24-May-95			1.8
19-Oct-94			3.3	25-May-95			2.0
20-Oct-94			1.8	26-May-95			5.1
4-Nov-94			12.7	29-May-95			16.3
19-Nov-94			2.0	30-May-95			7.9
6-Dec-94			1.0	2-Jun-95			1.0
27-Dec-94			1.3	3-Jun-95			6.4
30-Dec-94			1.0	10-Jun-95			13.5
				2-Jul-95			14.5
				14-Jul-95	16.5	33.0	
				18-Jul-95			3.3
				24-Jul-95	10.1	20.3	
				26-Jul-95	10.1	20.3	
				31-Jul-95			4.8
				1-Aug-95			2.7
				2-Aug-95			1.3
				19-Aug-95	25.4	50.8	
				9-Sep-95			2.8
				10-Sep-95			13.2
				12-Sep-95			13.5
				15-Sep-95			90.0
				19-Sep-95			55.0
				29-Sep-95			100.0
Total [mm]	**81.6**	**139.7**	**168.6**		**151.0**	**213.3**	**449.4**

Table 3. Cultural operations at two sites and growing seasons.

Cultural Practice	New Deal	Brownfield
Wheat Planted	3-Dec-93	15-Dec-93
	1-Dec-94	4-Dec-94
Wheat Terminated	4-May-94	28-Apr-94
	26-Apr-95	21-Apr-95
Cotton Planted	20-May-94	5-May-94
	15-May-95	12-May-95
Cotton Harvested	6-Nov-94	7-Nov-94
	10-Nov-95	2-Nov-94

times per month [46], and these were converted to volumetric water content using appropriate bulk density values provided by Baumhardt *et al.* [47]. On each sampling date, 6 - 10 soil samples per plot were taken in 0.30-m increments to a 1.2-m depth. Water used through E_{soil} and/or T during different crop stages was based on the water balance, inputs and outputs and included precipitation, irrigation and net change in soil water content. When no crops were on the ground, all the water loss was assumed to be through E_{soil}, and if a crop did exist, the water loss was assumed to be a combination of E_{soil} and T [48]. Specifically, the following quantities were determined: 1) water required to grow an adequate amount of residue; 2) water used by the wheat after being chemically terminated; 3) water lost through E_{soil}; 4) water used by each cotton crop; and 5) the effect of each cotton cropping system on cotton lint yields. Runoff was assumed to be negligible because of slopes < 0.1% and soil water content below the 1.2-m depth was not measured because of the existence of a caliche layer around the 1.4-m depth. Cotton lint yield was estimated by hand harvesting 10 m^2 of plot area in each replication of each treatment. Cotton yield components were determined and their relative contribution to total lint yield evaluated.

2.5. Additional Experiment

An additional experiment was done in New Deal, TX in 1996 to measure evaporative losses of cotton planted in a bare soil and wheat terminated residue, without any irrigation. To measure the loss of water by evaporation, gravimetric soil samples were taken in the top 0.60 m of the soil profile in about five-day increments starting on the day the wheat was terminated by applying glysophate, 15 May 1996, and ending on the 3 June 1996, when more than 75% of the cotton planted had emerged. Gravimetric soil water content was converted to a volume basis using the measured soil bulk density and these measurements were replicated six times. The bare soil and wheat residue treatment was replicated three times and each plot was the same size as previously used, *i.e.*, 156 × 183 m.

3. Results and Discussion

3.1. Wheat Water Use

The first research objective was to measure and compare the water used by the wheat to the soil water lost to evaporation from conventionally tilled bare soil. The amount of water used to establish a wheat residue and the amount of water lost to bare E_{soil} for Brownfield and New Deal are given in **Table 4**. During the wheat growth stage, all plots received the same irrigation amount. These results showed that in 1995, the wheat used more (P ≤ 0.05) water than the bare soil lost E_{soil} in both treatments and locations. At Brownfield, the wheat used 40% more water (P ≤ 0.05) than the bare soil lost to water evaporation in the loamy fine sand textured soil. At New Deal, the wheat used 36% more water (P ≤ 0.05) than the bare soil lost to evaporation in the clay loam soil. During the spring of 1995, establishing a wheat residue required more water compared to leaving the soil bare at both locations and on both soil textures.

The water depth in the upper 0.6 m of the soil profile at wheat planting and at wheat termination at both sites,

Table 4. Water balance (inputs and outputs) at two locations for the 1995-growing season.

Year	Brownfield				New Deal			
	Inputs [mm]		Outputs[#] [mm]		Inputs [mm]		Outputs[#] [mm]	
			Residue	Treatment			Residue	Treatment
	Irrigation[*]	Rain	Bare Soil	Residue	Irrigation[*]	Rain	Bare Soil	Residue
1995	53.0	62.0	94.5 (17)	132.5 (16)	89.0	55.0	105 (12)	142.5 (13)

[*]Irrigation treatments were not active during the wheat growth; [#]Calculated values of outputs are the mean of all measurements and the standard deviations are given in parenthesis.

Brownfield and New Deal, TX are given in **Table 5**. Even though all the plots within each location received the same amount of irrigation during the wheat growth period, our measurements demonstrated the variability in soil water content at each location. The coefficient of variation of the measured mean soil water content was > 25% (data not shown). This variability could be explained by the irrigation treatments applied during the previous season because overall soil water content could increase over the growing season in the highest irrigation treatments compared to the lower ones.

At both locations, the bare soil treatment resulted in an increase of the soil water content in the top 0.6 m by > 30% ($P \leq 0.05$) between the time when the wheat was terminated and planted (**Table 5**). However, the wheat residue treatments showed no gain in stored water at wheat termination compared to the amount of soil water at wheat planting. In the loamy fine sand, the wheat residue caused a reduction ($P \leq 0.05$) of 14% in soil water content. At Brownfield, the bare soil gained 71% ($P \leq 0.05$) soil water and at New Deal the gain was 37% ($P \leq 0.05$) from wheat planting to wheat termination. The statistical difference ($P \leq 0.05$) between the amount of soil water at wheat termination at Brownfield was 36 mm between the bare soil and wheat residue treatments. The statistical difference ($P \leq 0.05$) between soil water at wheat termination at New Deal was 40 mm of water in the top 0.6 m of soil. Therefore, there was no gain, 36 vs. 40 mm, in stored water by the wheat residue plots over the spring of 1995 at either location, while the bare soil was able to increase the water content in the top 0.6 m of soil by more than 30% ($P \leq 0.05$) over the spring months.

3.2. Wicking

The second objective was to determine if upward movement of water through the wheat residue, *i.e.*, wicking, occurred after the wheat was terminated and before the cotton was planted. The amount of water lost to evaporation from both the bare soil and terminated wheat plots between wheat termination and cotton emergence for 1995 at Brownfield and New Deal are shown as a bar graph in **Figure 1**. These results suggested, and as expected the evaporation of water was greater in the finer textured soil when compared to the coarser textured soil. In the fine loamy sand soil, the bare soil evaporated 44% more ($P \leq 0.05$) soil water than the wheat residue. However, in the clay loam soil the wheat residue evaporated 66% more ($P \leq 0.05$) soil water than the bare soil. This result suggested that there might be a soil textural effect on the amount of water that can evaporate from a bare soil and from a terminated wheat residue [19].

It was speculated that differences in water evaporation from the bare and from the wheat-covered soil were due to the different textures and the process of wicking. The soil at New Deal is a clay loam, *i.e.*, more conducive for upward capillary flow of water; while Brownfield has a loamy fine sand, whose large pores would decrease capillary flow [13]. Under wicking, wheat residue could lose more water to evaporation than bare soil because many roots act as capillaries and the surface area provided by the leaves of the terminated wheat contribute to the evaporation. A possible solution to this problem could be to mechanically sever the roots of the wheat plant after termination at a 0.2 - 0.3 m depth and in this way disrupt the capillary movement of water from the soil to the atmosphere.

The cumulative evaporation of soil water from the bare soil and terminated wheat, between wheat termination to cotton emergence is shown in **Figure 2**. The bare soil had an 18% larger ($P \leq 0.05$) amount of water evaporation loss compared to the wheat residue. This is a result that in part may be explained by the frequent irrigation and rain events that occurred during the study. These wetting events resulted in a bare soil that was wetter than the top of the wheat residue, leading to near free-water evaporation to occur at faster rate than wicking. However,

Table 5. Amount of water in the top 0.6 m of the soil for wheat residue at planting and termination at two locations.

Year	Brownfield Residue Treatment [mm]				New Deal Residue Treatment [mm]			
	Bare Soil		Residue		Bare Soil		Residue	
	Planting	Termination	Planting	Termination	Planting	Termination	Planting	Termination
1995	47.0	80.5	51.5	44.5	104.0	142.5	102.0	103.0

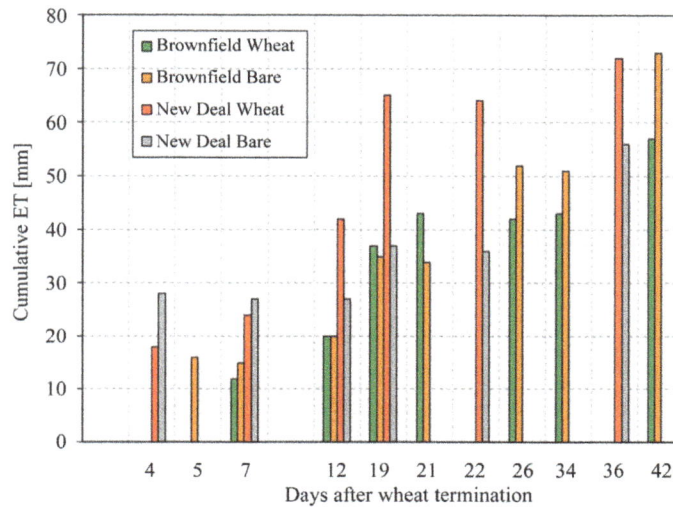

Figure 1. Cumulative measured evapotranspiration (ET, mm) as a function of days after wheat termination at Brownfield (BF) and New Deal (ND) from the wheat and bare soil residue treatment for the 1995-growing season.

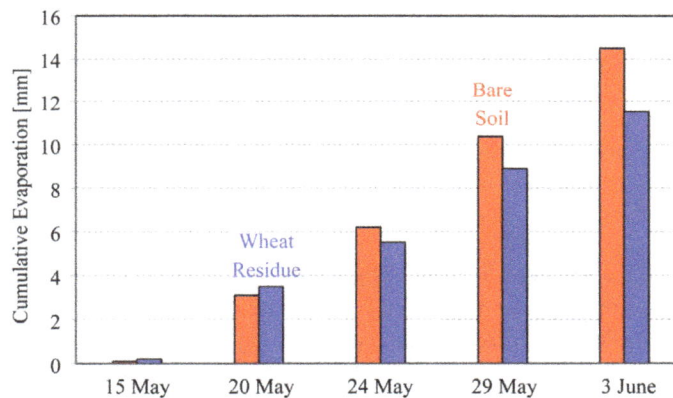

Figure 2. Cumulative measured evaporation in mm from a bare soil and wheat residue treatment for the time period between wheat termination (15 May 1996) to cotton emergence (3 June 1996) in New Deal TX.

the results from this study were not conclusive to determine the effects of wicking in a clay loam soil and the measurement of E_{soil}.

Part of the difficulty in determining differences of E_{soil} under field conditions is that the amount of water that may be lost due to wicking is of the order of 1 mm·d^{-1} and this loss takes place from mainly the surface 0.1 m layer of the soil profile. This rate of water evaporation represents a change of 0.01-m^3·m^{-3} volumetric water content on a daily basis, which is beyond the sensitivity of current instrumentation to measure volumetric soil water content [46]. Nevertheless, a wicking loss of 1 mm·d^{-1} for 20 days represents 20 mm of water, and this amount of water could determine the difference of having adequate water in the seedbed to germinate and to emerge the cotton crop planted in the residue [13].

The water used by the wheat and lost by E_{soil} affects the amount of water available for cotton growth. The amount of water content in the 1.2 m soil profile at cotton planting for both sites and growing seasons season is shown in **Table 6**. The bare soil treatments had more water at cotton planting than the wheat residue treatments at both locations except for the 40% ET_o plots at New Deal and the 50% ET_o plots at Brownfield for 1994. All the treatments in New Deal excluding the 40% ET_o irrigation treatment in 1994 showed a significant increase (P $\leq$ 0.05) in soil water for the bare soil treatments compared to the wheat residue treatments, while Brownfield did not have a significant increase (P $\leq$ 0.05) in soil water for the bare soil treatments at cotton planting. This stored water is important, particularly at the time when the cotton is planted as it necessary for seed germination and emergence impacting the vitality and vigor of the cotton seedling, which can later affect lint yields [49] [50].

Even though the wheat used more water in Brownfield in 1995 than the bare soil (**Table 5**), it appears that the wheat only extracted water from the top 0.6 m of soil because the water depth in the top 1.2 m of soil at cotton planting (**Table 6**) showed that the soil profile for each residue treatment contained about the same amount of water in the top 1.2 m of soil. However, at New Deal, capillary movement of water by a hydraulic gradient and by wheat roots may have caused a reduction in the amount of water in the top 1.2 m of soil for the wheat residue plots, allowing the bare soil to store more water at the time that cotton was planted (**Table 6**).

3.3. Cotton Water Use

The last and third objective of this study was to determine the effects bare and residue-covered soils have on the water available for cotton growth and development. Water inputs and water used by the cotton in both seasons and at both sites are given in **Table 7**. There were no significant differences (P $\leq$ 0.05) in the amount of water used by the cotton for any of the residue treatments at Brownfield (**Table 7**). However, the 80% ET_o bare soil treatments used at least 10% more (P $\leq$ 0.05) water to grow the cotton than the residue treatments for both years at New Deal (**Table 7**) and as previously discussed, could be accounted for sampling error (**Table 5**).

Cotton lint yields were used to evaluate the effects residues have on cotton growth and development. Cotton lint yields for both locations and growing seasons are given in **Table 8**. These cotton lint yield values were similar to cotton yields reported for surrounding areas [51] [52]. In 1994, both locations had significantly higher (P $\leq$ 0.05) lint yields (at least 8% higher) from the conventionally tilled treatments than the wheat residue plots for all irrigation treatments (**Table 8**). The only treatment with a loss in lint yield from the conventionally tilled treatments was in the in the low irrigation treatment at Brownfield in 1995 with a 14% decrease (P $\leq$ 0.05) in lint yields. The greatest gain in lint yield from the conventionally tilled treatments over the terminated wheat treatments occurred in the 40% irrigation treatment at New Deal in 1994 with a 42% gain (P $\leq$ 0.05). Therefore, our results suggested that the presence of residues had a significant (P $\leq$ 0.05) impact with a loss of lint yield for both years of the study. This result confirms our hypothesis that in a semiarid region the use of a small grain cover crop under irrigated conditions uses more water than it conserves compared to conventional tilled cotton, reducing cotton lint yields.

As expected and in climates where water is a limiting factor in crop production, the irrigation treatments at each location that received the greatest amount of water resulted in higher lint yields compared to the lower irrigation amounts (**Table 8**). For example, Bordovsky et al. [45] showed that higher irrigation amounts also gave higher yields, but also reported that at a 50% ET_o irrigation level, conventionally tilled and wheat residue cotton systems yielded about the same (P $\leq$ 0.05), while at a 100% ET_o irrigation level, the conventionally tilled cotton cropping system yielded 17% less (P $\leq$ 0.05) as compared to the wheat residue treatment at Lamesa, TX, which has a sandy loam soil.

Table 6. Amount of water in the top 1.2 m of the soil when cotton was planted at two locations and two years.

Year	Brownfield Residue Treatment [mm]		New Deal Residue Treatment [mm]	
	Bare Soil	Residue	Bare Soil	Residue
1994	207.0	201.5	296.0	299.0
1995	177.5	169.5	238.0	192.0

Table 7. Water balance (inputs and outputs), for two irrigation treatments at two locations and two years.

	Brownfield						
	Water Inputs [mm]			Water Outputs [mm]			
	Irrigation Treatment			Irrigation Treatment			
Year			Rain [mm]	50% ET_o		100% ET_o	
	50% ET_o	100% ET_o		Residue Treatment		Residue Treatment	
				Bare Soil	Residue	Bare Soil	Residue
1994	125	200	93	276	295	395	403
1995	41	81	375	315	314	406	403

	New Deal						
	Water Inputs [mm]			Water Outputs [mm]			
	Irrigation Treatment			Irrigation Treatment			
Year			Rain [mm]	40% ET_o		80% ET_o	
	40% ET_o	80% ET_o		Residue Treatment		Residue Treatment	
				Bare Soil	Residue	Bare Soil	Residue
1994	83	140	172	283	304	442	387
1995	93	130	382	541	539	620	558

Table 8. Average cotton lint yields for two locations, and two irrigation and residue treatments, for two years.

Brownfield – Cotton Lint Yield [kg·ha^{-1}]				New Deal Cotton Lint Yield [kg·ha^{-1}]			
Irrigation Treatment				Irrigation Treatment			
50% ET_o		100% ET_o		40% ET_o		80% ET_o	
Residue Treatment		Residue Treatment		Residue Treatment		Residue Treatment	
Bare Soil	Residue	Bare Soil	Residue	Bare Soil	Residue	Bare Soil	Residue
440 (20)[*]	392 (16)	933 (31)	816 (11)	561 (19)	394 (23)	674 (11)	599 (21)
380 (17)	444 (21)	594 (26)	546 (24)	436 (11)	401 (31)	603 (19)	559 (11)

[*]The standard error of the mean is in parenthesis. Mean separation was calculated using PROC MIXED with a t-type confidence interval of 0.05.

4. Conclusions

In this study, the cotton water use and lint yield of two cotton-cropping systems in the semiarid climate of the Texas High Plains was compared. These cropping systems were conventionally tilled and cotton planted into terminated wheat, on two sites and a low and high irrigation rate over two growing seasons. Our objective was to evaluate the effectiveness of planting cotton into a crop residue obtained by terminating a winter wheat crop planted in December. The purpose of the residue is mainly to provide a mechanical barrier to wind and thus protect cotton seedlings at a time when they are most vulnerable to sandblast damage due to high wind speeds. The selected sites, allowed us to compare the impact of soil texture, *i.e.*, coarse (Brownfield, TX) and fine (New Deal, TX). The residue treatments did not affect the total amount of water evaporated and transpired during the two seasons. However, at both locations the bare residue treatment had a higher lint yield than the cotton grown under the wheat residue. Wicking, the capillary movement of water through the wheat residue was only statistically significant at the site with the coarser texture (Brownfield, TX) and only in one (1994) of the two years.

These results suggest that the use of terminated wheat residue did not have a positive impact on increasing water storage for the subsequent cotton crop. Furthermore, terminated wheat residue adversely affected cotton lint yield on both years and sites and thus no advantages found in planting cotton into a terminated wheat residue in the semiarid Texas High Plains. Nevertheless, over a longer period of time, the use of a wheat residue may improve soil structure, which may lead to higher infiltration rate reducing runoff and storing more rainfall.

It was postulated that in a semiarid climate growing a small grain cover crop might use more water than it conserves when combined with a cotton crop. During the 1995 wheat growth period, the conservation tillage treatment used 20% more water than the conventional tillage treatment for all treatments at both locations supporting our hypothesis. It was further hypothesized that the use of residue would not increase cotton lint yields, and all the conventionally tilled treatments for two seasons had higher lint yields except for one plot in 1994 at Brownfield over the wheat residue plots. Therefore, for these two locations for the 1994 and 1995 cotton crop seasons, the use of terminated wheat residue did not improve the growing conditions and environment for the cotton crops compared to conventional tillage cropping systems. The significance of this work is that growers using this "conservation compliance" tillage system may not be using water optimally and may even be experiencing decreased cotton lint yields.

Acknowledgements

This work is a tribute to the many research contributions of Dr. Dan Krieg, Professor emeritus, Plant and Soil Science, Texas Tech University, to crop physiology and agronomy of dryland cropping systems. On the 18 June 2015 he lost his battle with cancer.

References

[1] USDA-NASS (2014) Texas Crop Production, USDA—National Agricultural Statistics Service, Issue No. PR-163-14, 10 December 2014.
http://www.nass.usda.gov/Statistics_by_State/Texas/Publications/Current_News_Release/2014_Rls/pr16314.pdf

[2] Lascano, R.J. (2000) A General System to Measure and to Calculate Daily Crop Water Use. *Agronomy Journal*, **92**, 821-832. http://dx.doi.org/10.2134/agronj2000.925821x

[3] Colaizzi, P.D., Gowda, P.H., Marek, T.H. and Porter, D.O. (2009) Irrigation in the Texas High Plains: A Brief History and Potential Reductions in Demand. *Irrigation & Drainage*, **58**, 257-274. http://dx.doi.org/10.1002/ird.418

[4] Lascano, R.J., Van Bavel, C.H.M., Hatfield, J.L. and Upchurch, D.R. (1987) Energy and Water Balance of a Sparse Crop: Simulated and Measured Soil and Crop Evaporation. *Soil Science Society of America Journal*, **51**, 1113-1121. http://dx.doi.org/10.2136/sssaj1987.03615995005100050004x

[5] Lascano, R.J. and Baumhardt, R.L. (1996) Effects of Crop Residue on Soil and Plant Water Evaporation in a Dryland Cotton System. *Theoretical Applied Climatology*, **54**, 69-84. http://dx.doi.org/10.1007/BF00863560

[6] Krieg, D.R. (1996) Physiological Aspects of Ultra Narrow Row Cotton Production. Proceedings of the Beltwide Cotton Conference, Nashville, TN. National Cotton Council, Memphis, 9-12 January 1996, 66.

[7] Darawsheh, M.K., Khah, E.M., Aivalakis, G., Chachalis, D. and Sallaku, F. (2009) Cotton Row Spacing and Plant Density Cropping Systems. I. Effects on Accumulation and Partitioning of Dry Mass and LAI. *Journal of Food, Agriculture & Environment*, **7**, 258-261.

[8] Lascano, R.J., Baumhardt R.L., Hicks, S.K. and Heilman, J.L. (1994) Soil and Plant Water Evaporation from Strip-Tilled Cotton: Measurement and Simulation. *Agronomy Journal*, **86**, 987-994. http://dx.doi.org/10.2134/agronj1994.00021962008600060011x

[9] Musick, J.T., Pringle, F.B., Harman, W.L. and Stewart B.A. (1990) Long-Term Irrigation Trends-Texas High Plains. *Applied Engineering Agriculture*, **6**, 717-724. http://dx.doi.org/10.13031/2013.26454

[10] Lascano, R.J. and Nelson, J.R. (2014) Circular Planting to Enhance Rainfall Capture in Dryland Cropping Systems at a Landscape Scale: Measurement and Simulation. In: Ahuja, L.R., Ma, L. and Lascano, R.J., Eds., *Practical Applications of Agricultural System Models to Optimize the Use of Limited Water*, Advances in Agricultural Systems Modeling, ASA, CSSA, SSSA, Madison, Volume 5, 85-111.
http://dx.do.org/10.2134/advagricsystmodel5.c4

[11] Bordovsky, J.P., Lyle, W.M. and Keeling, J.W. (1994) Crop Rotation and Tillage Effects on Soil Water and Cotton Yield. *Agronomy Journal*, **86**, 1-6. http://dx.doi.org/10.2134/agronj1994.00021962008600010001x

[12] Baker, J.T., McMichael, B., Burke, J.J., Gitz, D.C., Lascano, R.J. and Eprath, J.E. (2009) Sand Abrasion Injury and Biomass Partitioning in Cotton Seedlings. *Agronomy Journal*, **101**, 1297-1303.
http://dx.doi.org/10.2134/agronj2009.0052

[13] Koekkoek, E.J.W., Lascano, R.J., Hicks, S.K., Krieg, D.R. and Stroosnijder, L. (1995) Loss of Water through Terminated Wheat Plants: A Wick Effect. *Proceedings of the 1995 Annual Meetings*, St. Louis, 29 October-3 November 1995, 15.

[14] Van Rensburg, L.D. (2010) Advances in Soil Physics: Application in Irrigation and Dryland Crop Production, South African. *Journal of Plant and Soil*, **27**, 9-18. http://dx.doi.org/10.1080/02571862.2010.10639966

[15] Unger, P.W. (1994) Residue Management for Winter Wheat and Grain Sorghum Production with Limited Irrigation. *Soil Science Society American Journal*, **58**, 537-542. http://dx.doi.org/10.2136/sssaj1994.03615995005800020041x

[16] Nielsen, D.C., Unger, P.W. and Miller, P.R. (2005) Efficient Water Use in Dryland Cropping Systems in the Great Plains. *Agronomy Journal*, **97**, 364-372. http://dx.doi.org/10.2134/agronj2005.0364

[17] AG-CARES (2004-2008) AG-CARES Annual Report. Texas A & M AgriLife Research & Extension Center, Lubbock. http://lubbock.tamu.edu/ag-cares/

[18] Helms Farm Research Reports (2001-2005) Helms Farm Annual Reports. Texas A & M AgriLife Research & Extension Center, Lubbock. http://lubbock.tamu.edu/programs/disciplines/irrigation-water/helms-farm-research-reports/

[19] Jones, O.R., Hauser, V.L. and Popham, T.W. (1994) No-Tillage Effects on Infiltration, Runoff, and Water Conservation on Dryland. *Transactions of the ASAE*, **37**, 473-479. http://dx.doi.org/10.13031/2013.28099

[20] Howell, T.A., Steiner, J.L., Schneider, A.D. and Evett, S.R. (1995) Evapotranspiration of Irrigated Winter Wheat—Southern High Plains. *Transactions of the ASAE*, **38**, 745-759. http://dx.doi.org/10.13031/2013.27888

[21] Guerif, J., Richard, G., Durr, C., Machet, J.M., Recous, S. and Roger-Estrade, J. (2001) A Review of Tillage Effects on Crop Residue Management, Seedbed Conditions and Seedling Establishment. *Soil & Tillage Research*, **61**, 13-22. http://dx.doi.org/10.1016/S0167-1987(01)00187-8

[22] Hobbs, P.R., Sayre, K. and Gupta, R. (2008) The Role of Conservation Agriculture in Sustainable Agriculture. *Philosophical Transactions of the Royal Society B: Biological Sciences*, **363**, 543-555. http://dx.doi.org/10.1098/rstb.2007.2169

[23] Alvarez, R. and Steinbach, H.S. (2009) A Review of the Effects of Tillage Systems on Some Soil Physical Properties, Water Content, Nitrate Availability and Crops Yield in the Argentine Pampas. *Soil & Tillage Research*, **104**, 1-15. http://dx.doi.org/10.1016/j.still.2009.02.005

[24] Baumhardt, R.L., Keeling, J.W. and Wendt, C.W. (1993) Tillage and Residue Effects on Infiltration into Soils Cropped to Cotton. *Agronomy Journal*, **85**, 379-383. http://dx.doi.org/10.2134/agronj1993.00021962008500020038x

[25] Dao, T.H. (1993) Tillage and Winter Wheat Residue Management Effects on Water Infiltration and Storage. *Soil Science Society of America Journal*, **57**, 1586-1595. http://dx.doi.org/10.2136/sssaj1993.03615995005700060032x

[26] Hoogmoed, W.B. and Stroosnijder, L. (1984) Crust Formation on Sandy Soils in the Sahel. I. Rainfall and Infiltration. *Soil & Tillage Research*, **4**, 5-23. http://dx.doi.org/10.1016/0167-1987(84)90013-8

[27] Lascano, R.J., Vorheis, J.T., Baumhardt, R.L. and Salisbury, D.R. (1997) Computer-Controlled Variable Intensity Rain Simulator. *Soil Science Society of America Journal*, **61**, 1182-1189. http://dx.doi.org/10.2136/sssaj1997.03615995006100040025x

[28] Alberts, E.E. and Neibling, W.H. (1994) Influence of Crop Residues on Water Erosion. In: Unger, P.W., Ed., *Managing Agricultural Residues*, Lewis Publ., Chelsea, MI, 19-39.

[29] Savabi, M.R. and Stott, D.E. (1994) Plant Residue Impact on Rain Interception. *Transactions of the ASAE*, **37**, 1093-1098. http://dx.doi.org/10.13031/2013.28180

[30] Steiner, J.L. (1994) Crop Residue Effects on Water Conservation. In: Unger, P.W., Ed., *Managing Agricultural Residues*, Lewis Publ., Chelsea, MI, 41-76.

[31] Baumhardt, R.L., Schwartz, R., Howell, T.A., Evett, S.R. and Colaizzi, P. (2013) Residue Management Effects on Water Use and Yield of Deficit Irrigated Cotton. *Agronomy Journal*, **105**, 1026-1034. http://dx.doi.org/10.2134/agronj2012.0361

[32] Lascano, R.J. and Van Bavel, C.H.M. (1986) Simulation and Measurement of Evaporation from a Bare Soil. *Soil Science Society of America Journal*, **50**, 1127-1132. http://dx.doi.org/10.2136/sssaj1986.03615995005000050007x

[33] Farahani, H.J. and Ahuja, L.R. (1996) Evapotranspiration Modeling of Partial Canopy/Residue-Covered Fields. *Transactions of the ASAE*, **39**, 2051-2064. http://dx.doi.org/10.13031/2013.27708

[34] Klocke, N.L., Currie, R.S. and Aiken, R.M. (2009) Soil Water Evaporation and Crop Residues. *Transactions of the ASABE*, **52**, 103-110. http://dx.doi.org/10.13031/2013.25951

[35] Steiner, J.L. (1989) Tillage and Surface Residue Effects on Evaporation from Soils. *Soil Science Society American Journal*, **53**, 911-916. http://dx.doi.org/10.2136/sssaj1989.03615995005300030046x

[36] Enz, J.W., Brun, L.J. and Larsen, J.K. (1988) Evaporation and Energy Balance for Bare and Stubble Covered Soil. *Agricultural and Forest Meteorology*, **43**, 59-70. http://dx.doi.org/10.1016/0168-1923(88)90006-8

[37] Horton, R., Kluitenberg, G.J. and Bristow, K.L. (1994) Surface Crop Residue Effects on the Soil Surface Energy Balance. In: Unger, P.W., Ed., *Managing Agricultural Residues*, Lewis Publ., Chelsea, MI, 143-162.

[38] Heilman, J.L., McInnes, K.J., Gesch, R.W. and Lascano, R.J. (1992) Evaporation from Ridge-Tilled Soil Covered with Herbicide-Killed Winter Wheat. *Soil Science Society American Journal*, **56**, 1278-1286. http://dx.doi.org/10.2136/sssaj1992.03615995005600040045x

[39] Fryrear, D.W. and Bilbro, J.D. (1994) Wind Erosion Control with Residues and Related Practices. In: Unger, P.W., Ed., *Managing Agricultural Residues*, Lewis Publ., Chelsea, MI, 7-17.

[40] Evett, S.R. and Lascano, R.J. (1993) ENWATBAL.BAS: A Mechanistic Evapotranspiration Model Written in Compiled Basic. *Agronomy Journal*, **85**, 763-772. http://dx.doi.org/10.2134/agronj1993.00021962008500030044x

[41] Vorheis, J.T. (1997) Water Balance of Cotton Cropping Systems. Master's Thesis, Plant and Soil Science Department, Texas Tech University, Lubbock, 38 p.

[42] Ralston, J.T. (1997) Management Strategies for Dryland Cotton Production in West Texas. Master's Thesis, Plant and Soil Science Department, Texas Tech University, Lubbock, 47 p.

[43] Lyle, W.M. and Bordovsky, J.P. (1981) Low Energy Precision Application (LEPA) Irrigation System. *Transactions of the ASAE*, **24**, 1241-1245. http://dx.doi.org/10.13031/2013.34427

[44] Allen, R.G., Walter, I.A., Elliot, R., Howell, T.A., Itenfisu, D. and Jensen, M.E. (2005) The ASCE Standardized Reference Evapotranspiration Equation. ASCE-EWRI Task Committee Report, January 2005, 70 p. http://www.kimberly.uidaho.edu/water/asceewri/ascestzdetmain2005.pdf

[45] Bordovsky, J.P., Lyle, W.M., Lascano, R.J. and Upchurch, D.R. (1992) Cotton Irrigation Management with LEPA Systems. *Transactions of the ASAE*, **35**, 879-884. http://dx.doi.org/10.13031/2013.28673

[46] Evett, S.R. (2007) Soil Water and Monitoring Technology. In: Lascano, R.J. and Sojka, R.E., Eds., *Irrigation of Agricultural Crops*, 2nd Edition, ASA, CSSA, SSSA, Madison, 25-84.

[47] Baumhardt, R.L., Lascano, R.J. and Krieg, D.R. (1995) Physical and Hydraulic Properties of a Pullman and Amarillo Soil on the Texas South Plains. Technical Report D.R. No. 95-1, Texas A & M University Agricultural Research and Extension Center, Lubbock/Halfway.

[48] Bertuzzi, P., Bruckler, L., Bay, D. and Chanzy, A. (1994) Sampling Strategies for Soil Water Content to Estimate evapotranspiration. *Irrigation Science*, **14**, 105-115. http://dx.doi.org/10.1007/BF00193132

[49] Peng, S., Krieg, D.R. and Hicks, S.K. (1989) Cotton Lint Yield Response to Accumulated Heat Units and Soil Water Supply. *Field Crop Research*, **19**, 253-262. http://dx.doi.org/10.1016/0378-4290(89)90097-X

[50] Morrow, M.R. and Krieg, D.R. (1990) Cotton Management Strategies for a Short-Growing Season Environment: Water-Nitrogen Considerations. *Agronomy Journal*, **92**, 52-56. http://dx.doi.org/10.2134/agronj1990.00021962008200010011x

[51] USDA-NASS (1994) Agricultural Statistics-1994. http://www.nass.usda.gov/Publications/Ag_Statistics/agr4all.pdf

[52] USDA-NASS (1995) Agricultural Statistics-1995. Chapter II—Statistics of Cotton, Tobacco, Sugar Crops, and Honey. http://www.nass.usda.gov/Publications/Ag_Statistics/1995-1996/agr95_2.pdf

13

Long-Term Effects of Alternative Residue Management Practices on Soil Water Retention in a Wheat-Soybean, Double-Crop System in Eastern Arkansas

Ryan Norman[1], Kristofor R. Brye[1], Edward E. Gbur[2], Pengyin Chen[1], John Rupe[3]

[1]Department of Crop, Soil, and Environmental Sciences, University of Arkansas, Fayetteville, USA
[2]Agricultural Statistics Laboratory, University of Arkansas, Fayetteville, USA
[3]Department of Plant Pathology, University of Arkansas, Fayetteville, USA
Email: kbrye@uark.edu

Abstract

Soil water retention is a critical aspect of agricultural management, especially in areas such as the Lower Mississippi River Alluvial Valley that face potential water shortages in the near future. Previous studies have linked changes in soil water retention characteristics to agricultural management practices, especially as they affect the accumulation of soil organic matter (SOM). Therefore, the objective of this study was to determine the relationship between soil water potential and gravimetric soil water content in the top 7.5 cm as affected by nitrogen (N) fertilization/residue level (high and low), residue burning (burning and non-burning), tillage (conventional and no-tillage), and irrigation (irrigated and non-irrigated) after 12 complete cropping cycles in a wheat (*Triticum aestivum* L.)-soybean [*Glycine max* (L.) Merr.], double-crop production system in the Delta region of eastern Arkansas using soil wetting curves. The soil investigated was a Calloway silt loam (fine silty, mixed, active, thermic Glossaquic Fraglossudalf). The slope characterizing the relationship between the natural logarithm of the soil water potential and the gravimetric soil water content was only affected ($P < 0.05$) by the N-fertilization/residue-level treatment, while the intercept was unaffected by any field treatment. Averaged across tillage, burning, and irrigation, soil water contents under the high- exceeded those under low-N-fertilization/residue-level treatment at the same water potential, with the greatest differences observed at water contents > 0.12 g·g^{-1}. Understanding the ways in which alternative residue management practices affect soil water retention characteristics is an important component of conserving irrigation water resources.

Keywords

Water Retention, Tillage, Residue Burning

---

1. Introduction

Agronomic management practices that promote the formation of soil organic matter (SOM) and soil aggregation, such as reduced tillage and diversifying crop rotations, can increase plant available water in the soil [1] and likely have many more positive, long-term effects on soil water characteristics. For example, significant differences have been observed between soil water retention curves in the top 10 cm for native prairie (SOM = 22 $g \cdot kg^{-1}$) and cultivated agricultural soil (SOM = 10.8 $g \cdot kg^{-1}$) in eastern Arkansas [2]. Specifically, the native prairie soil contained a greater soil water content than the cultivated agricultural soil at the same water potential, indicating a possible correlation between increased SOM and water retention. Similarly, decreased soil water retention under conventional tillage (CT) management was reported compared to increased soil water retention and unsaturated hydraulic conductivity under no-tillage (NT) management in a continuous corn (*Zea mays* L.) study on Mollisols in Iowa [3]. Verkler *et al.* [4] reported slower soil dry down after wetting under non-burned management compared to burned residue management, as well as slower soil dry down under NT compared to CT when examining soil water content dynamics in a wheat (*Triticum aestivum* L.)-soybean [*Glycine max* (L.) Merr.], double-crop system on a silt-loam soil in eastern Arkansas after three years and four complete cropping cycles. Clearly, residue and field management practices influence soil water retention characteristics, which may be related to agricultural management effects on soil aggregation and SOM.

Increases in SOM have been associated with increased infiltration, greater hydraulic conductivity, and increased water retention [5]. Therefore, management practices such as tillage and nitrogen (N) fertilization that may affect the accumulation of SOM may also affect soil water retention characteristics. In a previous study of alternative residue management practice effects on near-surface soil properties in a wheat-soybean, double-crop production system on a silt-loam soil in eastern Arkansas, Amuri *et al.* [6] reported increasing soil carbon (C) and SOM over time in the top 10 cm across all treatment combinations over the course of six years and seven complete wheat-soybean cropping cycles following conversion to alternative management practices, where trends were likely due to the increase in crop residue returned to the soil as a result of converting from mono-culture soybean to a wheat-soybean double-crop system. Smith *et al.* [7] reported that the abundance of water-stable aggregates was significantly affected by tillage, irrigation, and N-fertilization treatments. Nitrogen fertilization promotes wheat biomass, which may eventually contribute to an increase in SOM and soil aggregation. Therefore, N fertilization, and other management practices that promote SOM and soil aggregation, may affect the relationship between soil water potential and the soil water content, hence soil water retention. For example, Bowman and Halvorson [8] reported significant increases in soil organic C (SOC), and therefore SOM, in the top 5 cm under increased N-fertilization management. Similarly, SOC and SOM increased at a greater rate under a high (134 $kg \cdot N \cdot ha^{-1} \cdot yr^{-1}$) than under low N-rate (<90 $kg \cdot N \cdot ha^{-1} \cdot yr^{-1}$) treatments in a wheat-containing rotation managed consistently for 10 yr near Akron, Colorado [9].

An understanding of traditional and alternative residue management practices on soil water retention is critical to determining the best management practices in highly productive agricultural regions, especially in areas such as the Delta region of eastern Arkansas that face potential water shortages in the future. Scott *et al.* [10] used a regression equation based on annual water use rate to determine that 75% of the Alluvial Aquifer, the shallowest aquifer underlying most of the Delta region of eastern Arkansas, will be depleted from large irrigation withdrawls by 2041. Therefore, the objective of this study was to examine the effects of N fertilization/residue level (high and low), residue burning (burning and non-burning), tillage (conventional and no-tillage), and irrigation (irrigated and non-irrigated) on the relationship between soil water potential and gravimetric soil water content in the top 7.5 cm after 12 complete cropping cycles in a wheat-soybean, double-crop production system in eastern Arkansas using soil wetting curves. It was hypothesized that water contents under NT will be greater than that under CT at the same water potential due to effects on aggregation and that the history of irrigation or dryland production will have little effect on soil water retention using the wetting curve approach. It was also hypothesized that the cumulative effects of 12 years of residue burning would render soil water retention in the burned treatment lower than that in the non-burned treatment due to the hydrophobic characteristics of the added ash. In addition, it was hypothesized that increased above and belowground biomass inputs would contribute to greater water retention in the high than in the low N-fertilization/residue-level treatment.

2. Materials and Methods

2.1. Site Description

A field study was initiated in Fall 2001 at the Lon Mann Cotton Branch Experiment Station (N34°44'2.26"; W90°45'51.56") [11] in the Lower Mississippi River Alluvial Valley near Marianna, AR. The study site lies within the Southern Mississippi Valley Loess [Major Land Resource Area (MLRA) 134] [12]. This region consists of a series of loess covered hills and alluvial terraces. Despite being prone to erosion, this region has been highly agriculturally productive for decades [12]. The field site is on a Calloway silt loam (fine silty, mixed, active, thermic Glossaquic Fraglossudalf) [13], which consists of 16% sand, 73% silt, and 11% clay in the top 10 cm [14]. The 30-yr mean air annual temperature of the region is 15.6°C and the 30-yr mean annual precipitation is 128 cm [15]. The 30-yr mean maximum and minimum air temperatures of the region are 32.8°C in July and 2.4°C in January, respectively [15]. This study follows a series of several previous studies at this same study site that have evaluated a variety of short- and long-term effects of alternative management practices effects on plant and soil properties [6] [7] [16].

2.2. Experimental Design

Between Fall 2001 and Spring 2005, the study consisted of a three-factor, split-strip-plot, randomized complete block experimental design with six replications of each of eight treatment combinations [11]. The three experimental factors were i) N fertilization/residue level (*i.e.*, a high N fertilization/residue level, achieved with a split application of N fertilizer, and a low N/fertilization/residue level, achieved with minimal to no N additions); ii) residue burning (*i.e.*, burning or non-burning); and iii) tillage (*i.e.*, CT or NT) [11]. However, an irrigation factor was introduced in 2005 by dividing the site into an irrigated half and a non-irrigated half [4]. Since 2005, the experimental area has consisted of 48, 3- × 6-m plots with six replications for every N-fertilization/residue-level-burning-tillage treatment combination and three replications for every N-fertilization/residue-level-burning-tillage-irrigation treatment combination (**Figure 1**) [6].

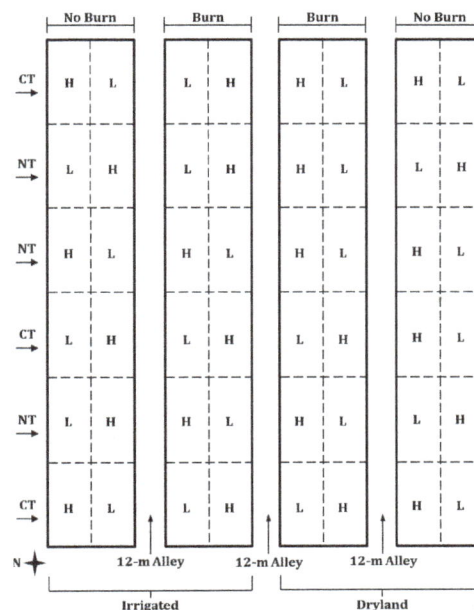

Figure 1. Experimental layout at the Lon Mann Cotton Branch Experiment Station in eastern Arkansas depicting 48, 3- × 6-m plots under residue-level [high (H) and low (L)], burn, tillage [conventional tillage (CT) and no-tillage (NT)], and irrigation treatments.

2.3. Field Management

Prior to the initiation of the study, the site was managed as a continuous, mono-cropped soybean system using CT [11]. Due to the consistent field management prior to beginning this long-term study in 2001, near-surface soil properties throughout entire study area were assumed uniform and any subsequent observed differences in measured soil and/or plant properties were assumed be to the result of imposed field treatments rather than inherent differences among plots [14].

To prepare for this study, initial field preparations in Fall 2001 involved disking twice followed by broadcast applications of N, phosphorous, potassium, and pelletized limestone at rates of 20, 22.5, 56, and 1120 kg·ha^{-1}, respectively, prior to wheat planting. Wheat was drill seeded with a 19-cm row spacing each fall thereafter. All plots were manually broadcast fertilized in early March 2002 through 2004 with urea (46% N) at the rate of 101 kg·N·ha^{-1}. High-N-fertilization/residue-level plots (n = 24) were manually broadcast fertilized in late March at approximately the late-jointing stage with an additional 101 kg·N·ha^{-1} to produce different levels of wheat residue. No N-fertilizer was applied in Spring 2005 due to a failure to establish wheat stands caused by prolonged wet soil conditions in Fall 2004. Since 2006, initial applications of 56 kg·N·ha^{-1} as urea were broadcast on high-N-fertilization/residue-level plots in approximately late February, followed by a split application of an additional 56 kg·N·ha^{-1} at the late-jointing stage in approximately late March. Since 2006, the low-N-fertilization/residue-level plots have received no N fertilization in order to achieve the desired residue-level difference.

Wheat was harvested using a plot combine in approximately early June each year. Wheat residue left behind the plot combine was uniformly spread by hand over each plot immediately following wheat harvest. Any remaining wheat stubble was mowed with a rotary mower to a height of ~3 cm from the soil surface in order to achieve a uniform residue-covered surface for soybean planting. Following mowing, the burning treatment was imposed on half of the plots by propane flaming. The residue-burning treatment was not able to be imposed in 2005, 2007, and 2012 due to the absence of a wheat stand in Spring 2005, prolonged wet soil conditions in Spring 2007, and overly weedy conditions in 2012. Imposition of the burning treatment was followed by imposing the tillage treatment each year. The CT plots were disked at least twice with a tandem disk to a depth of ~10 cm followed by seedbed smoothing with at least three passes of a soil conditioner. The CT practices used in this study are representative of widely used pre-soybean-planting tillage operations throughout the region.

A glyphosate-resistant soybean cultivar, maturity group 5.3 or 5.4, was drill-seeded with 19-cm row spacing at a rate of approximately 47 kg seed ha^{-1} in early to mid-June each year. Potassium fertilizer was applied according to recommended rates [17] when the previous year's soil test indicated potassium was needed. In 2002 through 2004, all plots were furrow-irrigated as needed, three to four times each soybean-growing season. A levee was created in 2005 to exclude furrow-irrigation water from the non-irrigated (*i.e.*, dryland) treatment, which received only natural rainfall. Weeds and insects were managed consistently throughout the entire study area as necessary based on University of Arkansas Cooperative Extension Service's recommendations, which generally consisted of herbicide and insecticide applications during both the wheat- and soybean-growing seasons [17]. Soybean were harvested with a plot combine from late October to early November each year. Each year within two weeks of soybean harvest, the subsequent wheat crop was sown into the soybean residue, which was left in place without any manipulations.

2.4. Plant Sample Collection and Processing

To verify achievement of the N-fertilization/residue-level treatment, immediately following wheat harvest and rotary mowing and before residue burning each year, aboveground residue was assessed. A residue sample was collected from within a 0.25-m^2 metal frame, oven-dried for 3 to 7 days at 55°C, and weighed to obtain an estimate of aboveground residue mass into which soybean were subsequently planted.

2.5. Soil Sample Collection and Processing

Consistent with prior annual soil sampling conducted at this field site [6] [7] [14] [16], at wheat harvest in May 2014, a single soil sample was collected from the top 10 cm in each plot using a 4.8-cm-diameter stainless steel core chamber beveled to the outside to minimize compaction. Soil samples were oven-dried for 48 hr at 70°C, weighed for bulk density determinations, ground to pass through a 2-mm mesh screen, and analyzed for selected soil chemical properties. Soil pH was determined potentiometrically in a 1:2 (w/v) soil-to-water suspension. Soil OM was determined by weight-loss-on-ignition after 2 hr at 360°C. Total soil C and N were determined by high-temperature combustion with an Elementar VarioMAX Total C and N Analyzer (Elementar Americas Inc., Mt. Laurel, NJ). All measured soil C was assumed to be organic given the lack of effervescence upon treatment with dilute hydrochloric acid [14]. The soil C: N ratio was calculated from measured C and N concentrations. Total C and N and SOM contents ($kg \cdot m^{-2}$) were calculated from measured concentrations ($g \cdot kg^{-1}$) and measured bulk densities in the top 10 cm from the May 2014 soil sampling.

To assess field treatment effects on the relationship between soil water potential and soil water content, 12, ~2-cm-diameter soil samples were collected from the top 7.5 cm of each plot at wheat harvest in May 2014 and combined into one composite soil sample per plot. Each sample was manually homogenized and air-dried for approximately 5 d, ground, and sieved to pass through a 2-mm mesh screen. Subsamples were weighed, oven-dried at 70°C for 48 hr, and reweighed to obtain the initial moisture content of the air-dried sample. Following the procedures of Brye [2], seven, 5 ± 0.01-g subsamples of air-dried soil from each of the 48 plots were added to small mixing cups. Drops of distilled water (*i.e.*, 2, 4, 6, 10, 12, 15, and 20 drops) were added to each of the seven mixing cups with an eyedropper and manually mixed with a spatula to achieve a range of gravimetric soil water contents. The moist soil in each mixing cup was transferred to small plastic instrument cups, 4 cm in diameter by 1 cm tall, and lightly packed to a uniform bulk density of ~0.7 $g \cdot cm^{-3}$. The plastic instrument cups were capped and allowed to equilibrate overnight to room temperature (*i.e.*, ~20°C). The next morning after over-night temperature equilibration, the water potential was subsequently measured with a WP4 Dewpoint PotentiaMeter (Decagon Devices, Inc., Pullman, WA), which was calibrated using a standard potassium chloride solution. After the water potential was recorded, each instrument cup was weighed, oven-dried at 70°C for 48 hr, then reweighed for gravimetric water content determinations. Measured water potentials were natural-logarithm transformed to facilitate statistical analyses.

In addition, to determine actual residue management effects on near-surface bulk density, a third set of soil cores were collected between 8 and 10 weeks after soybean planting in 2014, similar to previous annual bulk densities assessments [6] [7] [14] [16]. A single 4.8-cm-diameter soil core was extracted from the top 10 cm in each plot using a stainless steel core chamber and slidehammer. Samples were oven-dried at 70°C for 48 hr, and weighed for an additional bulk density measurement.

2.6. Statistical Analyses

An analysis of variance (ANOVA) was conducted using SAS (version 9.3, SAS Institute, Inc., Cary, NC) to evaluate the N-fertilization/residue-level treatment effect on aboveground residue mass for annual measurements conducted between 2007 and 2014. In addition, an ANOVA was also conducted using SAS, based on the strip-split-plot experimental layout of the field treatments (**Figure 1**), to evaluate the effects of N fertilization/residue level, burning, tillage, irrigation, and their interactions on soil pH, SOM, total N, and C contents, C:N ratio, and bulk density in the top 10 cm measured in 2014, which represents the cumulative effects of 12 consecutive years of consistent management. Due to practical limitations of the study area, the addition of the irrigation treatment since 2005 was superimposed on the burning treatment (**Figure 1**). Therefore, irrigation and burning treatments could not be simultaneously analyzed within this experimental design. Therefore, two separate ANOVAs were conducted, each excluding one of the confounding factors (*i.e.*, burning and irrigation). When appropriate, treatment means were also separated by least significant difference (LSD) at the 0.05 level.

To evaluate field treatment effects on soil water retention, an analysis of covariance (ANCOVA) was conducted using SAS to examine the long-term effects of N fertilization/residue level, burning, tillage, irrigation, and their interactions on the linear relationship between the natural logarithm (log) of the measured soil water potential and the gravimetric water content from the soil wetting-curve data. For the purposes of simultaneously evaluating all field treatments and their interactions, the experimental design was assumed to be completely random with three replications of each of 16 treatment combinations. The full ANCOVA model was reduced using a hierarchal principle to remove non-significant terms, and non-significant terms were only included the final model when they participated in higher-order, complex treatment interactions. When appropriate, treatment means for slopes and intercepts from the log-transformed relationships were separated by LSD at the 0.05 level.

3. Results and Discussion

3.1. Aboveground Residue Levels

Following the 13 wheat crops produced in this study between 2002 and 2014, the N-fertilization scheme used annually produced numerically greater aboveground residue amounts, into which the subsequent soybean crop was planted, in the high than in the low N-fertilization/residue-level treatment in 12 of the 13 years, with 2002 being the exception [6] [18]. In addition, significantly ($P < 0.05$) greater aboveground residue amounts were produced in the high than in the low N-fertilization/residue-level treatment in 10 of the 13 years, with 2002, 2004 [6], and 2010 being the exceptions [18]. Therefore, it is clear that the intended residue-level differences were achieved in more than the majority of the 13 years to justify the N-fertilization/residue-level treatment.

3.2. 2014 Soil Properties

After 13 complete wheat-soybean cropping cycles (*i.e.*, 2001 to 2014) and 12 years of consistent management, soil C and N contents, soil C:N ratio, and soil pH in the top 10 cm were affected ($P < 0.05$) by field treatments. When irrigation was excluded from the model, soil C content ($P = 0.038$) and soil C: N ratio ($P = 0.033$) differed between burn treatments in 2014. Averaged across tillage, N fertilization/residue level, and irrigation, soil C content averaged 1.22 and 1.42 kg·m^{-2}, while the soil C:N ratio averaged 9.2 and 10.2 under burning and non-burning, respectively. However, when burning was excluded from the model, soil N content ($P = 0.032$) and the C:N ratio ($P = 0.021$) differed between the N-fertilizer/residue-level treatments (**Table 1**). Averaged across tillage, burning, and irrigation, soil N content averaged 0.14 and 0.13 kg·m^{-2} under the high and low N-fertilization/residue-level treatments, respectively. Similarly, averaged across tillage, burning, and irrigation, the soil C: N ratio averaged 9.4 and 10.0 under the high and low N-fertilization/residue-level treatments, respectively.

In 2014, soil pH in the top 10 cm differed ($P = 0.021$) between irrigation treatments within N-fertilizer/residue-level treatments (**Table 1**) when burning was excluded from the model. Averaged across tillage and burning, soil pH was greater under irrigation regardless of N-fertilization/residue level, where soil pH averaged 7.26 under the high and 7.28 under the low N fertilization/residue level, than that under the dryland treatment, where soil pH averaged 6.67 under the low, which was greater than that under the high N fertilization/residue level (*i.e.*, pH averaged 6.48). However, all soil pH values, regardless of management, exceeded the minimum soil pH threshold of 6.0, below which soybean yield reductions can be expected on silt-loam soils in eastern Arkansas [19]. Therefore, the differences in soil pH among irrigation and N-fertilization/residue-level treatment combinations were agronomically non-significant with regards to soybean production on silt-loam soils in eastern Arkansas.

In contrast to other initial soil properties, after 13 complete wheat-soybean cropping cycles (*i.e.*, 2001 to 2014) and 12 years of consistent management, bulk density and SOM contents were unaffected ($P > 0.05$) by any of the field treatments in 2014 regardless of whether burning (**Table 1**) or irrigation (data not shown) were excluded in the statistical model. Therefore, bulk density averaged 1.21 g·cm^{-3} [standard error (SE) = 0.01] and SOM content averaged 2.9 kg·m^{-2} (SE = 0.06) across all field treatments in 2014.

Table 1. Analysis of variance summary of the effects of N-fertilization/residue level (residue level), tillage, irrigation, and their interactions on soil bulk density, pH, soil organic matter (SOM), total carbon (TC), total nitrogen (TN), and the soil C: N ratio in the top 10 cm from spring 2014 after 12 complete cropping cycles in a wheat-soybean, double-crop production system in eastern Arkansas.

Source of Variation	Bulk Density	pH	SOM	TC	TN	C:N
			P			
Residue level	ns[†]	ns	ns	ns	0.032	0.021
Tillage	ns	ns	ns	ns	ns	ns
Irrigation	ns	0.023	ns	ns	ns	ns
Residue level × tillage	ns	ns	ns	ns	ns	ns
Residue level × irrigation	ns	0.021	ns	ns	ns	ns
Tillage × irrigation	ns	ns	ns	ns	ns	ns
Residue level × tillage × irrigation	ns	ns	ns	ns	ns	ns

[†]Not significant (ns, $P > 0.05$).

3.3. Soil Water Retention

As was expected, the relationship between soil water potential and gravimetric water content across all data and treatment combinations followed a curvilinear pattern, where the water potential increased exponentially as gravimetric soil water content increased (**Figure 2**). Though soil properties, and water retention, were assumed to be uniform at the beginning of the study site in 2001, after 13 complete wheat-soybean cropping cycles (*i.e.*, 2001 to 2014) and 12 years of consistent management, the slope characterizing the linear relationship between the natural logarithm of water potential and gravimetric water content in the top 7.5 cm was affected ($P = 0.007$) by only the N-fertilization/residue-level treatment (**Table 2**; **Figure 3**), and was unaffected ($P > 0.05$) by tillage, burning, irrigation or any of their interactions (**Table 3**). The intercept characterizing the linear relationship between the natural logarithm of water potential and gravimetric water content was unaffected ($P > 0.05$) by any field treatment (**Table 2**; **Figure 3**). Averaged across tillage, burning, and irrigation, the slope for the low was greater (*i.e.*, more negative) than that for the high N-fertilization/residue-level treatment (**Table 3**). This result indicated that, on average, there was a greater change in water content per unit change in water potential associated with the soil from the low compared to the high N-fertilization/residue-level treatment. The greatest differences between high and low N-fertilization/residue treatments were observed at water contents > ~0.12 $g \cdot g^{-1}$ (**Figure 3**). Conversely, as soil water potential decreased, gravimetric soil water contents became increasingly similar under both N-fertilization/residue-levels treatments.

The results of this study on a silt-loam, loessial soil from the Delta region of eastern Arkansas were similar to the soil moisture characteristic curve results reported by Brye [2] using a similar wetting-curve approach. Brye [2] demonstrated that soil water contents in both native prairie and cultivated agricultural silt-loam soils in east-central Arkansas became increasingly similar as soil water potential approached permanent wilting point (*i.e.*, −1.5 MPa), regardless of field treatments imposed. Verkler *et al.* [4] also reported numerically greater maximum soil water contents at the 7.5 cm depth under the high compared with the low N-fertilization/residue-level treatment, although the differences were statistically non-significant. Management practices that increase the amount of crop residue returned to the soil, such as with greater above- and belowground biomass achieved with differential N fertilization, have been shown to increase infiltration, bulk density, and water storage capacity [20].

Though N fertilization/residue level did not significantly affect SOM or C contents in the top 10 cm in 2014 after 13 years of consistent residue management, one possible explanation for the significant effect of N fertili-

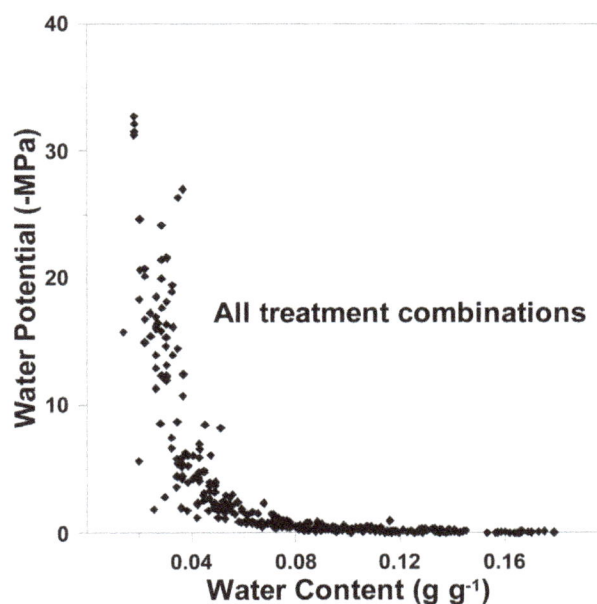

Figure 2. Raw data from all treatment combinations depicting the relationship between soil water potential and gravimetric soil water content from soil wetting curves for the top 7.5 cm in a wheat-soybean, double-crop system in eastern Arkansas after 13 years (*i.e.*, 2001 to 2014) of consistent residue management.

Table 2. Analysis of covariance summary of the effects of N-fertilization/residue level (residue level), burning (burn), tillage, irrigation, and their interactions on the linear relationship between the natural logarithm of soil water potential and gravimetric soil water content using a soil wetting curve approach from the top 7.5 cm after 12 complete cropping cycles in a wheat-soybean, double-crop production system in eastern Arkansas.

Source of variation[†]	P
Intercept term	
Residue level	ns[††]
Burn	ns
Tillage	ns
Irrigation	ns
Residue level × burn	ns
Residue level × tillage	ns
Residue level × irrigation	ns
Burn × tillage	ns
Burn × irrigation	ns
Tillage × irrigation	ns
Residue level × burn × tillage	ns
Residue level × burn × irrigation	ns
Residue level × tillage × irrigation	ns
Burn × tillage × irrigation	ns
Residue level × burn × tillage × irrigation	ns
Liner term	
Water content	<0.001
Residue level × water content	0.007
Burn × water content	ns
Tillage × water content	ns
Irrigation × water content	ns
Residue level × burn × water content	ns
Residue level × tillage × water content	ns
Residue level × irrigation × water content	ns
Burn × tillage × water content	ns
Burn × irrigation × water content	ns
Tillage × irrigation × water content	ns
Residue level × burn × tillage × water content	ns
Residue level × burn × irrigation × water content	ns
Residue level × tillage × irrigation × water content	ns
Burn × tillage × irrigation × water content	ns
Residue level × burn × tillage × irrigation × water content	ns

[†]Non-significant interactions ($P > 0.05$) were removed in the final model, except when non-significant terms participated in higher-order, complex treatment combinations. [††]Not significant (ns, $P > 0.05$).

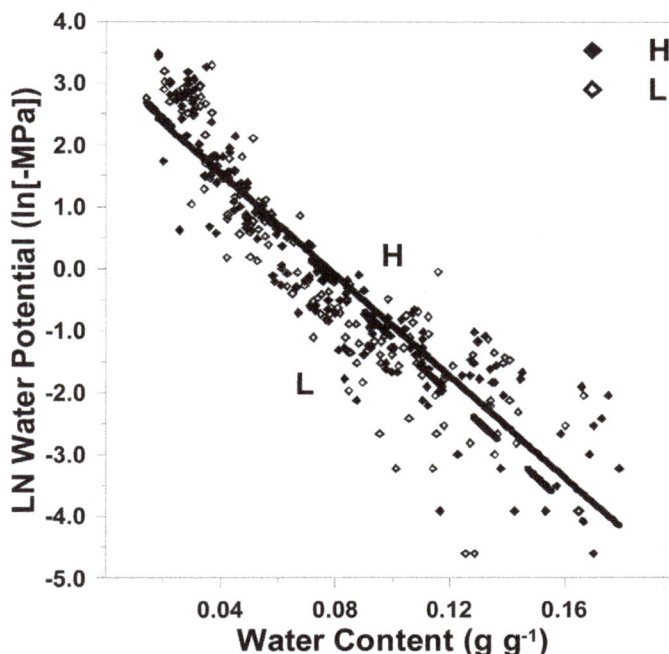

Figure 3. Influence of residue level [high (H, ——) and low (L, -----)] on the relationship between the natural logarithm (LN) of the water potential and the gravimetric soil water content from soil wetting curves for the top 7.5 cm in a wheat-soybean, double-crop system in eastern Arkansas after 13 years (*i.e.*, 2001 to 2014) of consistent residue management.

Table 3. Summary of estimated regression parameters for the N-fertilization/residue-level [High (H) and Low (L)] treatment for the linear relationship between the natural logarithm of soil water potential (−MPa) and gravimetric soil water content (g·g^{-1}) in the top 7.5 cm after 12 complete cropping cycles in a wheat-soybean, double-crop production system in eastern Arkansas. Coefficient estimates with the same lower case letter do not differ (*P* > 0.05).

Regression Term	Treatment	Coefficient Estimate	*P*[†]
Intercept	High	3.108 a	<0.001
	Low	3.357 a	0.002
Linear	High	−39.712 a	<0.001
	Low	−45.207 b	0.001

[†]*P* < 0.05 indicates coefficient estimate was significantly different from 0.

zation/residue level on the relationship between the natural logarithm of water potential and the gravimetric soil water content was that the high N-fertilization/residue-level treatment promoted increased soil structure development and SOM more than the low N-fertilization/residue-level treatment. While it was concluded in a previous study analyzing soil properties in the same plots used in the current study that N fertilization/residue level alone had no obvious, observable effects on the trend in SOM content in the top 10 cm over time between 2007 and 2014, N fertilization/residue level did affect (*P* < 0.05) the trend in SOM content over time as part of complex treatment combinations [18].

It is also possible that the N-fertilization/residue-level treatment may have impacted SOM content and soil aggregates in the top 7.5 cm differently than in the top 10 cm due to the greater accumulation of both above- and below-ground plant biomass concentrated near the soil surface. A previous study analyzing soil aggregation in the top 10 cm in the same plots used in the current study reported that the concentration of water-stable aggregates was 11% greater in the top 5 cm than in the 5 to 10 cm depth interval after 7 years of consistent management [7] suggesting that SOM and soil aggregates may be more concentrated in the 7.5-cm depth samples used for the current study than in the 10-cm depth samples used for previous studies [6] [18]. Therefore, it is possible that the N-fertilization/residue-level treatment affected the <2-mm-sized soil aggregates, which may have occluded SOM, in the top 7.5 cm, without clearly and obviously affecting SOM contained in the aggregate size classes larger than 2 mm in the top 10 cm. Such an increase in occluded SOM in smaller aggregates might account for an increase in soil water content [5] [21]. For example, Brye [2] reported greater soil water contents in the top 10 cm of a native prairie soil than soil water contents of cultivated agricultural soils at the same water potential and ascribed at least a partial explanation to the greater SOM content in the prairie soil (SOM = 22 $g \cdot kg^{-1}$) compared wo the cultivated agricultural soil (SOM = 10.8 $g \cdot kg^{-1}$) examined.

Somewhat surprisingly, neither burning nor tillage had an observable effect ($P > 0.05$; **Table 2**) on the relationship between the natural logarithm of water potential and the gravimetric soil water content after 13 years of consistent residue management and 12 complete cropping cycles. These results were in contrast with the original hypothesis that tillage would strongly affect the relationship between soil water potential and soil water content, such that soil water content under NT would exceed that under CT at the same water potential. Similarly, these results were in contrast to the original hypothesis that residue burning would render soil retention in the burned treatment lower than that in the non-burned treatment due to cumulative effects of hydrophobic ash additions. It is possible that any potential effects of burning and/or tillage on water retention characteristics were masked as a result of the sample preparation procedure (*i.e.*, grinding and sieving to ≤2 mm), which rendered the soil structure and natural aggregation and porosities highly disrupted from their in-situ, undisturbed state. As hypothesized, the cumulative effects of irrigation or dryland soybean production had no effect on resulting water retention characteristics as determined using soil wetting curves.

In contrast to the results of this study, other soil water retention studies have reported significant correlations between cultivation and near-surface soil water retention characteristics [2] [3] [5], likely due to the increased hydraulic conductivity and infiltration rates associated with relatively undisturbed NT compared to highly disturbed CT soils. Verkler *et al.* [4] reported that residue burning significantly affected maximum soil water contents during irrigation events, and that the mean maximum soil water content was 3% (v/v) greater under residue burning compared with non-burning. However, the water content measurements conducted by Verkler *et al.* [4] were made in-situ (*i.e.*, in undisturbed soil in the field), and are therefore fundamentally different from the oven-dried, ground, sieved, and rewetted soil samples used in the current study.

4. Conclusion

Following conversion to alternative residue and water management practices and after 13 consecutive years (*i.e.*, 2001 to 2014) of management, N fertilization/residue level, but not burning, tillage, or irrigation treatments, significantly affected the linear relationship between the natural logarithm of water potential and the gravimetric soil water content as determined by soil wetting curves. The wetting-curve procedure provided a simple, replicatable, and useful approach to evaluate long-term effects of various field treatment on soil water retention. It can be inferred from the results of this study that differences in N-fertilization/residue-level management can have a cumulative effect on soil water retention characteristics, possibly as a result of the increased soil aggregation and occluded SOM associated with increased crop residue inputs, both above-and belowground, under a high compared to a low residue-level management scheme. Consideration of soil water retention characteristics is vital to planning sustainable use of irrigation water, especially in areas such as the Delta region of eastern Arkansas that will face potential water shortages in the near future.

Acknowledgements

Field assistance provided by Matt Gregory at the Cotton Branch Experiment Station is gratefully acknowledged. This work was funded by the Arkansas Soybean Research and Promotion board.

References

[1]   Nielsen, D.C., Vigil, M.F., Anderson, R.L., Bowman, R.A., Benjamin, J.G. and Halvorson, A.D. (2002) Cropping System Influence on Planting Water Content and Yield of Winter Wheat. *Agronomy Journal*, **94**, 962-967. http://dx.doi.org/10.2134/agronj2002.0962

[2]   Brye, K.R. (2003) Long-Term Effects of Cultivation on Particle Size and Water-Retention Characteristics Determined using Wetting Curves. *Soil Science*, **168**, 459-468. http://dx.doi.org/10.1097/01.ss.0000080331.10341.36

[3]   Hill, R.L., Horton, R. and Cruse, R.M. (1985) Tillage Effects on Soil Water Retention and Pore Size Distribution of Two Mollisols. *Soil Science Society of America Journal*, **49**, 1264-1270.

http://dx.doi.org/10.2136/sssaj1985.03615995004900050039x

[4]   Verkler, T.L., Brye, K.R., Gbur, E.E., Popp, J.H. and Amuri, N. (2008) Residue Management and Water Delivery Effects on Season-Long Surface Soil Water Dynamics in Soybean. *Soil Science*, **173**, 444-455. http://dx.doi.org/10.1097/SS.0b013e31817b6687

[5]   Azooz, R.H. and Arshad, M.A. (1996) Soil Infiltration and Hydraulic Conductivity under Long-Term No-Tillage and Conventional Tillage Systems. *Canadian Journal of Soil Science*, **76**, 143-152. http://dx.doi.org/10.4141/cjss96-021

[6]   Amuri, N., Brye, K.R., Gbur, E.E., Popp, J. and Chen, P. (2008) Soil Property and Soybean Yield Trends in Response to Alternative Wheat Residue Management Practices in a Wheat-Soybean, Double-Crop Production System in Eastern Arkansas. *Journal of Integrative Biosciences*, **6**, 64-86.

[7]   Smith, S.F., Brye, K.R., Gbur, E.E., Chen, P. and Korth, K. (2014) Residue and Water Management Effects on Aggregate Stability and Aggregate-Associated Carbon and Nitrogen in a Wheat-Soybean, Double-Crop System. *Soil Science Society of America Journal*, **78**, 1378-1391. http://dx.doi.org/10.2136/sssaj2013.12.0534

[8]   Bowman, R.A. and Halvorson, A.D. (1998) Soil Chemical Changes after Nine Years of Differential N Fertilization in a No-Till Dryland Wheat-Corn-Fallow Rotation. *Soil Science*, **163**, 241-247. http://dx.doi.org/10.1097/00010694-199803000-00009

[9]   Halvorson, A.D., Reule, C.A. and Follett, R.F. (1999) Nitrogen Fertilization Effects on Soil Carbon and Nitrogen in a Dryland Cropping System. *Soil Science Society of America Journal*, **63**, 912-917. http://dx.doi.org/10.2136/sssaj1999.634912x

[10]   Scott, H.D., Ferguson, J.A., Hanson, L., Fugitt, T. and Smith, E. (1998) Agricultural Water Management in the Mississippi Delta Region of Arkansas. Arkansas Agricultural Experiment Station, Division of Agriculture, University of Arkansas, Fayetteville.

[11]   Cordell, M.L., Brye, K.R., Longer, D.E. and Gbur, E.E. (2006) Residue Management Practice Effects on Soybean Establishment and Growth in a Young Wheat-Soybean Double-Cropping System. *Journal of Sustainable Agriculture*, **29**, 97-120. http://dx.doi.org/10.1300/J064v29n02_08

[12]   Brye, K.R., Mersiovsky, E., Hernandez, L. and Ward, L. (2013) Soils of Arkansas. Arkansas Agricultural Experiment Station, University of Arkansas System Division of Agriculture, Fayetteville, AR. 136 p.

[13]   Natural Resources Conservation Service (NRCS), Soil Survey Staff (2012) Web Soil Survey: Soil Data Mart. http://websoilsurvey.nrcs.usda.gov

[14]   Brye, K.R., Cordell, M.L., Longer, D.E. and Gbur, E.E. (2006) Residue Management Practice Effects on Soil Surface Properties in a Young Wheat-Soybean Double-Crop System. *Journal of Sustainable Agriculture*, **29**, 121-150. http://dx.doi.org/10.1300/J064v29n02_09

[15]   National Oceanic and Atmospheric Administration (NOAA) (2002) Climatography of the United States No. 81, Monthly Station Normals of Temperature, Precipitation, and Heating and Cooling Degree Days 1971-2000: Arkansas. United States Department of Commerce, National Climatic Data Center, Asheville.

[16]   Verkler, T.L., Brye, K.R., Popp, J.H., Gbur, E.E., Chen, P. and Amuri, N. (2009) Soil Properties, Soybean Response, and Economic Return as Affected by Residue and Water Management Practices. *Journal of Sustainable Agriculture*, **33**, 716-744. http://dx.doi.org/10.1080/10440040903220724

[17]   University of Arkansas Cooperative Extension Service (UACES) (2000) Arkansas Soybean Handbook. Miscellaneous Publication 197. Cooperative Extension Service, University of Arkansas, Little Rock.

[18]   Norman, C.R. (2015) Long-Term Effects of Alternative Residue Management Practices on Near-Surface Soil Properties and Soybean Production in a Wheat-Soybean, Double-Crop System in Eastern Arkansas. M.S. Thesis, University of Arkansas, Fayetteville.

[19]   Slaton, N., Roberts, T. and Ross, J. (2013) Fertilization and Liming Practices. In: Extension Soybean Commodity Committee, Ed., *Arkansas Soybean Production Handbook*, University of Arkansas Research and Extension, Little Rock. http://www.uaex.edu/publications/pdf/mp197/chapter5.pdf

[20]   Shaver, T.M., Peterson, G.A., Ahuja, L.R., Westfall, D.G., Sherrod, L.A. and Dunn, G. (2002) Surface Soil Physical Properties after Twelve Years of Dryland No-Till Management. *Soil Science Society of America Journal*, **66**, 1296-1303. http://dx.doi.org/10.2136/sssaj2002.1296

[21]   Dao, T.H. (1993) Tillage and Winter Wheat Residue Management Effects on Water Infiltration and Storage. *Soil Science Society of America Journal*, **57**, 158-159. http://dx.doi.org/10.2136/sssaj1993.03615995005700060032x

14

Tillage and Irrigation Requirements of Sorghum (*Sorghum bicolor* L.) at Hamelmalo, Anseba Region of Eritrea

Ramesh P. Tripathi*, Isaac Kafil, Woldeselassie Ogbazghi

Department of Land Resources and Environment, Hamelmalo Agricultural College, Keren, Eritrea
Email: *rp.tripathi52@gmail.com, nrd70hnk@yahoo.com, wogbazghi@gmail.com

Abstract

Most Eritrean farmers do not adopt soil conservation measures and till even sloppy fields 2 - 4 times for planting sorghum (*Sorghum bicolor* L.) with a view to facilitate rainwater intake. Field experiments were conducted at Hamelmalo to optimize tillage and irrigation requirements of sorghum in loamy sand. Tillage treatments were conventional tillage (4 times) on existing slopes (CTf), conventional tillage on managed plots (terraced) with residue (CTm + R) and without residue (CTm − R), reduced tillage (single tillage 4 days after heavy rainfall) on managed plots with residue (RTm + R) and without residue (RTm − R) and no tillage (direct planting) on managed plots with residue (NTm + R) and without residue (NTm − R) randomized in four replications. Tillage in CTm and CTf was same. Experiment was repeated in year II along with a new experiment in split plot design with same tillage treatments in main plots and 4 irrigation treatments in subplots in 3 replications. Irrigation treatments were rainfed (I_0), 70 mm irrigation at 50% depletion of soil moisture in CTm − R from 1 m profile after end of monsoon (I_1), 70 mm irrigation 7 days after irrigation in I_1 (I_2), and 70 mm irrigation 7 days after irrigation in I_2 (I_3). Bulk density increased and infiltration rates decreased by harvesting due to tillage but changes were lower in residue plots of NT and RT than CT. Optimum soil moisture for emergence of sorghum was within 0.145 ± 0.002 m^3 m^{-3} at which soil strength was well below critical level for root growth. Soil strength in tilled layer due to intermittent wetting and drying following planting exceeded 2000 k Pa when dried below 0.143 m^3 m^{-3} moisture. Soil profile in CTf did not recharge by rainfall even by end of the rainy season, whereas it was fully wetted in level and terraced plots. Conservation measures resulted 80 - 150 mm of residual moisture per 2 m of soil profile at sorghum harvesting. Residual moisture was relatively more in residue and irrigated plots than in nonresidue and CTf plots. Soil bunding and levelling alone raised sorghum yields in RT + R to 2887 kg ha^{-1} under rainfed and 3980 kg ha^{-1} under 70 mm irrigation 21 days after last rainfall of the season (I_1). Corresponding yields in CTf were 501 kg ha^{-1} under rainfed and 1161 kg ha^{-1} under irrigation. Single

*Corresponding author.

preplanting tillage 4 days after heavy rainfall (RT) was as good as 2 - 4 tillage (CT) practiced by farmers. Sorghum yields in Hamelmalo could be about 2752 kg ha^{-1} by water use of 344 mm and 4009 kg ha^{-1} by 432 mm. Water use in CTf was lowest (208 mm) under rainfed.

Keywords

Rainwater Conservation, Residue Mulch, Semiarid, Soil Properties, Sorghum Yield, Tillage

---

1. Introduction

Farmers in Hamelmalo region generally till twice before and once after broadcasting of sorghum (*Sorghum bicolor* L.) and once again about 25 - 30 days from planting on slopes <2% - 35% [1]-[3]. Farmers partly practice contour tillage on existing slopes but do not adopt any other conservation measure to prevent runoff and soil loss. Sorghum yields were relatively higher (0.8 - 1 t ha^{-1}) in the initial years than now (0.2 - 0.6 t ha^{-1}) perhaps due to release of nutrients by rapid oxidation of organic matter by tillage and relatively improved rainwater intake in soils [1] [3]-[5]. High yields in the initial years encouraged farmers to over-till the land without questioning its actual need and consequent deterioration of soil structure, organic matter and soil biota, accelerated soil erosion and overall declining soil quality [6]-[12]. However, it was slowly realised that conventional tillage practice without any conservation measures on the fragile land slopes in the region was unsustainable in terms of production and resources conservation [2] [13]-[16]. The conventional tillage on steep slopes resulted significant loss of top soil (>150 t ha^{-1} y^{-1}) and consequently crop yields in various parts of Africa [1] [3] [13] [15] [17]. Many smallholder farmers' fields in Anseba region are severely affected by sheet and gully erosion, which is greatly accelerated by repeated tillage. Degree of soil degradation can be assessed from the fact that average sorghum yields even in good rainfall years were less than 0.2 - 0.6 t ha^{-1} [1] [18].

Temesgen *et al.* [5] reported that traditional tillage in Ethiopian highlands was precisely to improve rainwater infiltration and reduce runoff and evaporation. However, only 30% - 40% of rainwater infiltrated in the tilled plots on 1% - 6% slopes and about 10% - 25% in non-agricultural lands on slopes > 10% - 30% under the conventional practice of management at Hamelmalo farm [3]. Regardless of the tillage system, almost 100% rainwater could infiltrate only on level and properly bunded fields covered 100% by plant residue mulch [2] [9] [12] [19].

No-till system has been advocated for more than 3 decades as effective option to combat land degradation, increase rainwater conservation and raise yields in semiarid regions like that of Eritrea [12] [20]-[23]. Rockstrom *et al.* [24] observed that minimum tillage increased water productivity and crop yields due to better water harvesting and improved fertilizer use from applications along the ripped and sub-soiled planting lines. Improved field water harvesting through bunding, levelling and mulching under no-till system has also shown to reduce soil erosion, increase biomass production and soil organic matter [2] [25]-[29]. Slower organic matter mineralization in reduced-till than conventional-till systems and consequent nutrient releases was important for resource-constrained farmers of Hamelmalo who suffer most from the consequences of poor soil fertility and soil degradation [3] [30]. Limited field experiments to evaluate tillage requirement of crops to date have been conducted in Eritrea. Ministry of Agriculture initiated research on conservation agriculture in collaboration with FAO [31]. The results were highly encouraging. Another experiment on no-till system with sorghum and groundnut was initiated in 2008 at Hamelmalo Agricultural College in collaboration with Australia. Results substantiated the potential of reduced and no-till systems in resources conservation and raising crop yields. However, as suggested by Giller *et al.* [30], conservation tillage practices must be evaluated to quantify benefits and role of different competing uses of crop residue and other inputs to raise crop yields and arrest soil degradation under local ecological and socio-economic conditions. Reports also indicated decreased yields with conservation tillage, increased labour requirements when herbicides were not used, increased labour burden to women and lack of mulch due to poor productivity and priority given to feeding of livestock with crop residues [30]. Effort to raise crop yields through optimization of tillage and supplementary irrigation requirements have shown potential in resolving many social and soil conservation issues [2] [3].

A preliminary survey of area in 2007 showed that more than 90% of the agricultural land in sub Zoba

Hamelmalo has some form of degradation, including severe soil erosion, sparse vegetation, poor rainwater infiltration, structural degradation and compaction [2] [3]. Rocks and stone pieces outcropping in a large proportion of the land surfaces all around speak of the severity of erosion. More than 90% of the land in Zoba Anseba is uncultivated due to steep slopes, which contributes 70% - 90% of the rainfall as runoff [3]. Whereas much of the runoff goes to neighbouring country and Red sea, it also raises groundwater table along the rivers and their tributaries criss-crossing the area. The groundwater is of good quality and is traditionally being used by farmers for drinking and irrigating orchards and vegetable crops. Farmers generally practice irrigation at every 4 - 6 days interval. Sorghum and pearl millet are common rainfed crops raised during the monsoon season. Although rainfall is sufficient for a good crop, yields are poor due to severe water stress during the grain filling stage (September to October first week) at which supplementary irrigations may increase grain yield prospects significantly [3] [32]. Objective of this research was thus to optimize tillage and supplementary irrigation requirements of sorghum at Hamelmalo to obtain sustainable high yields.

2. Materials and Methods

2.1. Soil

Soils of the Hamelmalo region have developed from fluvial deposits. Experimental soil was loamy sand with 83% sand, 11% silt and 6% clay overlying a layer of sandy loam down to 1.3 m followed by sand (89%) with <20% cobbles and boulders forming a porous bed (**Table 1**). Soil organic matter was 0.65% in the surface layer, which reduced by 50% in lower layers. Soil pH ranged from 7.82 in the surface layer to 8.4 below 1.3 m and electrical conductivity ranged from 0.08 - 0.15 d Sm^{-1}. Available nutrients were low but Ca^{2+}, Mg^{2+} and Na^+ contents increased with depth. Average bulk density of the surface soil was 1.59 Mg m^{-3}, which increased to 1.69 Mg m^{-3} in 0.2 - 0.5 m layer but reduced underneath to 1.54 Mg m^{-3}. Field capacity moisture was 0.195 m^3 m^{-3} and groundwater table fluctuated from <2.5 m during rainy season to >5 m in dry season.

2.2. Experimental Details

Experiments were conducted during 2007 and 2009 at Hamelmalo Agricultural College farm on predominantly occurring loamy sand in the area. The Hamelmalo Agricultural College is located at 15°52'20.6"N and 38°27'57.6"E and 1280 m above msl in the Anseba region of Eritrea. It is about 12 km north of the Keren town on the Keren-Nakfa road adjacent to the river Anseba. Mean annual rainfall in the past 7 years was 488 mm with a minimum of 370 mm and a maximum of 663.1 mm. Total rainfall was 460 mm in 2007 and 390 mm in 2009. Highest mean monthly temperature occurred in May (35.7°C) and lowest in January (11.1°C).

In 2007, effect of 7 tillage treatments viz., conventional tillage farmers practice (CTf, slope < 3% - 6%), conventional tillage on managed plots with residue (CTm + R) and without residue (CTm − R), reduced tillage on managed plots with residue (RTm + R) and without residue (RTm − R) and no tillage on managed plots with residue (NTm + R) and without residue (NTm − R), randomized in four replications in 8 m × 6 m plots, were evaluated on rainfed sorghum. Managed plots refer to terraced or level plots with bunds all around. CTf refers to

Table 1. Important properties of the experimental soil.

Depth, m	Soil fractions, %			Texture	pH (1:5)	EC (1:5), dS m^{-1}	OM, %	N, %	P, mg kg^{-1}	Exchangeable cations, cmolc kg^{-1}			
	Sand	Silt	Clay							Ca^{2+}	Mg^{2+}	K^+	Na^-
0 - 0.2	83	11	6	Loamy sand	7.8	0.08	0.65	0.06	9.32	11.5	3	0.15	0.35
0.2 - 0.5	70	14	16	Sandy loam	8.2	0.08	0.42	0.05	3.71	15.0	5	0.10	0.47
0.5 - 0.3	61	20	19	Sandy loam	8.2	0.14	0.42	0.05	2.91	20.0	5	0.14	0.55
>1.3	89	7	4	Sand	8.4	0.15	0.32	0.04	3.61	29.0	8	0.11	0.51

3 preplanting tillage followed by another 25 days after planting. Tillage in CTm was same as in CTf. RT refers to single preplanting tillage 4 days after heavy rainfall and NT refers to direct planting on managed plots. Except in CTf, bunds of height 0.30 m were made around each plot to prevent runoff. Entire experimental area was surrounded by trench to avoid any run-on. Each plot was separated by 2 m passage, which was sloping towards the main drain. Residue of the previous crop, chopped to 0.1 m, was applied uniformly @ 2.5 t ha^{-1} before tillage operations. The residue cover after planting was 66%, 33% and 63% in NT + R, CTm + R and RT + R plots, respectively.

Sorghum variety PP290 Shambuko was planted on 11 July 2007 at a seed rate of 12 kg ha^{-1} in rows 0.6 m apart at an average depth of 0.03 m using a seeder developed for this purpose. Wherever necessary, plants were thinned to about 0.2 m distance within row in about 20 days after planting. Diammonium phosphate (DAP) was applied @ 100 kg ha^{-1} and urea @ 50 kg ha^{-1} as per recommendations of NARI, Halhale, Eritrea. Entire DAP was applied at the last tillage before sowing and urea was top-dressed around 22 and 45 days from planting following rainfall. Weeds were removed manually on 15 and 35 days from planting. Except smut no major pests and diseases were observed. All infested plants were removed and destroyed after weighing their dry biomass.

Results of the 2007 experiment showed that supplemental irrigations after cessation of monsoon in September may cause significant increase in sorghum yield in the region as dry period coincides with milk to grain development stages. The experiment was thus repeated in 2009 along with a new experiment in split plot design with same 7 tillage treatments in main plots and 4 irrigation levels in subplots in 3 replications. The irrigation treatments were rainfed (I_0), 70 mm irrigation at 50% depletion of soil moisture in CTm − R from 1 m profile (I_1), 70 mm irrigation 7 days after irrigation in I_1 (I_2), and 70 mm irrigation 7 days after irrigation in I_2 (I_3). The crop was planted on July 9, 2009 with similar seed rate and management levels. Plot size of the repeated experiment was same but that of new experiment in split plot design was 5 m × 4 m. The crop was harvested from the central 3 m × 3 m area of the plots. The panicles of mature plants were cut, dried, threshed manually, cleaned and the grain was weighed on a precision electronic balance from each treatment and reported in kg ha^{-1} at 14% moisture content. Biomass yield was reported separately after air drying.

2.3. Measurement of Soil Properties

Soil properties determined were bulk density by core method, infiltration rate using double ring infiltrometer and soil moisture content by gravimetric method at planting, about a month after planting, end of monsoon, harvesting and before irrigations. Sampling for moisture determination was done in 50 mm cylindrical soil cores in duplicate from centre of the 0.2 m layer down to 1 m in one of the replications. High bulk density and low infiltration rates due to intermittent wetting and drying after planting prompted us to determine soil strength in the tilled layer by penetrometer to optimize soil moisture content during germination and emergence. For this reason, series of soil strength measurements were made at different moisture contents following rainfall to establish the relationship between soil moisture content and soil strength.

2.4. Water Use

Crop water use (ET, mm) in n days was calculated from the soil moisture data using water balance equation appropriate for the experimental conditions.

$$ET = RFe + IR - RO - DR \pm \Delta S \qquad (1)$$

where RFe, IR, RO, DR and ΔS refer to effective rainfall (mm), irrigation depth (mm), runoff (mm), drainage (mm), and change in soil water storage or loss (mm). Estimation of DR was done from the following considerations [33].

$$DR = 0 \text{ for } S < Sp, \text{ and } DR = S - Sp \text{ for } S > Sp \qquad (2)$$

where Sp is potential water storage capacity, considered as soil moisture storage at field capacity. RO was zero from all the plots except CTf in which it was estimated using the rainfall-runoff relationship Equation (3) developed on 3% slope from measurements made in the model watershed [3].

$$Runoff = 0.52 \times rainfall - 2.0, \qquad (3)$$

Both rainfall and runoff are in mm.

3. Results and Discussion

3.1. Soil Properties

3.1.1. Bulk Density, Infiltration Rate and Soil Strength

Effect of tillage and crop residue on bulk density and infiltration rates at planting and harvesting were not significant (**Table 2**). But bulk density at harvesting increased by $\approx 6\%$ and infiltration rate decreased by $\approx 12\%$ from planting. The effect was lowest in NT and RT plots and maximum in conventional tillage farmers practice (CTf) plots followed by CTm plots. Lower effects in NT and RT plots might be due to lesser degradation of organic matter compared to CTm and CTf plots [6] [27] [34]. Presence of high exchangeable Ca^{2+} and Mg^{2+} in soil (**Table 1**) might have also contributed to formation of denser structure due to series of wetting and drying cycles before harvesting. Tripathi [35] and Tripathi *et al.* [36] reported that both tillage and residue effects may become significant after 4 - 8 years of continuous practice.

Although there was no appreciable swelling and shrinking yet soil strength was rapidly changing with drying (**Figure 1**). Soil strength was lost on wetting close to field capacity and was below 2000 kPa, critical for seedling emergence and root proliferation [2] [37], until soil drying to 0.143 m^3 m^{-3} moisture but exceeded 2500 kPa with further drying to 0.138 m^3 m^{-3} moisture at which root growth and water uptake are greatly affected [38]. Farmers also observed that emergence and early establishment of sorghum was better at higher water contents around seed zone [2] [3]. Isaac [2] reported that optimum soil moisture for emergence of sorghum was 0.145 $\pm$ 0.002 m^3 m^{-3} at which soil strength was well below the critical level. Since rainfall was low and erratic, CT was believed to optimize wetness in seed zone in the conventional sloppy fields [5] but crusting in such over till plots due to rapid drying following rainfall after planting cannot be ruled out. A number of reports also indicate that moisture conservation can be better optimized through mulching, terracing, levelling and bunding than tillage

Table 2. Bulk density and infiltration rates at planting and harvesting in different tillage and residue plots.

Treatments	Bulk density (Mg m^{-3})		Infiltration rate (mm h^{-1})	
	At planting	At harvesting	At planting	At harvesting
CTf	1.61	1.70	7.8	6.6
NT − R	1.60	1.69	8.1	7.1
NT + R	1.59	1.68	8.3	7.3
CTm − R	1.58	1.68	7.9	6.7
CTm + R	1.57	1.67	8.2	7.3
RT − R	1.60	1.68	8.0	7.2
RT + R	1.59	1.66	8.4	7.5
Mean	1.59	1.68	8.1	7.1
LSD (5%)	NS	NS	NS	NS

CTf = Conventional practice on existing slopes, NT = No tillage, CTm = CTf on managed plots, m = managed (terraced or level plots with bunds), −R = Without residue, +R = With residue, NS = Not significant.

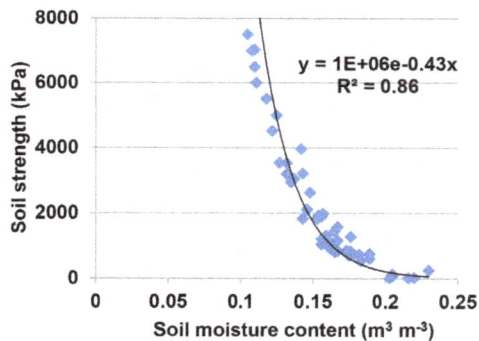

Figure 1. Soil strength as a function of soil moisture.

[3] [6] [34] [35] [39]-[41]. Optimization of tillage would not only reduce cost of cultivation but also facilitate soil and water conservation through improvements in soil structure [6] [35] [40].

3.1.2. Soil Moisture Content

At planting (July 11, 2007), soil moisture in 0 - 0.2 m layer was almost similar in all the plots and, therefore, moisture distribution only in CTf has been presented (**Figure 2**). Soil moisture at planting was about 4% below field capacity ($0.195 \ m^3 \ m^{-3}$) in 0 - 0.2 m layer and lower layers were still wetting (**Figure 2**). On 9 August (29 days from planting), all non-residue plots including CTf were still wetting in the lower layers and water content was much below field capacity but all residue plots were wetted beyond field capacity below 0.2 m depth. Total rainfall during June 27 to July 4 was 39.5 mm and that during July 10 to August 9 was 248 mm. Differences in soil moisture contents in CTf and NT-R (**Figure 2**) indicate the effect of levelling and bunding (terracing) and that in the residue plots was due to levelling, bunding and residue additions. Effect of residue on soil moisture was more in NT and RT plots perhaps because of its mulch action due to lesser incorporation in soil than in CT. Wetting in CTf and non-residue plots in lower layers continued until end of the rainy season (8 Sep) but developed uniform upward soil moisture gradient in all the plots before 23 September. The upward gradient in CTf was more because of drying of upper layers than by wetting of lower layers. Soil moisture distribution patterns (**Figure 2**) thus show that a) soil profile in the sloppy fields are not recharged by rainfall even by end of the rainy season, b) rainfall retention as soil water was unaffected by tillage and c) soil water contents in bunded and level plots of NT, RT and CT were always higher in residue than in nonresidue plots. Better capture of rainwater as soil water by crop residue than tillage has been reported by many researchers [25] [27] [34] [39] [41] [42]. Conventional tillage on sloppy fields (CTf) also encourages soil erosion compared to minimum tillage that improves water and fertilizer use [24].

In 2009, rainfall during June 28 to July 8 was 52.6 mm, which penetrated only down to 0.3 m in CTf plots but down to 1 m in the bunded and level plots (**Figure 3**). Except in CTf, wetness was greater and almost to the same depth in residue plots. In nonresidue plots, soil moisture was highest in RT – R and lowest in CTm – R.

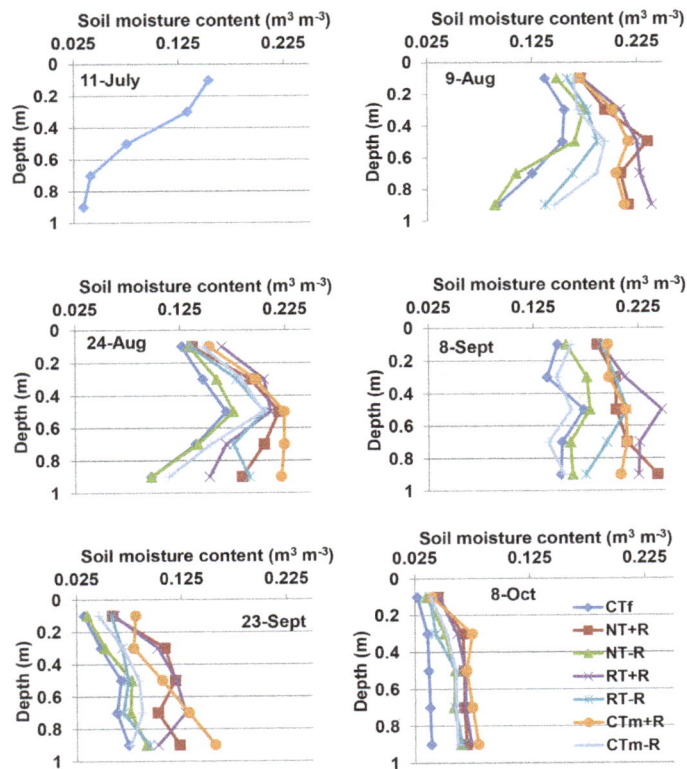

Figure 2. Soil moisture with depth in different tillage and residue plots on various dates during the crop season.

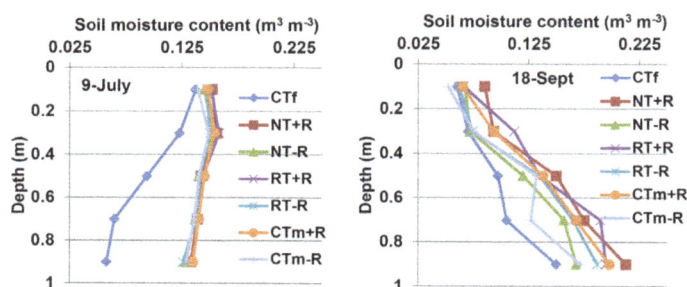

Figure 3. Soil moisture with depth in different tillage and residue plots on July 9, 2009 at planting and Sept 18, 2009 before irrigations.

Relatively more drying in CTm – R might be due to repeated tillage before planting. On September 18, surface 0.4 m soil profile in nonresidue plots dried to less than 7% moisture but wetness in the residue plots was close to field capacity below 0.6 m depth. Irrigations during September 20 to 1st week of October left more residual moisture that can be used by next crop in rotation. At harvesting, surface 0.1m layer dried almost to air dryness but, except in CTf, soil water was 40 - 80 mm m^{-1} of soil profile in lower layers (**Figure 4**). The residual moisture was relatively more in residue plots than in nonresidue plots and in I_3 (Irrigated 14 days after I_1) than in I_0. Results of 2009 confirm the observations of 2007 on tillage and residue effects on rainwater harvesting as soil water and residual moisture left in the soil profile after sorghum harvesting.

3.2. Yield

3.2.1. Tillage-Residue Effects

In 2007, highest yield of grain (2711 kg ha^{-1}) and stover (5812 kg ha^{-1}) under rainfed was observed in CTm + R, which was statistically at par with that in RT + R but significantly greater than in the remaining treatments (**Table 3**). It was lowest (1200 kg ha^{-1} grain and 3000 kg ha^{-1} stover) in CTf. All tillage treatments with residue performed significantly better than those without residue, perhaps due to better rainwater conservation (**Figure 2** and **Figure 3**). Yields were statistically equal in CTm − R and RT − R but significantly greater than in NT − R. This indicates superiority of RT over CT and NT. Similar results were also observed in 2009. Significantly lower yields under NT might be due to poor crop growth in the initial stages because of greater competition for nutrients and water with weeds compared to that under RT in which first flush of weeds that emerged in 4 days after rainfall were turned into the soil by tillage. The crop under NT also showed nitrogen deficiency, which recovered after weeding and urea applications. Two additional weedings were necessary in NT than in other plots. Results show that single preplanting tillage 4 days after heavy rainfall (RT) in well bunded and level plots that ensured rainwater conservation was sufficient to raise sorghum yields in the range of 2405 - 2797 kg ha^{-1} in Hamelmalo region. Crop yield improvements due to minimum tillage were also observed by Mbagwu [39], Steyn et al. [34] and Rockstrom et al. [24]. Despite potential benefits in crop yields, use of crop residue as mulch has been questioned by Giller et al. [30] in semiarid Africa due to its competing uses particularly as animal feed. The argument is valid if farmers are satisfied with current yield levels (1 - 2 t ha^{-1} sorghum stover) but if yields improve to the level obtained (>4 - 5 t ha^{-1}) through better management then at least half of the stover could be diverted back to the soil to which it belongs.

3.2.2. Tillage and Irrigation Effects

Mean grain yield due to tillage was highest in CTm + R (3823 kg ha^{-1}), which was at par with that in RT + R (3756 kg ha^{-1}) but significantly greater than in the remaining treatments (**Table 4**). Similarly mean yield due to irrigations was highest in I_1 (3548 kg ha^{-1}), which was as at par with that in I_2 (3492 kg ha^{-1}) but significantly greater than in I_0 (2416 kg ha^{-1}) and I_3 (3398 kg ha^{-1}). Interaction effects also show that grain yields in CTm + R and RT + R were at par, which further confirmed the superiority of reduced tillage. Except in I_1 and I_2 of CTm, grain yields due to residue were not significant (**Table 4**). Under rainfed (I_0), grain yield was lowest in CTf (501 kg ha^{-1}) and highest in CTm + R (2989 kg ha^{-1}), which was at par with that in RT + R (2887 kg ha^{-1}). Similar effects were observed on stover yields. First irrigation given on 21-day after last rainfall of the season (I_1) raised grain yield by 660 kg ha^{-1} in CTf and 1102 - 1408 kg ha^{-1} in the remaining treatments from that under rainfed.

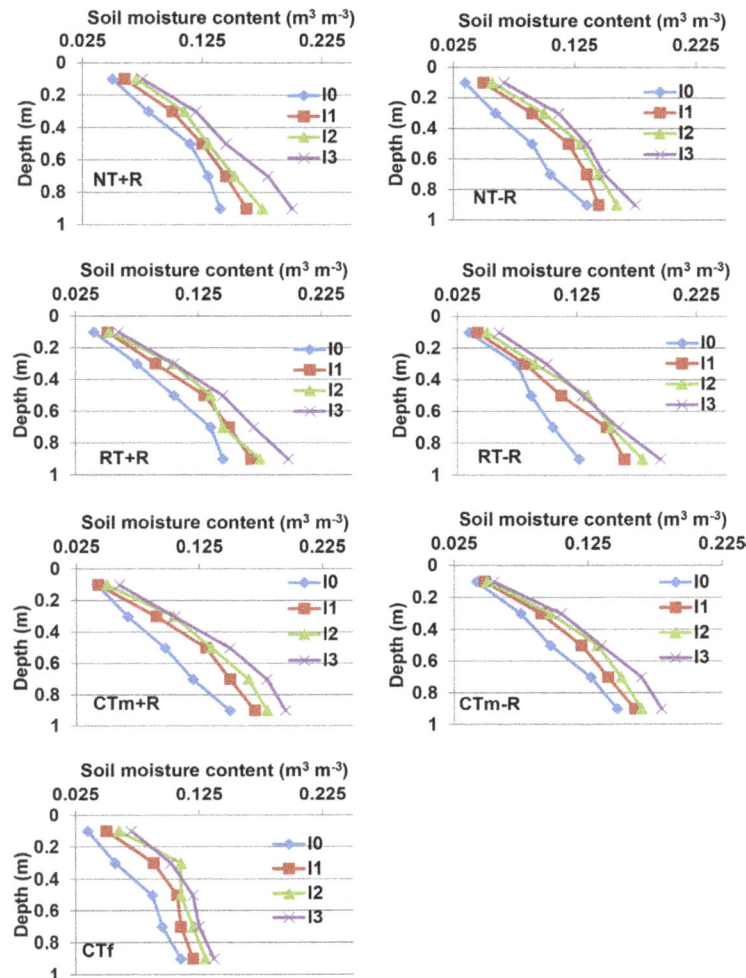

Figure 4. Soil moisture with depth at harvesting in different tillage, residue and irrigation plots.

Table 3. Grain and stover yields of rainfed sorghum in 2007 and 2009.

Treatments	Yield in 2007 (kg ha⁻¹)		Yield in 2009 (kg ha⁻¹)	
	Grain	Stover	Grain	Stover
CTf	1200	3000	629	1693
NT + R	2447	5474	2640	4975
NT − R	2085	5202	2450	4868
CTm + R	2711	5812	2880	5348
CTm − R	2440	5540	2676	5198
RT+R	2605	5704	2797	5301
RT − R	2405	5439	2590	5158
Mean	2270	5167	2409	4649
LSD (5%)	110	282	126	185

CTf = Conventional practice, NT = No tillage, CTm = CTf on well bunded and level plots, −R = Without residue, +R = With residue.

Table 4. Grain and stover yields of sorghum under different tillage and irrigations.

Tillage treatment, T	Grain yield (kg ha^{-1})					Stover yield (kg ha^{-1})				
	Irrigation levels, I					Irrigation levels, I				
	I_0	I_1	I_2	I_3	Mean	I_0	I_1	I_2	I_3	Mean
CTf	501	1161	983	879	881	1533	2266	2098	1983	1970
NT + R	2530	3880	3803	3731	3486	4973	5912	5872	5743	5625
NT − R	2397	3805	3657	3574	3358	4697	5815	5902	5823	5559
CTm + R	2989	4151	4156	3994	3823	5483	6310	6328	6328	6113
CTm − R	2802	3956	3893	3803	3614	5180	6008	6060.	6155	5851
RT + R	2887	3980	4108	4048	3756	5313	6285	6282	6207	6022
RT − R	2804	3906	3842	3754	3576	5188	5922	5957	5905	5743
Mean	2416	3548	3492	3398	3213	4624	5502	5500	5449	5269
Factors: LSD (P = 0.05)		T 122	I 64	T × I 185			T 174	I 96	T × I 271	

CTf = Conventional practice, NT = No tillage, CTm = CTf on well bunded and level plots, -R = Without residue, +R = With residue, I_0 = Rainfed, I_1 = 70 mm irrigation at 50% depletion of soil moisture from 1 m profile, I_2 = 70 mm irrigation 7 days after irrigation in I_1, I_3 = 70 mm irrigation 7 days after irrigation in I_2.

Similarly stover yields increased by 733 - 1118 kg ha^{-1}. The interaction effects justify the significance of reduced tillage (RT: one preplanting tillage 4 days after good rainfall) for optimum yields in well bunded and level plots. Maturity was delayed by 3 to 4 days in I_1 and 2 to 6 days in I_3.

Results thus show that soil bunding and levelling for rainwater conservation with single preplanting tillage can raise sorghum yields to more than 2405 kg ha^{-1} under rainfed and more than 3906 kg ha^{-1} under 70 mm irrigation 21 days after last rainfall of the season (**Table 4**). The three tillage practices (NT, RT, and CT) with residue showed lower weed incidence and produced better crop than without residue. Results also show that irrigations cannot bring significant increases in yield unless fields are level and bunded to arrest runoff and at least part of the residue is returned back to soil to maintain its quality.

3.3. Water Use and Production Function

Water use by rainfed sorghum was minimum (208 mm) in CTf and maximum (353 mm) in NT + R (**Figure 5**). Much of the irrigation water was also lost as runoff in CTf and, therefore, crop water use did not improve as much as in the other treatments. Sorghum crop at first irrigation was at grain filling stage and plants in CTf were almost drying due to inadequate soil water in the root zone but plants were perhaps within recoverable limits in other plots. Greater water use than irrigation amounts in the terraced plots (**Figure 5**) indicates improved upward flow of soil moisture into the root zone from lower layers and efficient root water extraction by mildly stressed plants [43].

A linear relationship between water use and grain yield (**Figure 6**) showed that yield increase due to single irrigation after cessation of monsoon was as important as rainfall during the crop season. Crop response to water availability in the root zone was thus crucial during September 20 to 1st week of October due to rapid drying. Irrigations raised not only the sorghum yield but also left more residual moisture for use by next crop in rotation. Production function (**Figure 6**) shows that farmers can raise rainfed sorghum yields from current <600 kg ha^{-1} [1] [18] to >2752 kg ha^{-1} with a water use of 344 mm through adoption of RT in terraced plots under dryland conditions of Hamelmalo. Grain yields may further increase to 4009 kg ha^{-1} by 70 mm irrigation about 21 days after cessation of the monsoon raising water use to 432 mm.

4. Conclusions

1) Single preplanting tillage 4 days after heavy rainfall (RT) was sufficient for optimum yield of sorghum in well bunded and level fields.

2) Soil bunding and levelling for rainwater conservation can raise rainfed sorghum yields to more than 2400 kg ha^{-1}.

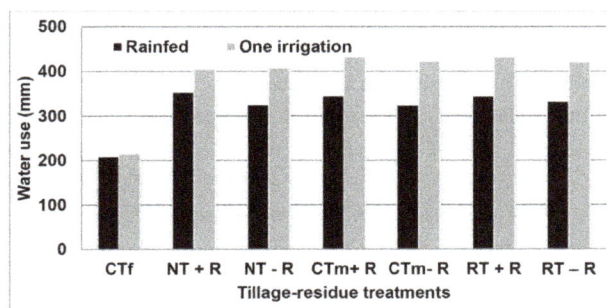

Figure 5. Water use by sorghum indifferent tillage, residue and irrigation plots.

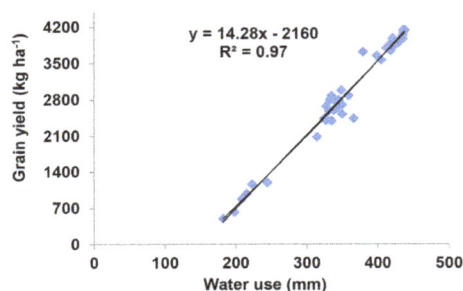

Figure 6. Sorghum yield as a function of water use.

3) Sorghum yields in Hamelmalo region can be raised to more than 3900 kg ha^{-1} by 70 mm irrigation applied 21 days after cessation of rainfall in September.

4) Grain yields would be better under residue than nonresidue conditions.

5) Crop response to water availability in the root zone is crucial for sorghum during September 20 to first week of October.

6) Sorghum root zone in the conventional cultivated sloppy fields is not recharged by rainfall even by end of the rainy season.

7) Irrigations cannot bring significant increases in yield unless fields are level and bunded.

Acknowledgements

Authors are grateful to the Centre for Development and Environment, Institute of Geography, University of Berne, Switzerland for providing financial support through Eastern and Southern Africa Partnership Programme to carry out this research.

References

[1]   MOA (Ministry of Agriculture) (2005) Area and Production by Zoba from 1992-2005. Asmara, Eritrea.

[2]   Isaac, K. (2008) Effect of Conservation Tillage Practices on Soil Properties, Growth and Yield of Sorghum (*Sorghum bicolour* L.) in the Semiarid Region of Eritrea. M.Sc. Thesis, Department of Land Resources and Environment, Hamelmalo Agricultural College, Eritrea.

[3]   Tripathi, R.P. and Ogbazghi, W. (2010) Development and Management of a Hilly Watershed in Hamelmalo Agricultural College Farm, as a Demonstration Site for Farmers and a Study Site for Students. Final Technical Report of the Project Financed by Eastern and Southern Africa Partnership Programme (ESAAP), Hamelmalo Agricultural College, Keren, 59 p.

[4]   Haile, A., Araiya, W., Omer, M.K., Ogbazghi, W. and Tewelde, M. (1995) Rehabilitation of Degraded Lands in Eritrea. Ministry of Agriculture, University of Asmara, International Development Center, Technical Paper No. 1.

[5]   Temesgen, M., Rockstrom, J., Savenije, H.H.C., Hoogmoed, W.B. and Alemu, D. (2008) Determination of Tillage Frequency among Smallholder Farmers in Two Semiarid Areas in Ethiopia. *Physics and Chemistry of the Earth*, **33**,

183-191. http://dx.doi.org/10.1016/j.pce.2007.04.012

[6]   Aina, P.O., Lal, R. and Roose, E.J. (1991) Tillage Methods and Soil and Water Conservation in West Africa. *Soil & Tillage Research*, **20**, 165-186. http://dx.doi.org/10.1016/0167-1987(91)90038-Y

[7]   Arnon, I. (1992) Agriculture in Drylands: Principles and Practices. Elsevier Science Publishers, B.V., Amsterdam.

[8]   Rockstrom J. and Jonsson, L.O. (1999) Conservation Tillage Systems for Dryland Farming: On-Farm Research and Extension Experiences. *East African Agricultural and Forestry Journal*, **65**, 101-114.

[9]   Derpsch, R. (2000) Frontiers in Conservation Tillage and Advances in Conservation Practice. FAO, Rome

[10] Karlen, D.L, Andrews, S.S. and Doran, J.W. (2001) Soil Quality: Current Concepts and Applications. *Advances in Agronomy*, **74**, 1-40. http://dx.doi.org/10.1016/S0065-2113(01)74029-1

[11] Tripathi, R.P., Sharma, P. and Singh, S. (2005) Tilth Index: An Approach to Optimize Tillage in Rice Wheat System. *Soil & Tillage Research*, **80**, 125-137. http://dx.doi.org/10.1016/j.still.2004.03.004

[12] Bollinger, A., Magid, J., Amado, T.J.C., Neto, F.S., Ribeiro, M.D.D., Calegari, A., Ralisch, R. and Neergaard, A.D. (2006) Taking Stock of the Brazilian "Zero-Till Revolution": A Review of Land Mark Research and Farmers' Practice. *Advances in Agronomy*, **91**, 47-110. http://dx.doi.org/10.1016/S0065-2113(06)91002-5

[13] Elwell, H.A. and Stocking, M.A. (1988) Loss of Soil Nutrients by Sheet Erosion Is a Major Hidden Farming Cost. *The Zimbabwe Science News*, **22**, 79-82.

[14] FAO (1994) Agriculture Sector Review for Eritrea. FAO, Rome.

[15] FAO (1998) Press Release 98/42. FAO: Conventional Tillage Severely Erodes Soil; New Concepts for Soil Conservation Required. http://www.fao.org/WAICENT/OIS/PRESS NE/PRESSENG/1998/pren 9842.htm

[16] Rockstrom, J. (2001) Green Water Security for the Food Makers of Tomorrow: Windows of Opportunity in Drought-Prone Savannahs. *Water Science & Technology*, **43**, 71-78.

[17] African Conservation Tillage Network, 2008. www.act-africa.org/

[18] FAO (2005) Global Information and Early Warning System on Food and Agriculture World Food Programme. Special report FAO/WFP Crop and Food Supply Assessment Mission to Eritrea.

[19] Rockstrom, J. and Falkenmark, M. (2000) Semiarid Crop Production from a Hydrological Perspective: Gap between Potential and Actual Yields. *Plant Science*, **19**, 319-346. http://dx.doi.org/10.1080/07352680091139259

[20] Laryea, K.B., Pathak, P. and Klaiji, M.C. (1991) Tillage Systems and Soils in the Semi-Arid Tropics. *Soil & Tillage Research*, **20**, 201-218. http://dx.doi.org/10.1016/0167-1987(91)90040-5

[21] Fowler, R. and Rockstrom, J. (2001) Conservation Tillage for Sustainable Agriculture: An Agrarian Revolution Gathers Momentum in Africa. *Soil & Tillage Research*, **61**, 93-107. http://dx.doi.org/10.1016/S0167-1987(01)00181-7

[22] Hobbs, P.R. (2007) Conservation Agriculture: What Is It and Why Is It Important for Future Sustainable Food Production? *The Journal of Agricultural Science*, **145**, 127-137. http://dx.doi.org/10.1017/S0021859607006892

[23] Hobbs, P.R., Sayre, K. and Gupta, R. (2008) The Role of Conservation Agriculture in Sustainable Agriculture. *Philosophical Transactions of the Royal Society B*, **363**, 543-555. http://dx.doi.org/10.1098/rstb.2007.2169

[24] Rockstrom, J., Kaumbutho, P., Mwalley, J., Nzabi, A.W., Temesgen, M., Mawenya, I., Barron, J. and Damgaard-Larsen, S. (2008) Conservation Farming Strategies in East and South Africa: Yields and Rain Water Productivity from On-Farm Action Research. *Soil & Tillage Research*, **103**, 23-32. http://dx.doi.org/10.1016/j.still.2008.09.013

[25] Lal, R. (1998) Soil Erosion Impact on Agronomic Productivity and Environment Quality. *Critical Reviews in Plant Sciences*, **17**, 319-464. http://dx.doi.org/10.1016/S0735-2689(98)00363-3

[26] Erenstein, O. (2002) Crop Residue Mulching in Tropical and Semi-Tropical Countries: An Evaluation of Residue Availability and Other Technological Implications. *Soil & Tillage Research*, **67**, 115-133. http://dx.doi.org/10.1016/S0167-1987(02)00062-4

[27] Scopel, E., Findeling, A., Guerra, E.C. and Corbeels, M. (2005) Impact of Direct Sowing Mulch-Based Cropping Systems on Soil Carbon, Soil Erosion and Maize Yield. *Agronomy for Sustainable Development*, **25**, 425-432. http://dx.doi.org/10.1051/agro:2005041

[28] Zingore, S., Manyame, C., Nyamugafara, P. and Giller, K.E. (2005) Long-Term Changes in Organic Matter of Woodland Soils Cleared for Arable Cropping in Zimbabwe. *European Journal of Soil Science*, **56**, 727-736. http://dx.doi.org/10.1111/j.1365-2389.2005.00707.x

[29] Farage, P.K., Ardo, J., Olsson, L., Rienzi, E.A., Ball, A.S. and Pretty, J.N. (2007) The Potential for Soil Carbon Sequestration in Three Tropical Dryland Systems of Africa and Latin America: A Modelling Approach. *Soil & Tillage Research*, **94**, 457-472. http://dx.doi.org/10.1016/j.still.2006.09.006

[30] Giller, K.E., Witter, E., Corbeel, M. and Tittonell, P. (2009) Conservtation Agriculture and Smallholder Farming in Africa: The Heretics' View. *Field Crops Research*, **74**, 1-12.

[31] NARI (National Agricultural Research Institute) Eritrea (2001-2003) Reports on Introduction of Conservation Agriculture in Eritrea, and Conservation Agriculture in Eritrea.

[32] Rockstrom, J., Barron, J. and Fox, P. (2002) Rainwater Management for Increased Productivity among Small Holder Farmers in Drought Prone Environments. *Physics and Chemistry of the Earth*, **27**, 949-959. http://dx.doi.org/10.1016/S1474-7065(02)00098-0

[33] Tripathi, R.P. and Mishra, R.K. (1986) Wheat Root Growth and Seasonal Water-Use as Affected by Irrigation under Shallow Water Table Conditions. *Plant and Soil*, **92**, 181-188. http://dx.doi.org/10.1007/BF02372632

[34] Steyn, J.T., Tolmay, J.P.C., Human, J.J. and Kilian, W.H. (1995) The Effects of Tillage Systems on Soil Bulk Density and Penetrometer Resistance of a Sandy Clay Loam Soil. *South African Journal of Plant and Soil*, **12**, 86-90. http://dx.doi.org/10.1080/02571862.1995.10634342

[35] Tripathi, R.P. (1992) Physical Properties and Tillage of Rice Soils in Rice-Wheat System. In: Pandey, R.K., Dwivedi, B.S. and Sharma, A.K., Eds., *Rice-Wheat Cropping System*, Project Directorate for Cropping Systems Research, Modipuram, 53-67.

[36] Tripathi, R.P., Sharma, P. and Singh, S. (2006) Soil Physical Response to Multi-Year Rice-Wheat Production in India. *International Journal of Soil Science*, **1**, 91-107. http://dx.doi.org/10.3923/ijss.2006.91.107

[37] Da Silva, A.P. and Kay, B.D. (1997) Estimating the Least Limiting Water Range of Soil from Properties and Management. *Soil Science Society of America Journal*, **61**, 877-883. http://dx.doi.org/10.2136/sssaj1997.03615995006100030023x

[38] Phillips, R.E. and Kirkham, D. (1962) Mechanical Impedance and Corn Seedling Growth. *Soil Science Society of America Proceedings*, **26**, 319-322. http://dx.doi.org/10.2136/sssaj1962.03615995002600040005x

[39] Mbagwu, J. (1990) Mulch and Tillage Effects on Water Transmission Characteristics of an Ultisol and Maize Grain Yield in SE Nigeria. *Pedologie*, **40**, 155-168.

[40] Lal, R. (1991) Tillage and Agricultural Sustainability. *Soil & Tillage Research*, **20**, 133-146. http://dx.doi.org/10.1016/0167-1987(91)90036-W

[41] Kronen, M. (1994) Water Harvesting and Conservation Techniques for Smallholder Crop Production Systems. *Soil & Tillage Research*, **32**, 71-86. http://dx.doi.org/10.1016/0167-1987(94)90034-5

[42] Bationo, A., Kihara, J., Vanlauwe, B., Waswa, B. and Kimetu, J. (2007) Soil Organic Carbon Dynamics, Functions and Management in West African Agro-Ecosystems. *Agricultural Systems*, **94**, 13-25. http://dx.doi.org/10.1016/j.agsy.2005.08.011

[43] Kramer, P.J. (1978) Plant and Soil Water Relationships: A Modern Synthesis. McGraw-Hill, Inc., New York, 482 p.

15

Effect of Mineral Fertilization and Irrigation on Sunflower Yields

Lucia Helena Garófalo Chaves[1*], Danila Lima Araujo[1], Hugo Orlando Carvallo Guerra[1], Walter Esfrain Pereira[2]

[1]Federal University of Campina Grande, Avenue Aprigio Veloso, Campina Grande, Brazil
[2]Federal University of Paraiba, Campus II, Areia, Brazil
Email: [*]lhgarofalo@hotmail.com

Abstract

Among the cultures used for the production of biofuels, the sunflower is one of the most important. Although some information exists, the water and nutritional needs of sunflower in the north east of Brazil are not well known. To fill knowledge gaps, an experiment was carried out to evaluate the effect of nitrogen (N), phosphorus (P), potassium (K) fertilization and available soil water (ASW) on sunflower yields. The sunflower cultivar Embrapa 122-V2000 was subjected to 44 treatments on a completely randomized design generated by the Baconian Matrix with four rates of N (0, 60, 80 and 100 kg·ha^{-1}), four rates of P$_2$O$_5$ (0, 80, 100 and 120 kg·ha^{-1}), four rates of K$_2$O (0, 80, 100 and 120 kg·ha^{-1}), and four available soil water (ASW) levels (55%, 70%, 85% and 100%) replicated three times. Urea was used as a source of N, triple super phosphate as P and potassium chloride as K. In all the experimental units was applied 2 kg·B·ha^{-1} as boric acid. The components of production evaluated were dry matter of the head, total number of achenes, total achenes' weight and 1000 achenes' weight. The results of this research showed that nitrogen had a significant effect on the dry matter of the head, total number of achenes and total achenes' weight. Phosphorus affected all production components and potassium affected the total number and the weight of achenes. With the exception of the 1000 achenes' weight, all the production components of the sunflower increased with the increased ASW level influenced significantly at 0.01 level of probably the total number of achenes. The highest rates of N, P and K (100, 120 and 120 kg·ha^{-1}, respectively) and 100% of available soil water produced the highest production.

Keywords

Water, Fertilizers, Yields, Oil Plants, Seeds

---

[*]Corresponding author.

1. Introduction

Sunflower (*Helianthus annuus* L.) occupies a prominent place among oilseed crops, as it contributes approximately 12% to global edible oil production. Water and nutrients play an important role in improving seed yield and oil quality of sunflowers [1]. Application of fertilizers substantially increases sunflower growth and yields, however, additions of nitrogen (N), phosphorus (P) and potassium (K) need to be optimized. In sunflowers, nutrient deficiency can result in up to 60% reduction in productivity [2]. The nutritional needs of sunflower are greater than many other crops like wheat, sorghum and corn, requiring higher amounts of N and other macronutrients [3].

The number of achenes per head is a reflection of action of N in critical early stages of flowering in sunflower development. The potential number of flowers is determined very early, and subsequently affects the number of achenes and head diameter [4]. Head diameter is one of the morphological characteristics most affected by addition of N, showing increases even with small N doses (25 kg·N·ha^{-1}). However, this increase in head diameter does not continue with further increases of N. Sachs *et al.* [5] observed an increase of achenes productivity with N doses up to 55 kg·N·ha^{-1}, 41 kg ha^{-1} of K_2O and 46 kg·ha^{-1} of P_2O_5. The achenes' oil content increased with the application of K_2O and P_2O_5 and the protein content decreased with increasing K_2O application.

Shortage of water, which is the most important component of life, limits plant growth and crop productivity, particularly in arid regions. Sunflower is commonly regarded as a plant that is tolerant to drought and it uses water efficiently. Nevertheless, the crop consumes a large amount of total water due to the fact that it produces high yields and a large vegetative bulk. It also has a long growing period coinciding with the warm months of spring and summer. Water stress on sunflower reduces plant height, root length, stomata number and causes early flowering, early maturity and seed yield reduction. Drought adversely influenced leaf area leaf area, days to maturity, leaf diameter, 1000-achene weight and achene yield per plant [6].

The present research aimed to investigate the yield of sunflower affected by the interaction between NPK fertilization treatments and irrigation regimes under semiarid Brazilian conditions.

2. Materials and Methods

The experiment was carried out from March to July 2011 under greenhouse conditions at the Agricultural Engineering Department of the Federal University of Campina Grande, Paraiba State, Brazil.

The sunflower cultivar used was the Embrapa 122-V2000. A total of 44 treatments on a completely randomized design generated by the Baconian Matrix (**Table 1**) with four doses of N (0, 60, 80 and 100 kg·ha^{-1}), four of P_2O_5 (0, 80, 100 and 120 kg·ha^{-1}), four of K_2O (0, 80, 100 and 120 kg·ha^{-1}), four available soil water (ASW) levels (55%, 70%, 85% and 100%) and three replicates resulting in 132 experimental units. Urea was used as a source of N; triple super phosphate as P and potassium chloride as K. In the soil in all the experimental units was applied, as pure solution, 2 kg·B·ha^{-1} as boric acid.

Each experimental unit consisted of a plastic pot filled with 32 kg of an Alfisol with the following attributes using the procedures recommended by [7]: sand = 553.40 g·kg^{-1}; silt = 117.30 g·kg^{-1}; clay = 329.30 g·kg^{-1}; pH (H_2O) = 6.6; Ca^{2+} = 1.45 cmol$_c$·kg^{-1}; Mg^{2+} = 1.65 cmol$_c$·kg^{-1}; Na = 0.17 cmol$_c$·kg^{-1}; K^+ = 0.21 cmol$_c$·kg^{-1}; H^+ + Al^{3+} = 0.79 cmol$_c$·kg^{-1}; organic matter = 22.2 g·kg^{-1}; Available P (Mehlich) = 8.1 mg·kg^{-1}.

Soil water content was monitored daily at three depth intervals: 0 - 10, 10 - 20 and 20 - 30 cm, using a Frequency Domain Reflectometry (FDR) segmented probe, inserted into the soil through an access tube installed in the pots. The volume of water required to maintain ASW for each treatment was calculated based the difference between field capacity and the permanent wilting point, and the FDR measurements. Irrigation was performed daily.

Five sunflower seeds were sown directly in the pots at a 2 cm depth. Twenty days after sowing, seedlings were thinned to one plant per pot.

When the experiment was finalized, plants were harvested and measured for dry matter (grams) of the head (DMH), total number of achenes (TNA), total weight (grams) achenes (TWA) and 1000 achenes weight (W1000A).

The results were analyzed statistically through the analyses of variance (ANOVA) described by [8], using SAEG software [9].

Table 1. Nitrogen (N), phosphorus (P_2O_5), potassium (K_2O) and available soil water levels generated by the Baconian matrix.

Treatments.	N	P_2O_5	K_2O	Water	Treatments	N	P_2O_5	K_2O	Water
	------kg·ha^{-1}-----			%		------ kg·ha^{-1}-----			%
1	0	0	0	55	23	0	0	0	85
2	0	80	80	55	24	0	80	80	85
3	80	80	80	55	25	80	80	80	85
4	100	80	80	55	26	100	80	80	85
5	60	0	80	55	27	60	0	80	85
6	60	100	80	55	28	60	100	80	85
7	60	120	80	55	29	60	120	80	85
8	60	80	0	55	30	60	80	0	85
9[*]	60	80	80	55	31[*]	60	80	80	85
10	60	80	100	55	32	60	80	100	85
11	60	80	120	55	33	60	80	120	85
12	0	0	0	70	34	0	0	0	100
13	0	80	80	70	35	0	80	80	100
14	80	80	80	70	36	80	80	80	100
15	100	80	80	70	37	100	80	80	100
16	60	0	80	70	38	60	0	80	100
17	60	100	80	70	39	60	100	80	100
18	60	120	80	70	40	60	120	80	100
19	60	80	0	70	41	60	80	0	100
20[*]	60	80	80	70	42[*]	60	80	80	100
21	60	80	100	70	43	60	80	100	100
22	60	80	120	70	44	60	80	120	100

[*]Reference level used by the sunflower growers of the region.

3. Results and Discussion

The ANOVA results are presented in **Table 2**. The results indicate that N, P and ASW had a significant effect on dry matter of the head (DMH) at 1% level of probability (**Table 2**) corroborating [10].

For plants treated with N the DMH increased with increasing N rate whose data were fitted to a quadratic regression model (**Figure 1(a)**). It is observed that the highest value of this variable (28 g) was registered for the highest dose of N (100 kg·ha^{-1}) 25% higher than the DMH found for the control treatment. The DMH was also influenced significantly by increasing P doses, and these results were fitted to a quadratic regression model (**Figure 1(b)**). The highest value of this variable (31 g) was obtained with the highest dose of P (120 kg·ha^{-1}) showing a superiority of 556% when compared with the reference level. The results corroborate [11] and [12] who found an increase of the DMH of sunflower cv. Embrapa 122/V-2000 with increasing N doses and ASW.

Potassium rates did not affect the DMH values (**Table 2**). These results disagree with those results found by [6] and [12] who found a positive significant effect of K on DMH.

The DMH increased linearly with increasing ASW from 18 to 24 g, for the lowest (55%) to the greatest ASW doses (100%), respectively (**Figure 2**). Guedes Filho *et al.* [11] observed an increase of 43% in DMH when

$$y = 0.0006x^2 + 0.0696x + 13.056$$
$$R^2 = 0.86$$

(a)

$$y = -0.001x^2 + 0.338x + 4.701$$
$$R^2 = 0.99$$

(b)

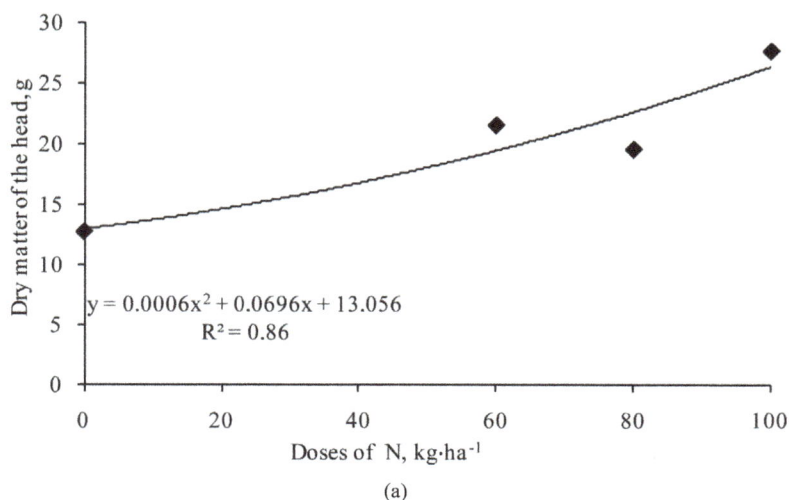

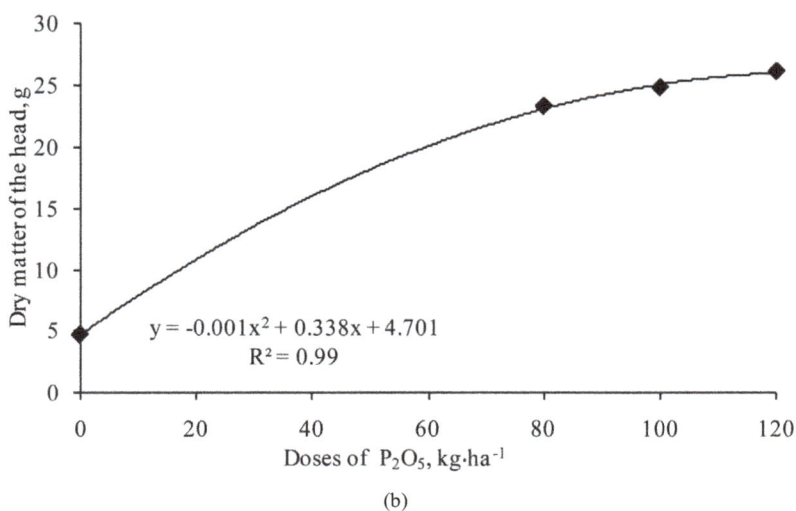

Figure 1. Regression curves for the relations N and P versus DMH.

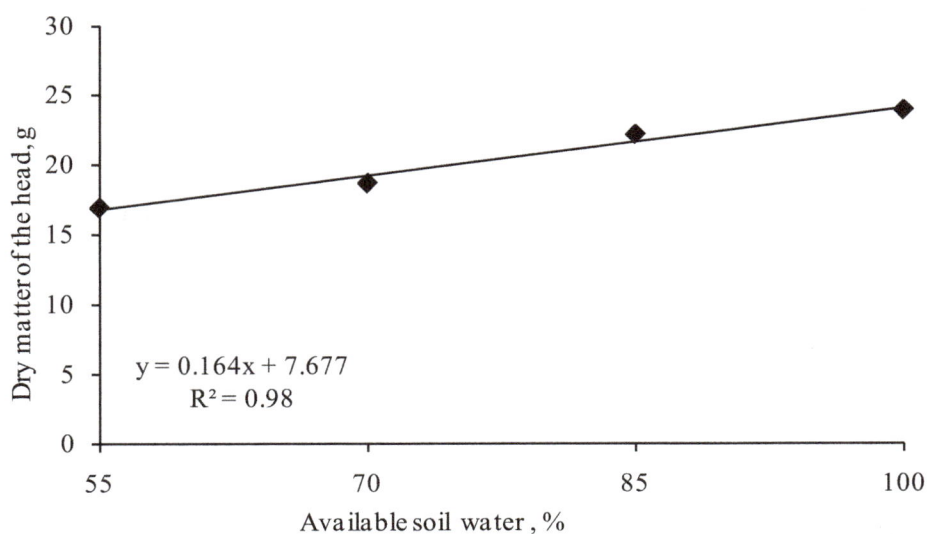

$$y = 0.164x + 7.677$$
$$R^2 = 0.98$$

Figure 2. Regression curve for the relation available soil water content × dry matter of the head.

Table 2. Analysis of variance for dry matter of the head (DMH), total achenes' number (TAN), total achenes' weight (TWA) and 1000 achenes' weight (W1000A) for different levels of N, P, K and ASW.

Source	DF	Mean square	Pr > F	Mean square	Pr > F	Mean square	Pr > F	Mean square	Pr > F
		DMH		TAN		TWA		W1000A	
N	3	141.76	<0.0001	77234.76	<0.0001	51.95	0.0013	618.67	0.0625
P	3	1682.16	<0.0001	423580.92	<0.0001	669.36	<0.0001	5869.42	<0.0001
K	3	5.40	0.7513	5677.88	0.0009	27.59	0.0338	270.98	0.3511
ASW	3	61.78	0.0047	50590.70	0.0020	46.66	0.0026	54.61	0.8804
N* ASW	9	19.08	0.1894	32852.11	0.0010	15.84	0.0924	249.73	0.4311
P* ASW	9	25.66	0.0593	14096.31	0.1659	24.62	0.0079	252.92	0.4211
K* ASW	9	18.56	0.2063	7348.53	0.6415	5.33	0.8070	250.49	0.4287
Contrast	DF								
N linear	1	109.05	0.0054	207053.43	<0.0001	142.18	0.0002	43.10	0.6760
N quadratic	1	177.65	0.0004	1227.04	0.7202	0.75	0.7745	229.97	0.3353
P linear	1	2417.57	<0.0001	556461.89	<0.0001	1031.08	<0.0001	8932.15	<0.0001
P quadratic	1	1139.87	<0.0001	318864.65	<0.0001	644.27	<0.0001	2925.41	0.0008
K linear	1	2.22	0.6850	207053.43	<0.0001	142.18	0.0002	43.10	0.6760
K quadratic	1	11.04	0.3664	1227.04	0.7202	0.75	0.7745	229.97	0.3353
ASW linear	1	124.20	0.0030	62694.06	0.0118	101.99	0.0012	130.48	0.4675
ASW quadratic	1	5.39	0.5275	87700.72	0.0031	17.67	0.1677	26.92	0.7411
N × ASW	1	51.84	0.0522	53772.80	0.0195	21.65	0.1272	555.46	0.1357
P × ASW	1	0.0007	0.9941	6512.35	0.4100	9.58	0.3085	359.46	0.2290
K × ASW	1	19.93	0.2258	3238.75	0.5608	6.20	0.4122	591.56	0.1238

ASW was increased from 55% to 100%. Similarly [13] observed that the level of ASW influenced the DMH linearly at 1% probability. The higher dry matter accumulation may be a reflection of a greater ion absorption in soil, since the increase in soil moisture on the development of sunflower crop can be significant in nutrient uptake by plants.

The total number of achenes (TAN) was significantly affected, at 1% level of probability, by the increase of N, P and K doses and available soil water. The interaction among N and ASW at 1% level of probability was also significant (**Table 2**) corroborating [12].

The increase of the number of achenes with the N doses was fitted to a linear model as observed in **Figure 3(a)**. The greatest TAN was obtained, thus, with the highest N doses.

The effect of P on the TAN was fitted to a quadratic model (**Figure 3(b)**) verifying the greatest achene number (416) at 80 kg·P·ha^{-1}. This kind of adjustment was also reported by [14] and [15]. The response of achene number to the K was also fitted to a quadratic model (**Figure 4(a)**) verifying the greatest achene number (506) for the 120 kg·K·ha^{-1}.

The increase in available soil water content increased, in a quadratic manner, the achene number from 301 to 448 when water increased from 55% to 100% ASW, an increase of 52.30% (**Table 2**). Otherwise, [11] found a quadratic effect producing above 600 achenes with 100% level available soil water.

The effect of ASW on the sunflower is similar to other studies [12] [16]-[18]. **Figure 4(b)** shows an increase of the TAN with the ASW in a quadratic model, obtaining the highest number of achenes for the highest level of water.

The interaction among N and the available soil water content on TAN was significant to the 1% level of probability (**Table 2**). **Figure 5** presents the cited interaction. It is observed that in general, the TAN increased

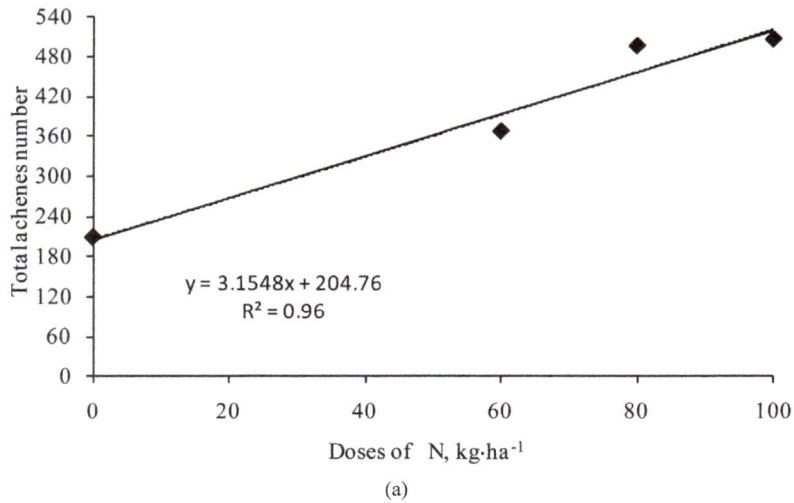

(a)

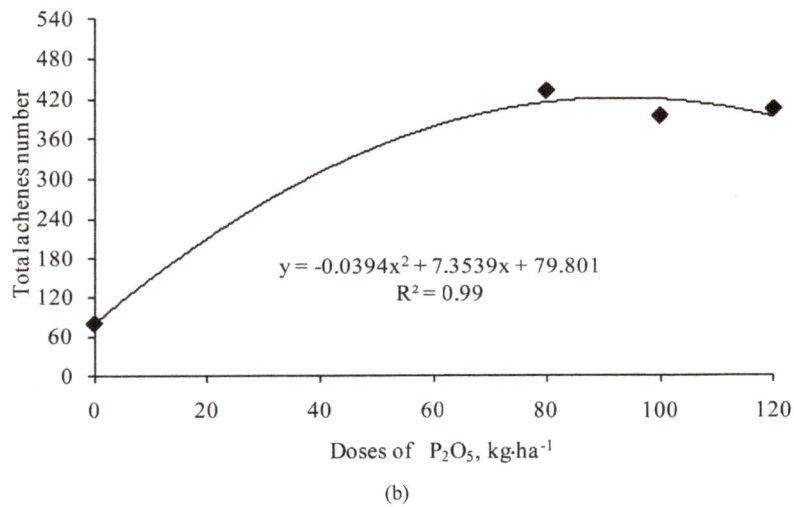

(b)

Figure 3. Regression curves for the relations N × total achenes' number and P × total achenes' number.

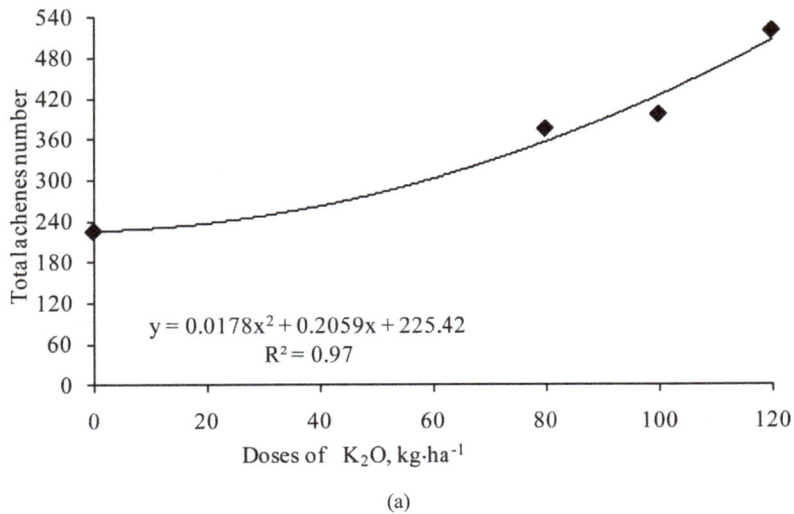

(a)

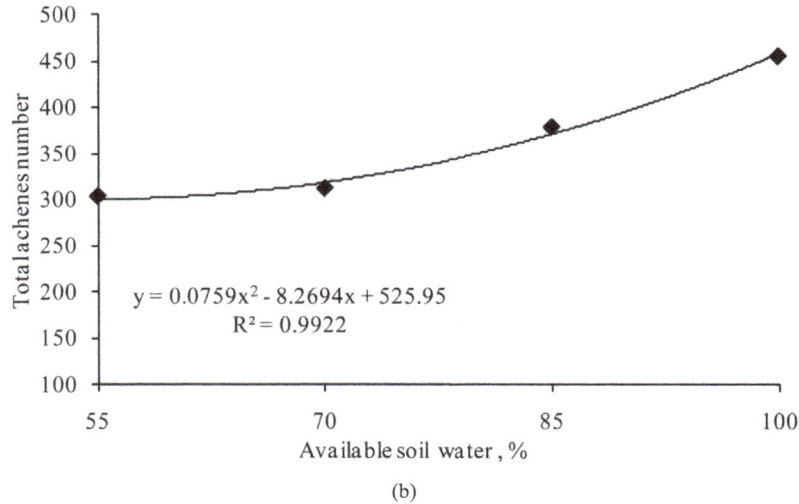

(b)

Figure 4. Regression curves for the relations K × TAN and ASW × TAN.

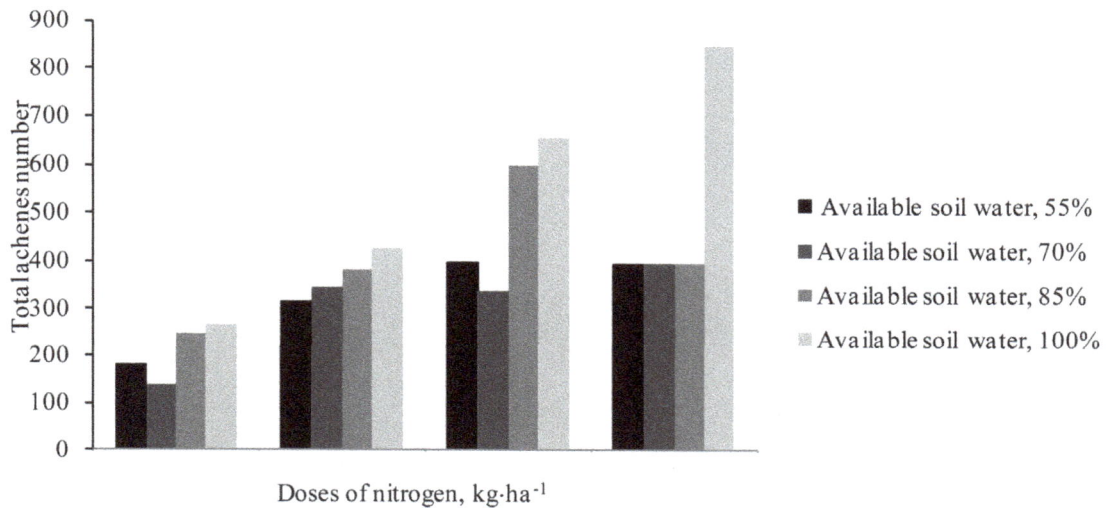

Figure 5. Interaction among N and the ASW on TAN.

with the N application and with the ASW. Thus the highest total number of achenes (844) was obtained with the highest N application (100 kg·ha^{-1}) and the highest ASW (100%).

The total achenes' weight was significantly influenced by increasing N, P, K rates and the ASW (**Table 2**). Total achenes' weight increased with N and P treatments, the data was adjusted to a linear and quadratic regression model, respectively (**Figure 6(a)** and **Figure 6(b)**, respectively). Analyzing **Figure 6(a)**, it is observed that the highest total achene' weight (16 g) was obtained with the highest N dose (100 kg·ha^{-1}). **Figure 6(b)** shows that the highest total achene' weight (16 g) was obtained with the highest P doses (96 kg·ha^{-1}). These highest weights were 136% and 728% superior to the control treatment, respectively.

The TWA increased linearly with increasing K (**Figure 7(a)**) verifying the greatest achene weight (15.67 g) for the 120 kg·P·ha^{-1}. The ASW treatments increased the total achenes weight linearly obtaining 8.94 and 15.22 g for the lowest and highest treatment, respectively. There was an increase of 70.25% between the lowest and highest water treatments.

The 1000 achenes' weight of sunflower was only affected by the P application, at the 1% level of probability. The regression was adjusted to a quadratic model (**Figure 8**), verifying an increase of the 1000 achenes' weight with P application and therefore, the greatest value for the treatment of 120 kg·P·ha^{-1} (60 g).

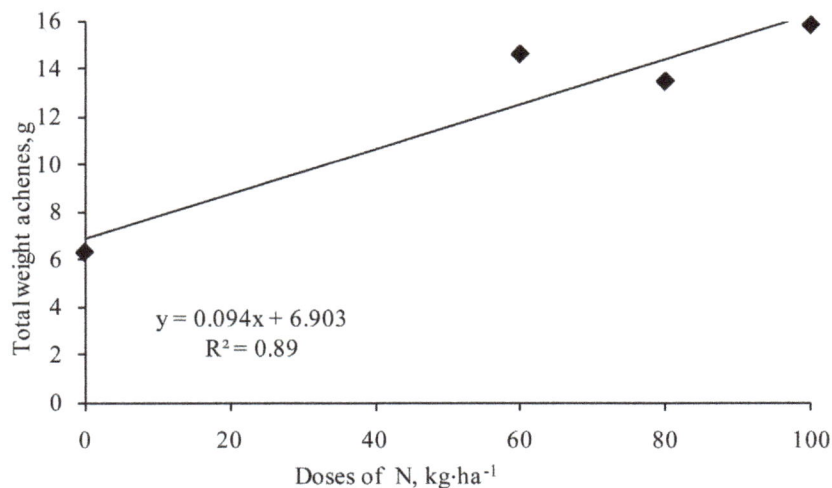

$$y = 0.094x + 6.903$$
$$R^2 = 0.89$$

(a)

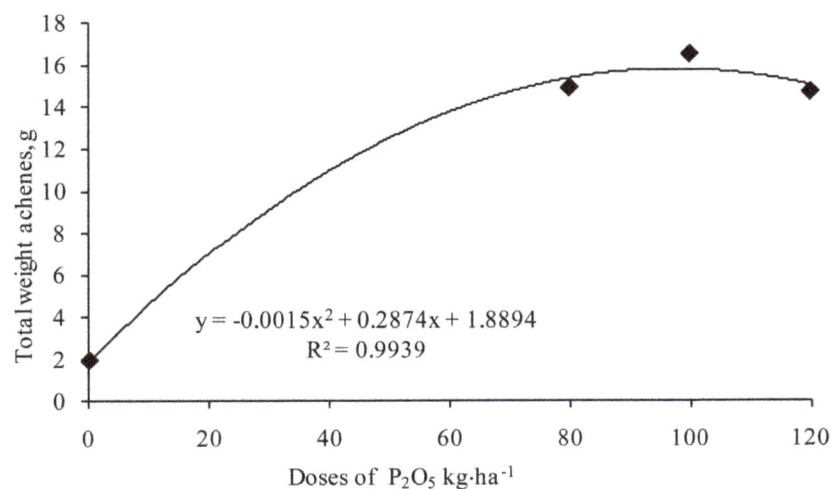

$$y = -0.0015x^2 + 0.2874x + 1.8894$$
$$R^2 = 0.9939$$

(b)

Figure 6. Regression curves for the relations Nx TWA and P × TWA.

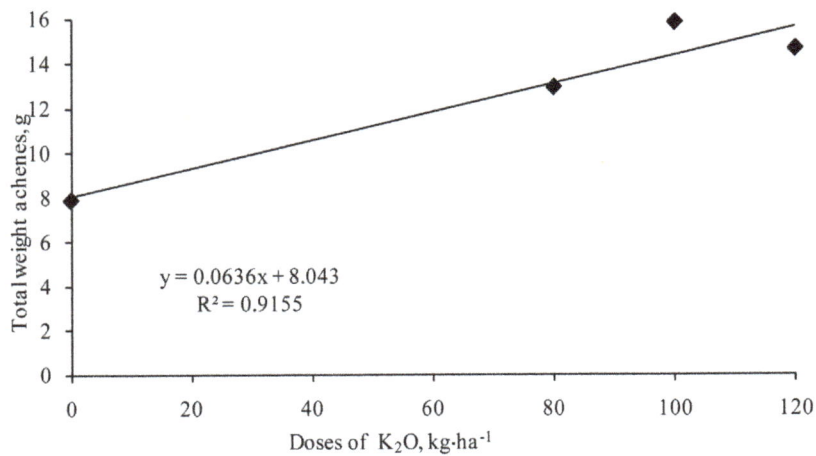

$$y = 0.0636x + 8.043$$
$$R^2 = 0.9155$$

(a)

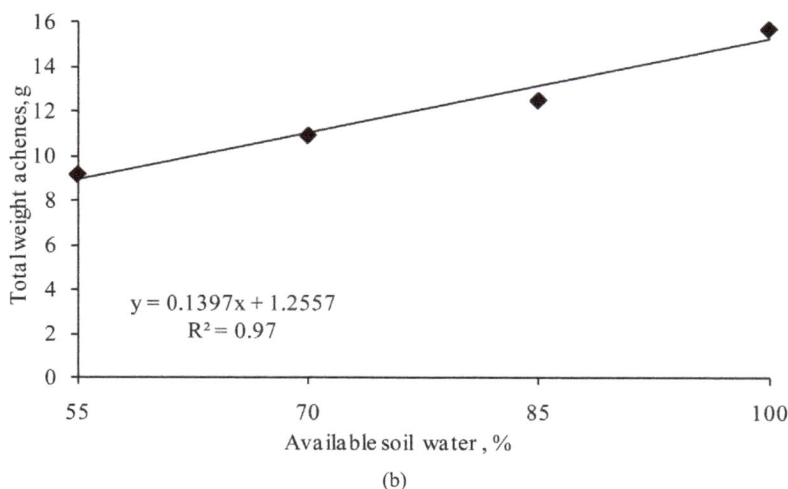

Figure 7. Regression curves for the relation K × total achenes' weight (a) and ASW × total achenes' weight (b).

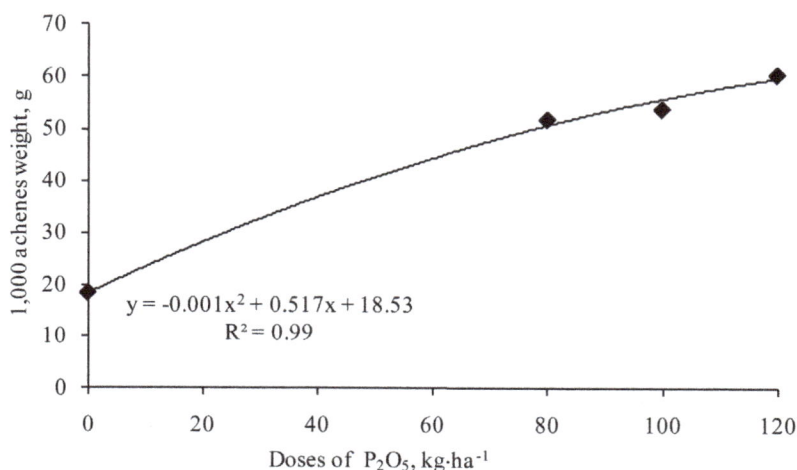

Figure 8. Regression curve for the relation P × 1000 achenes' weight.

4. Conclusions

The dry matter of the head, the total achenes' number and the total achenes' weight increased significantly with the nitrogen doses applied.

All the sunflower variables studied increased significantly with the phosphorus doses applied.

The total achenes' number and the total achenes' weight increased significantly with the potassium doses applied.

With the exception of the 1000 achenes' weight, whose effect was not significant, all the production components of the sunflower increased with the available water in the soil.

The interaction between N and the available soil water content was significant only for the TAN. It increased with the N application and with ASW.

The highest rates of N, P, and K (100, 120 and 120 kg·ha^{-1}, respectively) and the 100% available soil water produced the highest production.

Acknowledgements

Thanks to the Coordination of Improvement of Higher Education (CAPES) for the scholarship award to the second author at the Graduate School.

References

[1] Hussain, M.K., Rasul, E. and Ali, S.K. (2000) Growth Analysis of Sunflower (*Helianthus annuus* L.) under Drought Conditions. *International Journal of Agriculture and Biology*, **2**, 136-140.

[2] Biscaro, G.A., Machado, J.R., Tosta, M.S., Mendonça, V., Soratto, R.P. and Carvalho, L.A. (2008) Nitrogen Side Dressing Fertilization in Irrigated Sunflower under Conditions of Cassilândia-MS. *Ciência e Agrotecnologia*, **32**, 1366-1373. http://dx.doi.org/10.1590/S1413-70542008000500002

[3] Vigil, M.F. (2000) Fertilization in Dryland Cropping Systems: A Brief Overview Central Great Plains. Research Station-USDA-ARS.

[4] Zagonel, J. and Mundstock, C.M. (1991) Nitrogen Rates and Side-Dress Timing on Two Sunflower Cultivars. *Pesquisa Agropecuária Brasileira*, **26**, 1487-1492.

[5] Sachs, L.G., Portugal, A.P., Prudencio-Ferrreira, S.H., Ida, E.I. and Sachs, P.J. (2006) Efeito de NPK na produtividade e componentes químicos do girassol. *Semina: Ciências Agrárias*, **27**, 533-546. http://dx.doi.org/10.5433/1679-0359.2006v27n4p533

[6] Bakht, J., Shafi, M., Yousaf, M., Raziuddin R. and Khan, M.A. (2010) Effect of Irrigation on Physiology and Yield of Sunflower Hybrids. *Pakistan Journal of Botany*, **42**, 1317-1326.

[7] EMBRAPA, Centro Nacional de Pesquisa de Solos (1997) Manual de métodos de analise de solo. 2nd Edition, rev. atual, Embrapa, Rio de Janeiro. ISBN-85-85864-03-6.

[8] Ferreira, D.F. (2000) Sistema de análises de variância para dados balanceados. UFLA, Lavras (SISVAR 4. 1. Pacote computacional). http://www.scielo.br/scielo.php?script=sci_nlinks&ref=000117&pid=S0100-2945200900030002000012&lng=en

[9] Euclides, R.F. (1997) Manual de utilização do programa SAEG: sistema para análises estatísticas e genéticas. 2nd Edition, UFV, Viçosa. http://www.scielo.br/scielo.php?script=sci_nlinks&ref=000084&pid=S0100-204X20010002000180012&lng=en

[10] Silva, P.C.C., Couto, J.L. and Santos, A.F. (2010) Absorção dos íons amônio e nitrato e seus efeitos no desenvolvimento do girassol em solução nutritiva. *Revista de Biologia e Ciência da Terra*, **10**, 97-104.

[11] Guedes Filho, D.H., Chaves, L.H.G., Campos, V.B., Santos Júnior, A. and Oliveira, J.T.L. (2011) Production of Sunflower and Biomass Depending on Availables Oil Water and Nitrogen Levels. *Iranica Journal of Energy and Environment*, **2**, 313-319.

[12] Chaves, L.H.G., Guerra, H.O.C., Campos, V.B., Pereira, W.E. and Ribeiro, P.H.P. (2014) Biometry and Water Consumption of Sunflower as Affected by NPK Fertilizer and Available Soil Water Content under Semiarid Brazilian Conditions. *Agricultural Sciences*, **5**, 668-676.

[13] Oliveira, J.T.L., Chaves, L.H.G., Campos, V.B.J., Santos Júnior, A. and Guedes Filho, D.H. (2012) Fitomassa de girassol cultivado sob adubação nitrogenada e níveis de água disponível no solo. *Revista Brasileira de Agricultura Irrigada*, **6**, 23-32. http://dx.doi.org/10.7127/rbai.v6n100077

[14] Sadiq, S.A., Shahid, M., Jan, A. and Noor-Ud-Din, S. (2000) Effect of Various Levels of Nitrogen, Phosphorus, Potassium (NPK) on Growth, Yield, Yield Components of Sunflower. *Pakistan Journal of Biological Sciences*, **3**, 338-339. http://dx.doi.org/10.3923/pjbs.2000.338.339

[15] Gerendás, J., Abbadi, J. and Sattelmacher, B. (2008) Potassium Efficiency of Safflower (*Carthamus tinctorius* L.) and Sunflower (*Helianthus annuus* L.). *Journal of Plant Nutrition and Soil Science*, **171**, 431-439. http://dx.doi.org/10.1002/jpln.200720218

[16] Kakar, A.A. and Soomro, A.G. (2001) Effect of Water Stress on the Growth, Yield and Oil Content of Sunflower. *Pakistan Journal of Agricultural Science*, **38**, 73-74.

[17] Bajehbaj, A.A. (2010) Effects of Water Limitation on Grain and Oil Yields of Sunflower Cultivars. *Journal of Food, Agriculture and Environment*, **8**, 98-101.

[18] El Naim, A.M. and Ahmed, M.F. (2010) Effect of Irrigation Intervals and Inter-Row Spacing on Yield, Yields Components and Water Use Efficiency of Sunflower (*Helianthus annuus* L). *Journal of Applied Sciences Research*, **6**, 1446-1451.

16

Productivity and Dry Matter Accumulation of Sugarcane Crop under Irrigation and Nitrogen Application at Rio Verde GO, Brazil

Alefe Viana Souza Bastos*, Renato Campos de Oliveira, Nelmício Furtado da Silva, Marconi Batista Teixeira, Frederico Antonio Loureiro Soares, Edson Cabral da Silva

Instituto Federal Goiano (IFgoiano), Rio Verde, Brazil
Email: *alefe_viana@hotmail.com, renatocampos52@hotmail.com, nelmiciofurtado@gmail.com, marconibt@gmail.com, fredalsoares@hotmail.com, edsoncabralsilva@gmail.com

Abstract

Dry matter production and productivity of stem currently are being widely studied in sugarcane, reinforcing the study in question, which aims to assess the accumulation of dry matter of the aerial segment and the productivity of stems of sugarcane crops within the first cycle, at different levels of water replacement (WR) with and without nitrogen fertilization, through a subsurface drip irrigation system. The assay was conducted in the experimental area of the Federal Institut Goiano, Campus Rio Verde, GO, Brazil, in a dystroferric Rhodic Hapludox soil, cerrado phase (savannah), and comprised experimental splits of three furrows with an 8-meter long double row. Experimental design consisted of randomized blocks in a 5×2 factorial array, with four replications. Evaluated factors comprised five levels of WR (100%, 75%, 50%, 25% and 0% of field capacity), with and without the application of nitrogen (0 and 100 kg·ha^{-1} urea). Harvest occurred in May 2013 and stem productivity (SP), productivity of pointers (PP), productivity of straw (PS), harvest index (HI), dry matter of stem (DMS), dry matter of pointers (DMP), the relationship between dry matter of pointer and dry matter of stem (DMP/DMS) and total dry matter of the aerial segment (TDM) were determined. The variables SP, PP, DMS and DMP had a linear growth in proportion to WR increase, whereas HI and DMP/DMS adjusted to a quadratic model. Nitrogen fertilization affected positively the variables SP, HI, DMS and DMP/DMS and occurred interaction to TDM; also increasing the productivity stem and the harvest index.

Keywords

Sugarcane Ratoon, Water Replacement, Nitrogen Fertilizer

---

*Corresponding author.

1. Introduction

Sugarcane is one of the most important crops in Brazil due to its great socio-economic relevance in the production of several products, mainly biofuel used worldwide. Brazil is currently the biggest world producer of sugarcane with a total production of crushed sugarcane for the 2014-15 harvest estimated at 642.1 millions of tons, cultivated in approximately 9004.5 thousands of hectares in the producing states [1].

According to Silva *et al.* [2], for maximum yield, sugarcane needs adequate moisture, directly proportional to the transpired water, throughout its vegetal growth. Thus, irrigation becomes indispensable for greater productivity.

Useful information may be found in the literature on responses and benefits of nitrogen in sugarcane plants. Several authors report that responses to the nutrient are generally more frequent in sugarcane ratoon than in the sugar plant and that the crop absorbs between 50% and 60% of the N amount applied to the soil [3]-[5]. However, for higher productivity, the interaction of the plant's genetic factors, climate, soil and management must be taken into account [6].

The dry matter of the aerial segment is also relevant. In mechanized sugarcane harvest, part of the phytomass, mainly comprising leaves and pointers (leaves at the tip of the stem), is placed on the soil as a soil cover or chaff. Chaff is not only a protection but provides nutrients and improves the soil's physical and chemical properties [7] [8] and, at the same time, may enhance the temporary immobilization of nitrogen [6]. The production of dry matter is one of the most studied variables in assays on agriculture. According to Faroni *et al.* [9], sugarcane phytomass comprises all vegetal matter ranging from the aerial segment to the radicular system (rhizomes and roots).

In the medium and long terms, soil may accumulate organic C and N when sugarcane is managed without burning, even though, in the short term, the amount of residues with high C/N may demand more mineral N due to microbial immobilization. Several studies have shown that the best N dose should be 60 kg·ha^{-1}, which is higher than that used in burnt sugarcane [10]. Current analysis may provide important information on nitrogen fertilization associated with irrigation and may contribute towards recommendations for nitrogen fertilizers and the rational use of water in sugarcane crops in the soil-climate conditions of the Brazilian savannah.

The method of harvesting sugarcane without the use of fire, mandatory by environmental legislation, implies in the management of nitrogen fertilization concerning doses, sources and methods of application. Non-burnt sugarcane produces residues of dry leaves and pointers on the soil which may vary between 10 and 20 t·ha^{-1} dry matter, a C/N ratio above 100 and N contents between 40 and 80 kg·ha^{-1} [11] [12]. Consequently, when straw decomposes, some N rates may be absorbed by the subsequent stubbles and others may supply the N stock in the soil [12] [13].

Current analysis quantifies the dry matter of the aerial segment of the plant and evaluates the productivity of the sugarcane stem, pointers and straw in sugarcane ratoon at different levels of water replacement (WR) with and without nitrogen fertilization with surface drip irrigation system.

2. Material and Methods

2.1. Experimental Conditions

Current study was conducted in the experimental area of the Instituto Federal Goiano, campus Rio Verde GO Brazil, at 17°48'28"S and 50°53'57"W, mean altitude 720 m. The region's climate may be classified as tropical Aw, following Köppen [14], with a rainy season between October and May and a dry season between May and September. Mean annual temperature varies between 20°C and 35°C, with yearly rainfall between 1500 and 1800 mm; land relief is moderately undulated (6% declivity).

The study was carried in a soil classified as Dystropherric Typic Rhodic Hapludox soil [15], and Dystroferric Red Latosol (LVdf), medium texture, cerrado (savannah) phase. For characterization physical and chemical of soil (**Table 1**), a representative sub-sample was sieved to 2 mm and analyzed according to methodology described in EMBRAPA, [16].

Sugarcane, variety RB 85 - 5453 characterized by toughness and fastness, was planted in March 2011. This variety was chosen due to adaptability to the savannah and for having large acreage in the region. Experimental splits with three furrows and double rows (W-like planting) were established for current experiment. Spacing comprised 1.80 m between the double rows and 0.4 m between the 8-meter rows, totaling 35.2 m^2 per split. Planted area comprised two linear meters of the central row of each split.

The first ratoon cycle started in May 2012, when it was found in third day after the cut the first buds; the treatments were maintained under the same conditions as those of the sugarcane crop; after 390 days from the

Table 1. Physical, water and chemical characteristics of the soil in the experiment area, at layers 0 - 0.20 and 0.20 - 0.40 m deep.

Physical and water characteristics									
Layer	Granulometry (g·kg^{-1})			θ_{CC}	θ_{PMP}	Ds	PT	Texture classification	
(m)	Sand	Silt	Clay	m^3·m^{-3}		g·cm^{-3}	cm^3·cm^{-3}		
0 - 0.20	458.3	150.2	391.5	51.83	30.5	1.27	0.55	Loamy-clayey	
0.20 - 0.40	374.9	158.3	466.8	55	31.33	1.28	0.51	Clay	

Chemical characteristics										
Layer	pH	MO	P	K	Ca	Mg	Al	H + Al	S	CTC	V
(m)	in H$_2$0	(g·kg^{-1})	(mg·dm$^{-3)}$)				(mmol·dm^{-3})				(%)
0.00 - 0.20	6.2	63.42	7.06	2.04	20.40	16.80	0.0	57.75	41.80	99.55	41.99
0.20 - 0.40	6.6	44.47	2.65	4.09	14.40	13.20	0.0	44.55	31.69	76.24	41.57

θ_{CC}, field capacity (10 KPa); θ_{PMP}, permanent wilting site (1.500 KPa); Ds, soil density; PT, total porosity; pH in distilled water. P and K, extractor Mehlich^{-1}. M.O—organic Matter. V—Saturation per base.

first cutting, the first ratoon (on which current study focuses) was harvested (May 2013).

2.2. Experimental Design

The experimental design comprised 5 × 2 randomized blocks, with four replications; treatments consisted of a combination of 5 levels of water replacement (100%, 75%, 50%, 25% and 0% of field capacity), with or with the application of N-fertilizer (0 and 100 kg·ha^{-1}; urea).

2.3. Irrigation Management

Irrigation system consisted of a drip tube 20 cm deep, between two furrows, featuring the following characteristics: drip tube (model Dripnet PC 16150), with thin walls; pressure at 1.0 bar; nominal discharge 1.0 L·h^{-1}; spacing between drips equivalent to 0.50 m.

Irrigation was conducted by 0.1 kPa sensitive digital tension meter and tension probes were installed at depths 0.20, 0.40, 0.60 and 0.80 m, at 0.15, 0.30, 0.45 and 0.60 m distance from the drip tube. Critical tension 40 kPa and depth 0.40 m were employed to assess irrigation needs. The physical and hydric characteristics of the soil were determined by the water retention curve in the soil according equation van Genuchten [17], was developed to convert measured Ψ_m into water contents in the soil (θ), minimizing the sum of the squares of the deviations, with software RETEC, to obtain the parameters of adjustment used in Equation (1):

$$\theta = \frac{0.5643}{\left[1+\left(0.2933\left|\Psi_m\right|\right)^{1.4937}\right]^{0.330522}} \tag{1}$$

where

θ = contents of water in the soil (g·g^{-3});

Ψ_m = matrix potential (mca).

Daily results of water content in the soil determined applied water volume for each level of water replacement. The 100% treatments were based on the rise of soil moisture for field capacity.

Treatments received 120 kg·ha^{-1} P$_2$O$_5$ (source, simple superphosphate) and 80 kg·ha^{-1} K$_2$O (source, potassium oxide), following the soil's chemical analysis and recommendations by Sousa & Lobato [18]. The Phosphate fertilizers were applied in the furrows in its entirety in planting, whereas nitrogen and potassium (parceled out in four times) were placed on the coverage in treatment 0% of water replacement with irrigation water throughout the cane ratoon development cycle in water replacement treatments. Ten equal N applications, equivalent to 10 kg, were applied throughout the crop cycle, with a total of 100 kg·ha^{-1}.

At the end of the experiment, the water supply in the soil was measured to determine the volume of water supplied. Rates amounted to 0, 115, 230, 345 and 460 mm for water replacement of 0%, 25%, 50%, 75% and 100% of soil moisture. Effective rainfall (1015.33 mm) during the experiment was also calculated (**Figure 1**). A ten-day water balance for sugarcane management was estimated, based on climatological data during the experimental period, according to method by Thornthwaite & Mather [19]. Reference Evapotranspiration (ET_0) was calculated by Penman-Monteith equation [20].

2.4. Variables Review

The entire area of each split was harvested. Stems were weighed to determine productivity (SP, $t \cdot ha^{-1}$); pointers and straw were weighed to determine productivity of pointers (PP, $t \cdot ha^{-1}$) and straw (PS, $t \cdot ha^{-1}$), respectively. Rates were obtained by the proportional relationship of the harvested area and calculated for one hectare. Weighing was done by a 100 kg clock-type scale, subdivided into 500 g. Methodology derived from several studies on sugarcane crop [22] [23].

So that the dry matter of the stem (DMS, $t \cdot ha^{-1}$) and the dry matter of the pointer (DMP, $t \cdot ha^{-1}$) could be calculated, the samples were harvested and dried in a forced air-circulation buffer at 65°C - 70°C and then weighted in a precision balance with maximum capacity 3200 g and 0.01 g precision.

Total dry matter of the aerial segment (TDM, $t \cdot ha^{-1}$) was calculated by the sum of DMS and DMP, whilst the relationship between the dry matter of the pointer and stem (DMP/DMS) was calculated by dividing DMP by DMS, as described above.

Harvest Index (HI) measured the efficiency of the conversion of products synthetized in material of economic relevance, following Marafon [24], by Equation (2). In the case of mature culture, HI is defined as the ratio between dry matter mass of the economically profitable fraction (economic DM), or rather, the stems of the sugarcane crop, and TDM of the aerial segment.

$$HI = \frac{DM_{economic}}{TDM} \tag{2}$$

where

HI = harvest Index;
$DM_{economic}$ = economic Dry Matter ($t \cdot ha^{-1}$);
TDM = total dry matter of the aerial segment ($t \cdot ha^{-1}$).

2.5. Statistical Analysis

Data were submitted to analysis of variance by F-test at 0.05 probability; in significant cases, analysis of linear

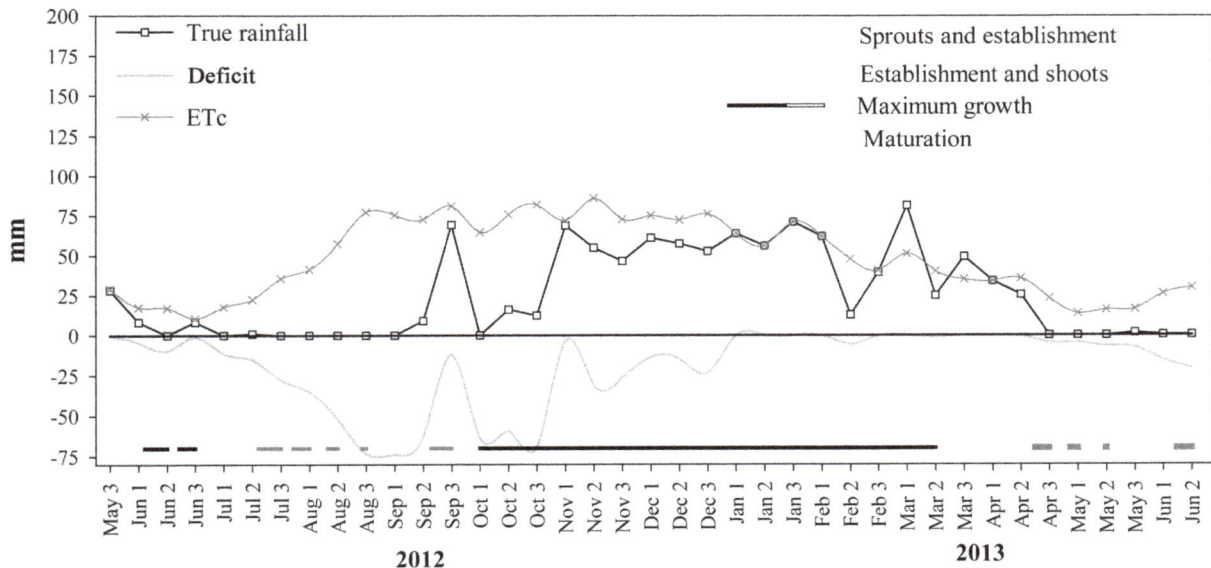

ETc—Evapotranspiration of crop. Crop phases (adapted from doorenbos and Kassam [21]): Sprouts and Establishment (Kc = 0.6); Establishment and shoots (Kc = 0.9 to 1.1); Maximum growth (Kc = 1.3); Maturation (Kc = 0.7 to 0.9).

Figure 1. Ten-day water balance in Rio Verde during the experimental period (sugarcane ratoon). Source: FESURV Meteorological Station—Universidade de Rio Verde, Rio Verde GO., Brazil.

and quadratic polynomial regression was undertaken to calculate water replacement levels of; in the case of N application, means were compared by Tukey's test at 0.05 probability. Statistical software SISVAR-ESAL was employed [25].

3. Results and Discussion

Table 2 shows that water replacement affected significantly all the variables under analysis, with the exception of PS. In N applications, there were significant differences for stem productivity (SP) and harvest index (HI), but no significant interaction between the factors studied for the assessed variables. There was an increase of 17.85 t·ha^{-1} for SP as a response for the application of N-fertilizer when compared to N-less treatments, where a 2.5% increase occurred for HI. As reported in current study and in several research works, sugarcane ratoon responds favorably to nitrogen application. Souza [26], reports that treatment without nitrogen fertilization had an increase of 5.04 and 7.41 t·ha^{-1}, when compared to 120 and 240 kg·ha^{-1} N doses, or rather, lower than the increase reported in current study. The result can be attributed to differences between cultivars, or even by soil type and/or nitrogen source applied.

Rípoli *et al.* [27] reported that sugarcane produces between 9% and 14% of straw (dry matter base), with approximately 10 t·ha^{-1}. Although not affected by evaluated factors, straw production in current analysis ranged between 15.92 and 17.16 t·ha^{-1}, higher than rates by the former authors.

The variables stem productivity and pointer productivity had a linear behavior for WR (**Figure 2**). There was a 22.6 t·ha^{-1} increase in stem production for a 25% increase of WR, with the highest productivity at 233.4 t·ha^{-1} stems for 100% WR (**Figure 2(a)**). The above WR is equivalent to a water depth of 1475.33 mm (100% WR plus true rainfall). The above rates were higher than those by Haddad *et al.* [28] who studied stem productivity and sugar yields of sugarcane according to water depth. They also registered productivity rise in proportion to increase in water depth, with maximum rate at 125.9 t·ha^{-1} for the application of 1854.4 mm. Difference in productivity may have been caused by better water distribution throughout the sugarcane cycle coupled to different fertilizations, management, climate and varieties. However, results were similar to those given by Andrade Júnior *et al.* [29] and Barbosa *et al.* [30] with subsurface drip irrigation.

There was a 2.1 t·ha^{-1} increase of pointer productivity for every 25% WR increase, with highest productivity at 48.8 t·ha^{-1} at 100% water depth (**Figure 2(b)**). Increase in the productivity of pointers enhanced stem productivity, which may be due to the source-drain ratio, where the pointer is the energy source for the stems (drain), and to the unit increase (UI) of WR. In fact, UI was three times greater for stem productivity than UI for point productivity, and thus the greater development of the pointer caused a greater production of energy used by the sugarcane stems.

Table 2. Summary of the analysis of variance for stem productivity (SP), straw productivity (PS), productivity of pointer (PP) and harvest index (HI) of sugarcane at different levels of water replacement (WR), with or without N application.

Variation source	DF	Mean square			
		SP	PS	PP	HI
Water replacement (WR)	4	10541.56**	4.44^{ns}	104.34**	26.73**
Nitrogen (N)	1	3186.40**	15.44^{ns}	1.38^{ns}	39.76*
Interaction WR × N	4	96.98^{ns}	12.71^{ns}	30.73^{ns}	1.55^{ns}
Blocks	3	43.43^{ns}	25.75^{ns}	27.12^{ns}	4.95^{ns}
Residue	27	182.68	14.16	12.02	6.68
CV (%)		7.18	22.76	7.77	3.20
Nitrogen (N)		Means (t·ha^{-1})		Mean (%)	
With		197.06 a	17.16 a	44.46 a	81.90 a
without		179.21 b	15.92 a	44.83 a	79.90 b
LSD		8.76	2.44	2.23	1.68
SE		3.02	0.84	0.77	0.58

**Significant at 0.01 probability by F-test; *Significant at 0.05 probability by F-test; ^{ns} not significant at 0.05 probability by F-test; Means followed by the same letter in the columns do not differ statistically at 0.05 probability by Tukey's test; LSD (least significant difference); SE (standard errors); DF (Degrees of freedom); SP (stem productivity); PS (straw productivity); PP (productivity of pointer); HI (harvest index).

Results for the different levels of water replacements show that the harvest index (HI) had a quadratic behavior (**Figure 3(a)**). HI results increased up to 70.22% at an estimated rate of 82.63; they decreased in proportion to water replacement increase. According to Doorenbos and Kassam [31], 80% may be adopted for sugarcane crops since it is close to obtained results. In fact, rates varied between 77.94 and 82.63, following water replacements.

Table 3 shows that variables DMS, DMP and DMP/DMS were significantly affected by WR. Application of N caused a significant effect on variables DMS and DMP/DMS, with an increase of 8.16 t·ha^{-1} and 13.64% for the respective variables, when compared to non-application of N. Total dry matter (TDM) of the aerial segment was the only variable which significantly differed with regard to WR × N.

Cantarella [32] assessed sugarcane response to N application in a three-year period and verified that the production of the dry matter of the stem was significantly higher in urea-fertilized crops (100 kg·ha^{-1} N). Moreover, the residual effect of fertilization was also confirmed in the following year. Results demonstrate that the application of N in sugarcane crops increases the dry matter of the stem and final productivity.

Variables DMP and DMS exhibited a linear behavior for WR (**Figure 4**). The amount of dry matter accumulated in the pointer (DMP) due to water replacement varied between 10.96 and 14.77 t·ha^{-1}. There was a 0.35% increase in DMP for every 1% increase in WP, or rather, a gain of 0.038 t·ha^{-1} (**Figure 4(a)**). Therefore, the accumulation of dry matter in the pointer increases in proportion to water increase. Oliveira *et al.* [33] reported rates between 7.5 and 12.0 t·ha^{-1} for DMP in irrigated sugarcane varieties, similar to that in current analysis.

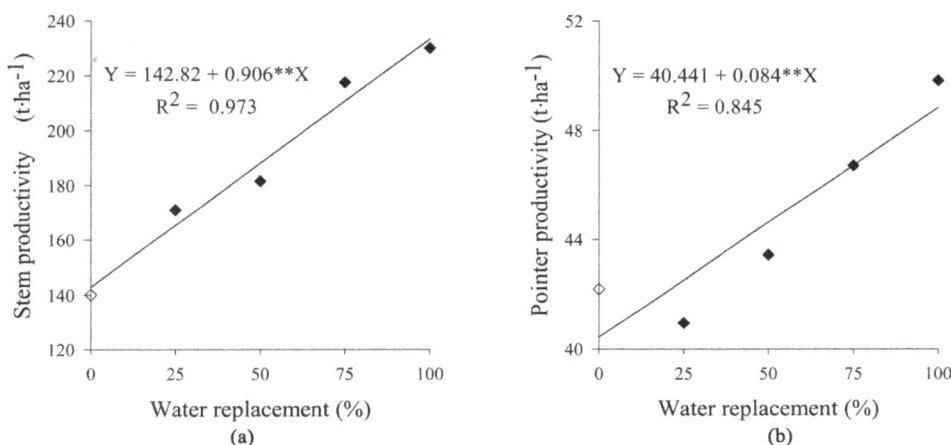

$**$ and $*$ respectively significant at p < 0.01 and 0.05, according to F-test.

Figure 2. Productivity of sugarcane stems (a) and pointers (b), due to water replacement.

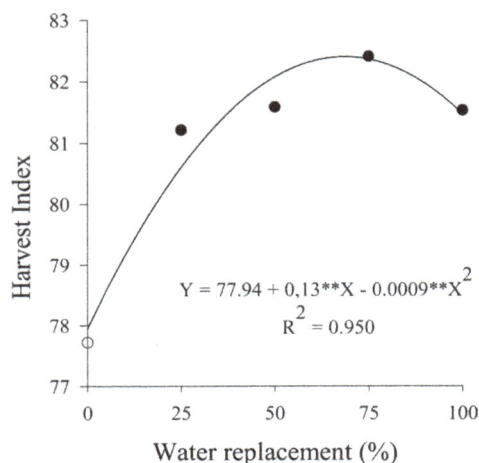

$**$ and $*$ significant respectively at p < 0.01 and 0.05, according to F-test.

Figure 3. Harvest Index of sugarcane according to water replacement.

Table 3. Summary of the analysis of variance for dry matter of stem (DMS), dry matter of pointer (DMP), DMP/DMS ratio and total dry matter from the aerial segment (TDM), at different levels of water replacement, with and without the application of Nitrogen.

Variation source	DF	Mean square			
		DMS	DMP	DMP/DMS	TDM
Water replacement (WR)	4	1004.48**	21.67**	0.0068**	1292.47ns
Nitrogen (N)	1	666.26**	0.26ns	0.0096*	693.06ns
WR × N interaction	4	71.61ns	1.28ns	0.0004ns	86.22*
Blocks	3	20.52ns	6.37ns	0.0012ns	45.75ns
Residues	27	32.77	2.18	0.0017	32.08
CV (%)		10.29	11.49	17.51	8.27
Nitrogen (N)		Means (t·ha⁻¹)		Means	Means (t·ha⁻¹)
With		59.73 a	12.94 a	0.25 a	72.67 a
Without		51.57 b	12.78 a	0.22 b	64.35 b
LSD		3.71	0.96	0.03	3.67
SE		1.28	0.33	0.009	1.27

**Significant at 0.01 probability by F-test; *Significant at 0.05 probability by F-test; ns not significant at 0.05 probability by F-test; Means followed by the same letter in the columns do not differ statistically at 0.05 probability by Tukey's test; LSD (least significant difference); SE (standard errors); DF (degrees of freedom); DMS (dry matter of stem); DMP (dry matter of pointer); TDM (total dry matter from the aerial segment).

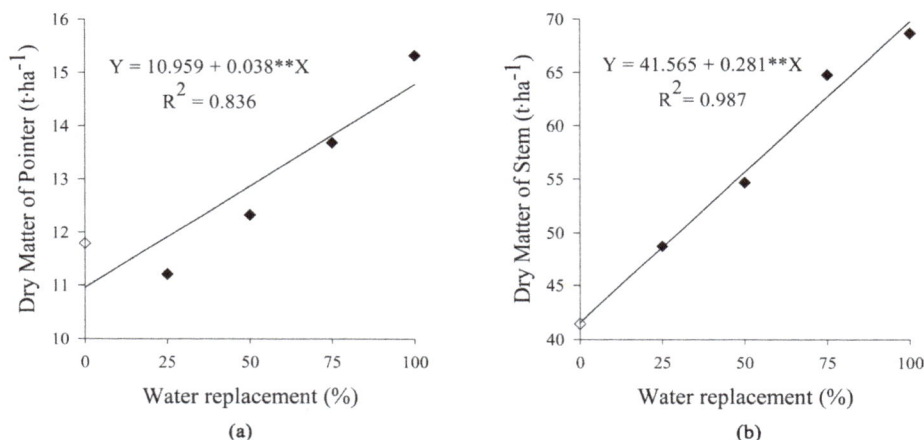

** and *respectively significant at $p < 0.01$ and 0.05, according to F-test.

Figure 4. Dry matter of sugarcane pointer (a) and dry matter of stem (b) according to water replacement.

Every 1% increase in WR, the increment for DMS is of 0.28 t·ha⁻¹ (**Figure 4(b)**), corroborating results by Farias *et al.* [34] who reported that water supply through irrigation provided an increase in the accumulation rate of the sugarcane stem's dry matter. Irrigated water supply for crops increases the production of dry matter of stem and pointer, underscoring that DMS production is 1.94 times greater than DMP production when there is a 1% WR increase by irrigation.

Therefore, irrigation increases the production of dry matter in the stem and, therefore, in sugarcane productivity. The above justifies the practice, especially when one observes in **Figure 1** that true rainfall supplied the requirements of the crop (ETc) only during approximately 70 days within the entire cycle.

DMP/DMS ratio (**Figure 5**) had a quadratic behavior: all rates decreased till 66.6%; from this point, rates increased due to WR. Relationship between pointer and stem revealed that the decrease of this index caused a proportional increase in the production of stems and a decrease in the production of pointers. Machado *et al.* [35] reported that after 100 days the dry matter of leaves and of pointer provided 70% of the plant's biomass and that, after this period, there was a progressive decrease in the accumulation of dry matter, or rather, 9% of dry matter accumulated by the plant at the end of the cycle.

Figure 6 shows total dry matter from the aerial segment behavior in sugarcane ratoon: when nitrogen was applied, productivities were estimated at 53.13; 62.88; 72.63; 82.38; 92.13 $t \cdot ha^{-1}$ respectively for water replacement 0%, 25%, 50%, 75% and 100%; without the application of nitrogen, productivities were 51.91; 58.11; 64.31; 70.51; 76.71 $t \cdot ha^{-1}$, respectively for the same WR percentages. All rates were higher in treatments with nitrogen, or rather, increase in water irrigation depth provided a proportionate increase in the difference between rates. In other words, the amount of available water for the plant affected the absorption of nitrogen.

Figure 6(a) demonstrates that the evolution of the WR × N interaction increased linearly the productivity of total dry matter of the aerial segment. With regard to the evolution of WR for each N level, a 0.39 and 0.25 $t \cdot ha^{-1}$ increase occurred in the production of TDM in treatments with and without N for every 1% increase in WR. Consequently, irrigated sugarcane crops were more efficient in increasing TDM production when N-fertilizer was applied.

Above results disagree with those given by Arantes [36] who evaluated the production capacity of sugarcane cultivars with and without irrigation and reported but the cause of different results can be well attributed the cultivars used in the study of Arantes, since none were the same used in the current study. That there occurred no significant effect on the accumulation of total dry matter in each evaluation period.

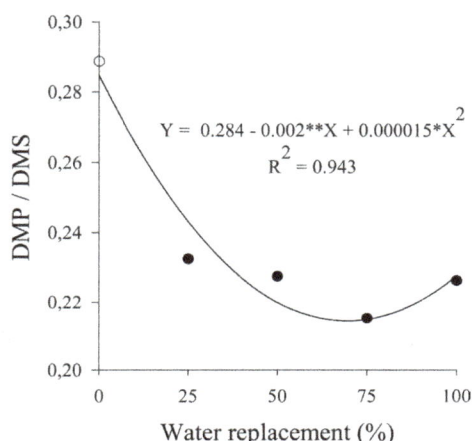

$$Y = 0.284 - 0.002^{**}X + 0.000015^{*}X^{2}$$
$$R^{2} = 0.943$$

** and *significant respectively at $p < 0.01$ and 0.05, according to F-test.

Figure 5. Harvest index of sugarcane according to water replacement.

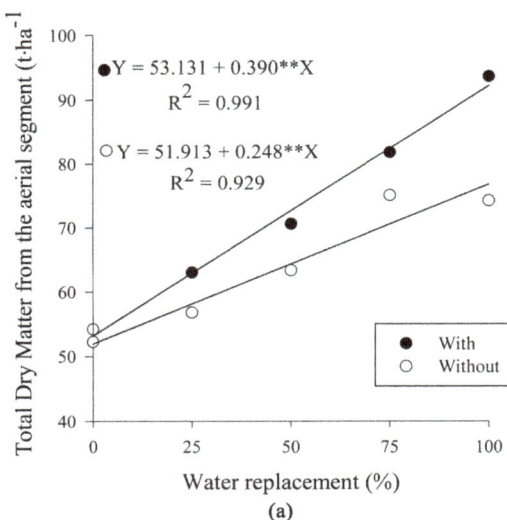

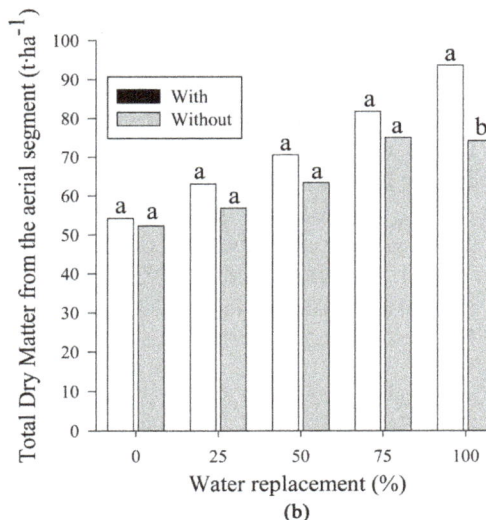

$$Y = 53.131 + 0.390^{**}X$$
$$R^{2} = 0.991$$

$$Y = 51.913 + 0.248^{**}X$$
$$R^{2} = 0.929$$

(a)   (b)

* and *respectively significant at $p < 0.01$ and 0.05, according to F-test.

Figure 6. Total dry matter of the sugarcane aerial segment due to water replacement for each application of nitrogen (a) and nitrogen application for each water replacement (b).

There was only a significant difference for TDM in WR 100% with a 20.1% increase, according to the supply of the nutrient (**Figure 6(b)**). As a rule, there was a 12.93% gain in the application of N at WR levels (**Table 3**).

Otto *et al.* [37] assessed the phytomass of roots and the aerial segment of sugarcane related to N fertilization and reported that splits which did not receive N fertilization showed great energy expenditure in the growth of the roots. This fact established a competition for photoassimilates in the growth of the aerial segment and also explained the low TDM rates in treatments without N-fertilizers in current assay.

4. Conclusions

Increase in water replacements significantly raised the productivity of dry matter in stems and pointers by the sugarcane.

There was an increase in harvest index up to water replacement 70.22%; for the DMP/DMS ratio, harvest index decreased up to water replacement 66.6%.

There was an increase of 90.61 $t \cdot ha^{-1}$ in the productivity of stems and 8.40 $t \cdot ha^{-1}$ in the productivity of pointers between treatments water replacement 0% and water replacement 100%.

Nitrogen fertilization significantly increased the productivity of stems, harvest index, dry matter of stem and DMS/DMP ratio.

Nitrogen fertilization increased productivity by 8.16 $t \cdot ha^{-1}$ for dry matter and 20.85 $t \cdot ha^{-1}$ for stems; in the case of DMP/DMS, increase reached 13.64% and 2.5% for harvest index.

Total dry matter of the aerial segment was affected by RH × N interaction: highest rates occurred in water replacement 100% with the application of nitrogen. Straw productivity was not affected by any of the factors evaluated.

Acknowledgements

The authors would like to thank the Brazilian National Council for Scientific and Technological development (CNPq) for funding current assay. Thanks are also due to the Brazilian Coordination for the Updating of Higher Education Personnel (CAPES), to the Foundation for Research in the State of Goiás (FAPEG), to Financing Agent for Studies and Projects (FINEP) and to the Federal Institute of Goiás, Campus Rio Verde GO Brazil.

References

[1] (2015) CONAB—Companhia Nacional do Abastecimento. Acomp. da Safra Bras. de Cana-de-açúcar, V. 1—Safra 2014/15, N. 3—Terceiro Levantamento, Brasília, dez. 2014. 28 p. http://Www.Conab.Gov.Br

[2] Silva, C.T.S., Azevedo, H.M., Azevedo, C.A.V., Neto, J.D., Carvalho, C.M. and Filho, R.R.G. (2009) Crescimento da Cana-de-açúcar Com E Sem Irrigação Complementar Sob Diferentes Níveis de Adubação de Cobertura Nitrogenada e Potássica. *Revista Brasileira de Agricultura Irrigada*, **3**, 3-12. http://dx.doi.org/10.7127/rbai.v3n100012

[3] Albuquerque, G.A.C. and Marinho, M.L. (1983) Adubação na Região Norte-Nordeste. In: Orlando Filho, J., Coord., *Nutrição de Adubação da Cana-de-açúcar no Brasil*. Instituto do Açúcar E do Álcool, Piracicaba, 267-286.

[4] Maeda, A.S., Buzetti, S. and Bolonhezi, A.C. (2009) Adubação Nitrogenada e Potássica na Qualidade e Produtividade da Cana-Soca. *Stab*, **27**, 40-45.

[5] Teodoro, I.T., Neto, J.D., Souza, J.L., Lira, G.B., Brito, K.S., Sá, L.A., Santos, M.A.L. and Sarmento, P.L.V.S. (2013) Isoquantas de Produtividade da Cana-de-açúcar em Função de Níveis de Irrigação e Adubação Nitrogenada. *Irriga*, **18**, 387-401. http://dx.doi.org/10.15809/irriga.2013v18n3p387

[6] Cantarella, H., Trivelin, P.C.O. and Vitti, A.C. (2007) Nitrogênio e Enxofre na Cultura da Cana-de-açúcar. In: Yamada, T., Abdalla, S.R.S. and Vitti, G.C., Eds., *Nitrogênio e Enxofre na Agricultura Brasileira*, Ipni Brasil, Piracicaba, 355-412.

[7] Christoffoleti, P.J., Carvalho, S.J.P. de, López-Ovejero, R.F., Nicolai, M., Hidalgo, E. and Silva, J.E. (2007) Conservation of Natural Resources in Brazilian Agriculture: Implications on Weed Biology and Management. *Crop Protection*, **26**, 383-389. http://dx.doi.org/10.1016/j.cropro.2005.06.013

[8] Aquino, G.S. de and Medina, C. de C. (2014) Produtividade e Índices Biométricos e Fisiológicos de Cana-de-Açúcar Cultivada Sob Diferentes Quantidades de Palhada. *Pesquisa Agropecuária Brasileira*, **49**, 173-180. http://dx.doi.org/10.1590/S0100-204X2014000300003

[9] Faroni, C.E., Trivelin, P.C.O., Silva, P.H., Bologna, I.R., Vitti, A.C. and Franco, H.C.J. (2007) Marcação de Fitomassa de Cana-de-açúcar Com Aplicação de Solução de Uréia Marcada Com 15n. *Pesquisa Agropecuária Brasileira*, **42**,

851-857. http://dx.doi.org/10.1590/S0100-204X2007000600012

[10] Thorburn, P.J., dart, I.K., Biggs, I.J., Baillie, C.P., Smith, M.A. and Keating, B.A. (2002) The Fate of Nitrogen Applied to Sugarcane by Trickle Irrigation. *Irrigation Science*, In Press.

[11] Trivelin, P.C.O., Victoria, R.L. and Rodriguês, J.C.S. (1995) Aproveitamento Por Soqueira de Cana-de-açúcar de Final de Safra do Nitrogênio da Aquamônia ^{15}n E Ureia ^{15}n Aplicado ao Solo em Complemento à Vinhaça. *Pesquisa Agropecuária Brasileira*, **30**, 1375-1385.

[12] Vitti, A.C., Franco, H.C.J., Trivelin, P.C.O., Ferreira, D.A., Otto, R., Fortes, C. and Faroni, C.E. (2011) Nitrogênio Proveniente da Adubação Nitrogenada e de Resíduos Culturais na Nutrição da Cana-Planta. *Pesquisa Agropecuária Brasileira*, **46**, 287-293. http://dx.doi.org/10.1590/S0100-204X2011000300009

[13] Basanta, M.V., dourado Neto, D., Reichardt, K., Bacchi, O.O.S., Oliviera, J.C.M., Trivelin, P.C.O., Timm, L.C., Tominaga, T.T., Correchel, V., Cássaro, F., Pires, L.F. and Macedo, J.R. (2003) Management Effects on Nitrogen Recovery In a Sugarcane Crop Grown in Brazil. *Geoderma*, **116**, 235-248. http://dx.doi.org/10.1016/S0016-7061(03)00103-4

[14] Castro Neto, P. (1982) Notas de Aula Prática do Curso de Agrometeorologia. Lavras, Esal, 45 p.

[15] Soil Survey Staff (2006) Keys to Soil Taxonomy. Tenth Edition, United States Department of Agriculture, Natural Resources Conservation Services.

[16] Embrapa Solos (2013) Empresa Brasileira de Pesquisa Agropecuária. Centro Nacional de Pesquisa de Solos. Sistema Brasileiro de Classificação de Solos. Rio de Janeiro.

[17] Van Genuchten M.T., Leij, F.J. and Yates, S.R. (2009) RETC, Code for Quantifying the Hydraulic Functions of Unsaturated Soils: Version 6.02. University of California, Riverside.

[18] Sousa, D.M.G. and Lobato, E. (2004) Cerrado: Correção do Solo E Adubação. 2nd Edition, Embrapa Informação Tecnológica, Brasília, 416 p.

[19] Thornthwaite, C.W. and Mather, J.R. (1955) The Water Balance, Laboratory of Climatology. *Centerton*, **8**, 1-14.

[20] Monteith, J.L. (1973) Principles of Environmental Physics. Edward Arnold, London, 241 p.

[21] Doorenbos, J. and Kassam, A.H. (1994) Efeito da Água no Rendimento das Culturas. Ufpb, Campina Grande, 306 p. (Irrigação e Drenagem, Estudos Fao, 33).

[22] Silva, T.G.F., de Moura, M.S., Zolnier, S., Soares, J.M., Vieira, V.J.deS. and Júnior, W.F.G. (2011) Demanda Hídrica E Eficiência do Uso da Água da Cana-de-açúcar Irrigada no Semi-Árido Brasileiro. *Revista Brasileira de Engenharia Agrícola e Ambiental*, **15**, 1257-1265. http://dx.doi.org/10.1590/S1415-43662011001200007

[23] Barros, A.C., Coelho, R.D., Marin, F.R., Polzer, D.L. and Netto, A.deO.A. (2012) Utilização do Modelo Canegro Para Estimativa de Crescimento da Cana-de-açúcar Irrigada E Não Irrigada Para As Regiões de Gurupi—To e Teresina—Pi. *Irriga*, **17**, 189-207. http://dx.doi.org/10.15809/irriga.2012v17n2p189

[24] Marafon, A.C. (2012) Análise Quantitativa de Crescimento em Cana-de-açúcar: Uma Introdução ao Procedimento Prático. Se. 1ª Ed., Embrapa Tabuleiros Costeiros, Aracaju.

[25] Ferreira, D.F. (2011) Sisvar: A Computerstatisticalanalysis System. *Ciência E Agrotecnologia*, **35**, 1039-1042.

[26] de Souza, J.P.S.P. (2012) Resposta da Cana-de-açúcar a Adubação Nitrogenada com Ureia Convencional e Revestida em Solo de Cerrado na Fazenda Água Limpa. Faculdade de Agronomia e Medicina Veterinária,Universidade de Brasília, Monografia (Agronomia), 34 p.

[27] Rípoli, T.C.C. and Rípoli, M.L.C. (2004) Biomassa de Cana-de-açúcar: Colheita, Energia e Ambiente. Usp/Esalq, Piracicaba, 302 p.

[28] Haddad, G.S.V., Mantovani, E.C., Sediyama, G.C., da Costa, E.L. and delazari, F.T. (2012) Produtividade de Colmos E Rendimento de Açúcares da Cana-de-açúcar em Função de Lâminas de Água. *Irriga*, **17**, 234-244. http://dx.doi.org/10.15809/irriga.2012v17n2p234

[29] de Andrade Jr., A.S., Bastos, E.A., Ribeiro, V.Q., Duarte, J.A.L., Braga, D.L. and Noleto, D.H. (2012) Níveis de Água, Nitrogênio e Potássio por Gotejamento Subsuperficial em Cana-de-açúcar. *Pesquisa Agropecuária Brasileira*, **47**, 76-84. http://dx.doi.org/10.1590/S0100-204X2012000100011

[30] Barbosa, E.A.A., Arruda, F.B., Pires, R.C.M., da Silva, T.J.A. and Sakai, E. (2012) Cana-de-açúcar Fertirrigada com Vinhaça e Adubos Minerais Via Irrigação Por Gotejamento Subsuperficial: Ciclo Dda Cana-Planta. *Revista Brasileira de Engenharia Agrícola e Ambiental*, **16**, 952-958. http://dx.doi.org/10.1590/S1415-43662012000900005

[31] Doorenbos, J. and Kassam, A.H. (1979) Yield Response to Water. FAO Irrigation and Drainage Paper, No. 33. FAO, Rome, 172 p.

[32] Cantarella, H. (2012) Avaliação de Resposta a N em Cana-de-açúcar Não Adubada Por Três Anos. (Relatório), Instituto Agronômico Centro ee Solos e Recursos Ambientais, Agrisus, Pesquisa Agronômica, 719/10.

[33] de Oliveira, E.C.A., de Oliveira, R.I., de Andrade, B.M.T., Freire, F.J., Júnior, M.A.L. and Machado, P.R. (2010)

Crescimento e Acúmulo de Matéria Seca em Variedades de Cana-de-açúcar Cultivadas Sob Irrigação Plena. *Revista Brasileira de Engenharia Agrícola e Ambiental*, **14**, 951-960. http://dx.doi.org/10.1590/S1415-43662010000900007

[34] de Farias, C.H.A., Fernandes, P.D., de Azevedo, H.M. and Dantas Neto, J. (2008) Índices de Crescimento da Cana-de-açúcar Irrigada e de Sequeiro no Estado a Paraíba. *Revista Brasileira de Engenharia Agrícola e Ambiental*, **12**, 356-362. http://dx.doi.org/10.1590/S1415-43662008000400004

[35] Machado, E.C., Pereira, A.R., Fahl, J.I., Arruda, J.V. and Cione, J. (1982) Índices Biométricos de Duas Cultivares de Cana-de-açúcar. *Pesquisa Agropecuária Brasileira*, **17**, 1323-1329.

[36] Arantes, M.T. (2012) Potencial Produtivo de Cultivares de Cana-de-açúcar Sob os Manejos Irrigado e Sequeiro. Dissertação (Mestrado em Agronomia), Universidade Estadual Paulista Faculdade de Ciências Agronômicas, Botucatu, 65 p.

[37] Otto, R., Franco, H.C.J., Faroni, C.E., Vitti, A.C. and Trivelin, P.C.O. (2009) Fitomassa de Raízes e da Parte Aérea da Cana-de-açúcar Relacionada à Adubação Nitrogenada de Plantio. *Pesquisa Agropecuária Brasileira*, **44**, 398-405. http://dx.doi.org/10.1590/S0100-204X2009000400010

17

Effect of Cultivar, Irrigation and Nitrogen Fertilization on Chickpea (*Cicer arietinum* L.) Productivity

Kico Dhima[1*], Ioannis Vasilakoglou[2], Stefanos Stefanou[1], Ilias Eleftherohorinos[3]

[1]Department of Agricultural Technology, Technological Educational Institute of Thessaloniki, Thessaloniki, Greece
[2]Department of Agricultural Technology, Technological Educational Institute of Thessaly, Larissa, Greece
[3]School of Agriculture, Faculty of Agriculture, Forestry and Natural Environment, Aristotle University of Thessaloniki, Thessaloniki, Greece
Email: *dimas@cp.teithe.gr

Abstract

A 2-year field study was conducted in northern Greece to investigate the effect of nitrogen fertilization and irrigation on productivity of three Greek chickpea varieties ("Amorgos" "Serifos", "Andros"). Chickpea, grown under irrigation regime (30 + 30 mm of water) and fertilized with 50 kg·N·ha^{-1} before planting and with 40 kg·N·ha^{-1} at blossom growth stage, produced more total dry biomass and seed yield as compared with that grown under non-irrigated conditions and fertilized with 50 kg·N·ha^{-1} before planting only. In particular, irrigation and nitrogen fertilization at blossom growth stage increased total dry weight of chickpea by 18.3% and 18.5%, respectively, as compared with that of non-irrigated and fertilized with N before planting. The corresponding increase of seed yield was 30.5% and 20%, respectively. The total dry biomass of "Amorgos" was 10% and 13% greater than that of "Serifos" and "Andros", while its respective seed yield increase was 5% and 16%. Finally, the quantum yield of photosystem II of chickpea was not affected by irrigation or fertilization. These results indicated that nitrogen fertilization at blossom growth stage combined with irrigation increased seed yield of all chickpea varieties, whereas the same treatments did not have any effect on plant quantum yield of photosystem II.

Keywords

Chickpea (*Cicer arietinum* L.), Irrigation, Nitrogen Fertilization, Dry Biomass, Seed Yield, Quantum Yield of Photosystem II

---

*Corresponding author.

1. Introduction

Chickpea (*Cicer arietinum* L.) is one of the most important grain legumes as it ranks third in the world after dry bean (*Phaseolus vulgaris* L.) and field pea (*Pisum sativum* L.) [1]. Its good source for protein, complex carbohydrates, fibre, vitamins and minerals [2] [3] makes this legume an important component of human diet in developing world [4] [5]. Generally, its protein quality is higher than that of many other legumes [6].

Chickpea is a significant contributor to agricultural sustainability due to its nitrogen fixation ability and for this reason is considered a good rotational crop [1]. Its presence improves soil health by promoting microbial population and activity [7] [8]. Although chickpea fixes nitrogen from atmosphere, there is strong evidence that nitrogen fertilization increases seed yield, seed protein and amino acids [9]-[12]. However, its requirements for nitrogen fertilization are lower than other crops to obtain higher yield and improved seed quality [13].

Chickpea is grown on a wide range of environments, from the subtropics (India and North-eastern Australia) to arid and semi-arid environments of Mediterranean climatic regions (Mediterranean basin and Southern Australia). Although several researchers [14]-[16] have reported that this crop can grow under environmental stress conditions such as drought, high temperatures and poor soils, Krishnamurthy *et al.* [17] found that drought stress reduced plant growth and yield by reducing leaf surface area and rate of photosynthesis. In addition, Leport *et al.* [18] showed that early water stress reduced total biomass and seed yield of chickpea, while Saraf *et al.* [19] stated that moisture excess or deficit caused significant yield reduction on this crop.

Taking into consideration the partial drought tolerance of chickpea along with its increasing grain legume value in sustainable agricultural systems [20], this crop can play an important role in the traditional semi-arid areas of Mediterranean basin. Furthermore, chickpea may be an important food security crop in the semi-arid and dry environments of northern Greece and it can serve as an important winter rotational crop in this region.

Based on the above findings that indicate the great importance of water and nitrogen availability on the potential growth and yield of chickpea [21], the objective of this research is to assess the effect of nitrogen fertilization at blossom growth stage on productivity of three Greek chickpea varieties ("Amorgos", "Serifos", "Andros") grown under irrigated or non-irrigated conditions.

2. Materials and Methods

2.1. Experimental Sites

Two chickpea field experiments were conducted in 2007/08 (Year 1) and 2009/10 (Year 2) at the Technological and Educational Institute Farm of Thessaloniki in northern Greece (22°44'10"E, 40°37'06"N, 0 - 1 m altitude). The experiments were established on a sandy loam (Typic Xeropsamment) soil with the following physico-chemical characteristics: sand 644 g·kg^{-1}, silt 280 g·kg^{-1}, clay 76 g·kg^{-1}, organic C content 5 g·kg^{-1} and pH (1:2 H_2O): 7.6. Soil analysis (0 - 30 cm soil depth) conducted before crop planting showed that initial nitrogen content ranged from 86 to 90 and 75 to 83 mg·kg^{-1} of soil in year 1 and year 2, respectively. The previous crop in year 1 and 2 was wheat and barley, respectively, whereas the mean monthly temperature and total monthly rainfall data recorded near the experimental locations are shown in **Figure 1**.

2.2. Treatments and Experimental Design

Three Greek chickpea varieties ("Amorgos" "Serifos" and "Andros") were planted by hand in 40-cm rows to achieve an approximately desired density of 500,000 plants ha^{-1}. The selected chickpea varieties are among the most commonly grown ones in Greece. The planting was performed November 02, 2007 (year 1) and November 14, 2008 (year 2). Two days before crop planting, 50 kg·N·ha^{-1} and 25 kg·P·ha^{-1} as diammonium thiophosphate (20-10-0) were dispersed uniformly and incorporated into the soil of the experimental area. In addition, 40 kg·N·ha^{-1}, as ammonium nitrate (33.5-0-0), were applied in half of the plots (namely as fertilized plots) at blossom growth stage. Weed control was achieved by pre-emergence applied pendimethalin at 1.65 kg·ai·ha^{-1} and by hand weeding.

A split-split-plot arrangement of treatments was employed for both experiments in a randomized complete block design. The three chickpea varieties were arranged as main plots, whereas the irrigation and nitrogen fertilization regimes were arranged as subplot and sub-subplot, respectively. The main plots (12 m × 3 m) were separated by a 3 m wide alley. Each main plot was divided into two irrigation subplots (5 m × 3 m) that were separated by a 2 m wide alley. Half of the subplots were irrigated two times (30 + 30 mm of water) during the

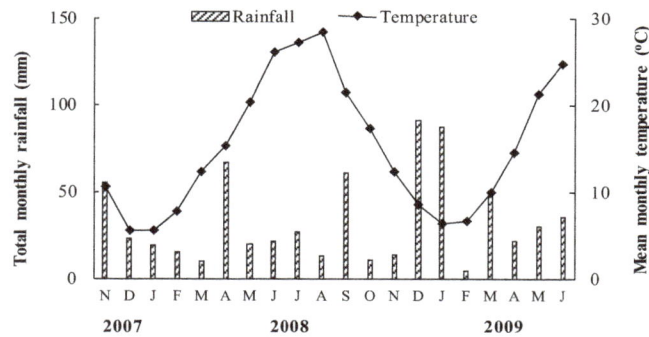

Figure 1. Mean monthly temperature and total monthly rainfall data.

blossom growth stage of crop and the other half was not irrigated. Drip irrigation was performed in rows located 1.5 m apart. The first irrigation was performed one week before nitrogen fertilization and the other two days later. Each subplot was divided into two (2 m × 3 m) sub-subplots. Half of the sub-subplots were fertilized with 40 kg·N·ha^{-1} at chickpea blossom growth stage and the other half was not fertilized. Each sub-subplot included five chickpea rows. There were four replicates for each combined treatment (variety × nitrogen fertilization × irrigation).

2.3. Measurements

Chickpea plants found in the two central rows of each plot were counted at 3 weeks after seeding (WAS). The quantum yield of photosystem II (Y) was also measured at early and late blossom growth stage of chickpea. In particular, the measurements of the chlorophyll fluorescence parameters were made using a chlorophyll fluorometer (MINI-PAM, Miniaturised Pulse-Amplitude-Modulated photosynthesis yield analyzer, Company Walz, Effeltrich, Germany). The measurement of light intensity was of 0.15 μmol·m^{-2}·s^{-1}, with a frequency of 0.6 kHz and a saturation pulse intensity of 16,000 μmol·m^{-2}·s^{-1} for 0.8 s. Two measurements per plant were made on the upper leaves of five marked plants in the center of each sub-subplot to determine fluorescence at steady-state (Fs) and the maximum fluorescence after saturation flash (Fm). Quantum yield of photosystem II (Y) was calculated using the equation: Y = (Fm' – Fs)/Fm. The average of the ten measurements per sub-subplot was used for further data analysis.

At harvest, total dry biomass, pod number, seed yield (at 14% seed moisture), and 1000-seed weight of chickpea were determined by hand-harvesting the chickpea plants of the two central rows (each 3 m long) of each sub-subplot. The plants from each sub-subplot were cut at ground level and, before determination of their total dry biomass and yield components, the samples were air-dried under shade conditions for 3 days and then oven-dried at 65°C for 24 h to constant weight.

2.4. Statistical Analysis

A combined across year analysis of variance (ANOVA) was performed for chickpea total dry biomass, pod number, seed yield, and 1000-seed weight data using a split-split plot factorial design (chickpea variety × irrigation × nitrogen fertilization). Also, a multivariate analysis of variance (MANOVA) was made for the obtained chickpea quantum yield of photosystem II (Y) data.

The Statistical Package for the Social Sciences [22] and the MSTAT program [23] were used for the MANOVA and ANOVA, respectively, whereas the Tukey's Honestly Significant Difference test procedures were used to detect and separate mean treatment differences at $P = 0.05$.

3. Results

3.1. Chickpea Emergence and Physiological Parameters

Chickpea emergence had been completed within three weeks after planting and the obtained crop density averaged 50 plants m^{-2} (data not shown), which reflects the desired plant density by the farmers.

Quantum yield of photosystem II (Y), determined at early and late blossom growth stage, did not indicate any significant differences due to chickpea varieties, irrigation or nitrogen fertilization (**Table 1**). However, all chickpea varieties, averaged across irrigation and fertilization regimes, indicated slightly greater yield of photosystem II (Y) at the early blossom growth stage than at the late one (**Table 1**).

3.2. Chickpea Total Dry Biomass

The chickpea total dry biomass indicated significant differences due to nitrogen fertilization ($P < 0.05$) and chickpea varieties ($P < 0.05$). In particular, chickpea (averaged across year, fertilization and varieties), grown in irrigated plots, provided 18% more total dry biomass than grown in non-irrigated plots, whereas 18.5% more dry biomass was obtained for chickpea (averaged across year, irrigation and varieties) grown in fertilized plots with 40 kg·N·ha^{-1} applied at blossom growth stage as compared with those grown in non-fertilized plots (fertilized only with 50 kg·N·ha^{-1} before crop planting) (**Table 2**). Also, "Andros" (averaged across year, irrigation and fertilization) produced 13% and 3% less total dry biomass than "Amorgos" and "Serifos", respectively (**Table 2**).

The three chickpea varieties, grown under any irrigated and fertilization conditions, produced more total dry biomass in year 2 than in year 1 (**Table 3**). "Amorgos", "Serifos" and "Andros" (averaged across year and fertilization) produced 21.4, 20% and 15% more dry biomass under irrigation regime than under non-irrigation, whereas their respective total dry biomass (averaged across year and irrigation), grown in fertilized plots, was 20.5, 19% and 18% higher than that grown in non-fertilized plots. "Amorgos", "Serifos", "Andros", averaged over years and grown in fertilized plots, produced 20%, 20% and 13% more dry biomass, respectively, than in non-fertilized plots, whereas their respective increase in irrigated and fertilized plots was 19%, 15% and 19% (**Table 4**). In non-irrigated and fertilized plots, '"Amorgos", "Serifos" and "Andros" produced 20.5, 22.1% and 24.1% more total dry biomass, respectively, than those grown under non-irrigated and non-fertilized plots (**Table 3**). Finally, in irrigated plots and fertilized plots, their respective increase of total dry biomass was 19%, 16.3% and 19.1%.

3.3. Chickpea Seed Yield

Pod number m^{-2} and 1000-seed weight were not significantly affected by variety, irrigation and nitrogen fertilization (data not shown), but seed yield was significantly affected by year ($P < 0.001$), nitrogen fertilization ($P < 0.001$), chickpea variety ($P < 0.001$), and by the interaction between irrigation × nitrogen fertilization ($P < 0.05$).

Table 1. Quantun yield of photosystem II (Y) as affected by irrigation, nitrogen fertilization and chickpea varieties. Means are averaged across two growing seasons (2007/08 and 2008/09).

Treatments	Y			
	Early blossom		Late blossom	
Irrigation				
With	0.630	a	0.600	a
Without	0.625	a	0.611	a
Nitrogen				
50 kg N[1]	0.633	a	0.613	a
50 kg N + 40 kg N[2]	0.622	a	0.598	a
Variety				
Amorgos	0.627	a	0.609	a
Serifos	0.633	a	0.601	a
Andros	0.623	a	0.606	a
CV%	7.7		6.7	

Means within each column followed by different letter indicate significant difference; according to Tukey's Honestly Significant Difference test at $P = 0.05$. [1]Fertilization before crop planting. [2]Fertilization at blossom growth stage.

Table 2. Total dry biomass and seed yield of chickpea as affected by irrigation, nitrogen fertilization and variety. Means are averaged across two growing seasons (2007/08 and 2008/09).

Treatments	Total dry biomass		Seed yield	
	t ha^{-1}			
Irrigation				
With	12.06	a	4.16	a
Without	9.85	b	2.89	b
Nitrogen				
50 kg N[1]	12.07	a	3.91	a
50 kg N + 40 kg N[2]	9.84	b	3.13	b
Variety				
Amorgos	11.84	a	3.79	a
Serifos	10.69	b	3.20	b
Andros	10.34	b	3.58	a
CV%	16.4		15.2	

Means within each column followed by different letter indicate significant difference; according to Tukey's Honestly Significant Difference test at $P = 0.05$. [1]Fertilization before crop planting. [2]Fertilization at blossom growth stage.

Table 3. Total dry biomass of three chickpea varieties as affected by irrigation and nitrogen fertilization during 2007/08 (year 1) and 2008/09 (year 2).

Chickpea variety	Irrigation (mm)							
	0				60			
	Nitrogen fertilization (kg·N·ha^{-1})							
	50[1]		50 + 40[2]		50		50 + 40	
Year 1	t ha^{-1}							
Amorgos	8.55	cdef	10.34	abcdef	13.36	abcd	14.73	a
Serifos	7.81	f	8.37	def	11.06	abcdef	12.28	abcdef
Andros	7.69	f	8.10	ef	10.64	abcdef	12.17	abcdef
Year 2								
Amorgos	10.12	abcdef	13.12	abcd	10.19	abcdef	14.33	ab
Serifos	8.89	cdef	13.06	abcde	10.57	abcdef	13.52	abc
Andros	9.87	abcdef	12.34	abcdef	9.35	bcdef	12.53	abcdef
CV, %	16.4							

Means with different letter indicate significant difference according to Tukey's Honestly Significant Difference test at $P = 0.05$. [1]Fertilization before crop planting. [2]Fertilization at blossom growth stage.

In particular, chickpea, averaged across year, fertilization and varieties, produced 30.5 more seed yield in irrigated than in non-irrigated plots, whereas 20% more seed yield, averaged across year, irrigation and varieties, was obtained when it was grown in fertilized than in non-fertilized plots (**Table 2**). Also, "Amorgos" (averaged across year and fertilization) produced 16% and 6% greater seed yield than "Serifos" and "Andros". Furthermore, the three chickpea varieties produced more seed yield in year 2 than in year 1, which was higher under any irrigated and fertilization regime (**Table 4**). "Amorgos", "Serifos" and "Andros" (averaged across year and fertilization) produced 30%, 31%, and 33% more seed yield under irrigated than under non-irrigated plots,

Table 4. Seed yield of three chickpea varieties as affected by irrigation and nitrogen fertilization during 2007/08 (year 1) and 2008/09 (year 2).

Chickpea variety	Irrigation (mm)							
	0				60			
	Nitrogen fertilization (kg ha^{-1})							
	50[1]		50 + 40[2]		50		50 + 40	
Year 1				t ha^{-1}				
Amorgos	2.19	h	3.13	defgh	3.73	cderf	4.90	abc
Serifos	2.23	gh	2.30	fgh	3.19	defgh	3.70	cdefg
Andros	2.28	fgh	2.56	efgh	3.44	cdefgh	4.03	bcde
Year 2								
Amorgos	3.31	defgh	3.90	bcde	3.58	cdefgh	5.61	a
Serifos	2.60	efgh	3.51	cdefgh	3.50	cdefgh	4.58	abcd
Andros	3.16	defgh	3.51	cdefgh	4.39	abcd	5.25	ab
CV, %				15.2				

Means with different letter indicate significant difference according to Tukey's Honestly Significant Difference test at $P = 0.05$.
[1]Fertilization before crop planting. [2]Fertilization at blossom growth stage.

whereas their respective seed yield (averaged across year and irrigation) was 27%, 18% and 13% higher in fertilized as compared with non-fertilized plots. Finally, "Amorgos", "Serifos", "Andros", averaged over years and grown in fertilized plots, produced 22.5, 14% and 11% more seed yield, respectively, than in non-fertilized plots, whereas their respective increase in irrigated and fertilized plots was 30, 15.5% and 15.5% (**Table 4**).

4. Discussion

The recorded greater total dry biomass and seed yield of chickpea varieties grown in irrigated and fertilized plots than in non-irrigated and non-fertilized plots agree with results reported by Islam et al. [24] who found that application of 15 kg·N·ha^{-1} before crop planting and 15 kg·N·ha^{-1} applied with irrigation at flower initiation growth stage increased chickpea yield by 51% as compared with that obtained in non-fertilized and non-irrigated plots. Also, Bakhsh et al. [25] found 36% increase in total dry weight due to irrigation. The chickpea dry biomass reduction in non-irrigated plots could be attributed to lower CO_2 accumulation in biochemical reactions of photosynthesis and therefore to lower carbohydrates production [26] [27].

The significantly greater seed yield of chickpea grown in year 2 could be mainly attributed to higher and more uniformly distributed rainfall and temperature prevailing during November 2008 to May 2009, as compared with the respective rainfall and temperature in year 1 (**Figure 1**).

The increased chickpea seed yield by 30.5% and 20% in irrigated and fertilized plots, respectively, as compared with that in non-irrigated and non-fertilized plots, is in agreement with the results reported by Pawar et al. [28] who found that seed yield was increased with irrigation and the increase was not affected by the growth stage of chickpea at the irrigation time. Also, Bakhsh et al. [25] found that seed yield and most of their yield components were improved by 17% with irrigation, while Rabieyan et al. [29] reported that the effect of biofertilizer (biosuper) application on water deficit stress was lower than that of complete irrigation. In addition, the combined effect of nitrogen + biofertilizer application on water deficit stress was higher than that recorded after their separate applications. In general, the application of nitrogen + biofertilizer under complete irrigation regime increased seed weight, pod weight per plant and 1000-seed weight of chickpea [30]. Furthermore, the combination of inoculation, fertilization (20 kg·N·ha^{-1}) and irrigation increased more seed yield of chickpea as compared with that grown under the combination of fertilization (60 kg·N·ha^{-1}) and irrigation [31]. In addition, the nitrogen fertilization or the inoculation resulted in higher seed yield and protein ratio of chickpea grown under irrigated than under non-irrigated conditions. Also, Zaman and Malik [32] reported that maximum seed yield, dry matter, pods per plant, seeds per pod and 1000-seed weight of chickpea were obtained with two irrigations, while Ali *et al.* [33] found that the rate of fertilization had a significant effect on seed yield of chickpea but it was differentiated among genotypes. Finally, the lower seed yield of chickpea grown in non-irrigated and non-fertilized plots could be attributed to drought stress that reduces plant growth and yield by reducing leaf surface area and rate of photosynthesis [17]. On the contrast to our results, Mohamed [34] found that the application of 90 or 180 $kg \cdot N \cdot ha^{-1}$ did not have any effect on seed yield and yield components of chickpea. The lack of irrigation or rainfall water to increase the efficient use of nitrogen applied could be the reason for these differences.

5. Conclusion

Although chickpea is considered one of the most tolerant food legumes to environmental stresses such as drought, high temperatures and poor soils, the results of this study indicate that the three chickpea varieties produce higher seed yield under irrigated and nitrogen fertilized conditions. These findings support strongly the evidence that the application of 90 $kg \cdot N \cdot ha^{-1}$ (50 $kg \cdot N \cdot ha^{-1}$ before crop planting + 40 $kg \cdot N \cdot ha^{-1}$ at blossom growth stage) in combination with irrigation (30 + 30 mm of water) results in higher chickpea seed yield, which increases its profitability and makes this crop more important as food security crop for smallholder farmers in the semi-arid Mediterranean environments.

References

[1]   FAO (2004) Production Yearbook 2003. Food and Agricultural Organization of the United Nations, Rome, Vol. 58.

[2]   Ali, N. and Kumar, S. (2006) Pulse Production in India. *Yojana*, 13-15.

[3]   Gupta, Y.P. (1988) Nutritive Value of Pulses. In: Rawanujam, B.S. and Jain, H.K., Eds., *Pulse Crop*, Oxford IBH Publishing Co. Pvt. Ltd., New Delhi, 561-601.

[4]   Siddique, K.H.M., Brinsmead, R.B., Knight, R., Knights, E.J., Paull, J.G. and Rose, I.A. (2000) Adaptation of Chickpea (*Cicer arietinum* L.) and Faba Bean (*Vicia faba* L.) to Australia. In: Knight, R., Ed., *Cool-Season Food Legumes*, Kluwer, Adelaide, 289-303.

[5]   Singh, K.B. (1997) Chickpea (*Cicer arietinum* L.). *Field Crops Research*, **53**, 161-170.
       http://dx.doi.org/10.1016/S0378-4290(97)00029-4

[6]   Singh, N., Kaur, M. and Sandhu, K.S. (2005) Physicochemical and Functional Properties of Freeze-Dried and oven Dried Corn Gluten Meals. *Drying Technology*, **23**, 1-14. http://dx.doi.org/10.1081/DRT-200054253

[7]   Kader, M.A. (2002) Effects of *Azotobacter inoculants* on the Yield and Nitrogen Uptake by Wheat. *Journal of Biological Science*, **2**, 259-261. http://dx.doi.org/10.3923/jbs.2002.259.261

[8]   Nishita, G. and Joshi, N.C. (2010) Growth and Yield Response of Chickpea (*Cicer arietinum* L.) to Seed Inoculation with *Rhizobium* sp. *Nature Science*, **8**, 232-236.

[9]   BARC (Bangladesh Agricultural Research Council) (1997) Fertilizer Recommendation Guide, Farmgate, Dhaka, 69.

[10]   Chaudhari, R.K., Patel, T.D., Patel, J.B. and Patel, R.H. (1998) Response of Chickpea Cultivars to Irrigation, Nitrogen and Phosphorus On Sandy Clay Loam Soil. *International Chickpea Newsletter*, **5**, 24-26.

[11]   Gupta, N. and Singh, R.S. (1982) Effect of Nitrogen, Phosphorus and Sulphur Nutrition on Protein and Amino Acids in Chickpea (*Cicer arietinum* L.). *Indian Journal of Agricultural Research*, **16**, 113-117.

[12]   Khan, H., Haqqani, A.M., Khan, M.A. and Malik, B.A. (1992) Biological and Chemical Fertilizer Studies in Chickpea Grown under Arid Conditions of Thal (Pakistan). *Sarhad Journal of Agriculture*, **8**, 321-327.

[13]   Jain, L.K. and Singh, P. (2003) Growth and Nutrient Uptake of Chickpea as Influenced by Phosphorus and Nitrogen. *Crop Research*, **25**, 401-413.

[14]   Siddique, K.H.M., Loss, S.P., Regan, K.L. and Jettner, R.L. (1999) Adaptation and Seed Yield of Cool Season Grain Legumes in Mediterranean Environments of South-Western Australia. *Australian Journal of Agricultural Research*, **50**, 375-387. http://dx.doi.org/10.1071/A98096

[15]   Thangwana, N.M. and Ogola, J.B.O. (2012) Yield and Yield Components of Chickpea (*Cicer arietinum*): Response to Genotype and Planting Density in Summer and Winter Sowings. *Journal of Food, Agriculture & Environment*, **10**, 710-715.

[16]   Singh, K.B. (1993) Problems and Prospects of Stress Resistance Breeding in Chickpea. In: Sing, K.B. and Saxena, M.C., Eds., *Breeding for Stress Tolerance in Cool-Seasons Food Legumes*, Wiley, Chichester, 17-35.

[17]   Krishnamurthy, L., Kashiwagi, J. and Vpadhayaya, M.D. (2003) Genetic Diversity of Drought Avoidance Root Traits in the Mini-Core Germplasm Collection of Chickpea. *International Chickpea and Pigeon Pea News Letters*, **10**, 21-29.

[18] Leport, L., Turner, N.C., Davies, S.L. and Siddique, K.H.M. (2006) Variation in Pod Production and Abortion among Chickpea Cultivars under Terminal Drought. *European Journal of Agronomy*, **24**, 236-246. http://dx.doi.org/10.1016/j.eja.2005.08.005

[19] Saraf, C.S., Baldev, B., Ali, M. and Slim, S.N. (1990) Improved Cropping Systems and Alternative Cropping Practices, In: *Chickpea in the Nineties: Proceedings of the 2nd Inter Workshop on Chickpea Improvement*, ICRISAT Center, India, Patancheru, 502-524.

[20] Rajala, A., Hakala, K., Makela, P., Muurinen, S. and Peltonen-Sainio, P. (2009) Spring Wheat Response to Timing of Water Deficit through Sink and Grain Filling Capacity. *Field Crops Research*, **114**, 263-271. http://dx.doi.org/10.1016/j.fcr.2009.08.007

[21] Saccardo, F., Crinò, P. and Giordano, I. (2001) Cece (*Cicer arietinum* L.). In: *Leguminose e agricoltura sostenibile*, Edagricole, Bologna, 555-590.

[22] Statistical Package for the Social Sciences (SPSS) (1998) SPSS Base 8.0 User's Guide and SPSS Applications Guide. SPSS, Chicago.

[23] MSTAT-C (1988) MSTAT-C, a Microcomputer Program for the Design, Arrangement, and Analysis of Agronomic Research. Michigan State University East Lansing, East Lansing.

[24] Islam, M.S., Kawochar, M.A., Karim, M.F., Ali, M.H. and Islam, M.K. (2010) Response of Chickpea to Integrated Nitrogen and Irrigation Management. *Journal of Experimental Biosciences*, **1**, 41-45.

[25] Bakhsh, A., Malik, S.R., Aslam, M., Iqbal, U. and Haqqani, A.M. (2007) Response of Chickpea Genotypes to Irrigated and Rain-Fed Conditions. *International Journal of Agriculture & Biology*, **9**, 590-593.

[26] Hopkins, W.G. and Hüner, N.P. (2004) Introduction to Plant Physiology. 3rd Edition, John Wiley & Sons, Inc, Hoboken.

[27] Potts, D.L., Stanley Harpole, W., Goulden, M.L. and Suding, K.N. (2008) The Impact of Invasion and Subsequent Removal of an Exotic Thistle, *Cynara cardunculus*, on CO_2 and H_2O Vapor Exchange in a Coastal California Grassland. *Biological Invasions*, **10**, 1073-1084. http://dx.doi.org/10.1007/s10530-007-9185-y

[28] Pawar, H.K., Khade, K.K. and More, V.D. (1992) Studies on Crop Sequences under Irrigation Constrain. *Journal of Maharashtra Agricultural Universities*, **17**, 299-301.

[29] Rabieyan, Z., Yarnia, M. and Kazemi-e-Arbat, H. (2011) Effects of Biofertilizers on Yield and Yield Components of Chickpea (*Cicer arietinum* L.) under Different Irrigation Levels. *Australian Journal of Basic and Applied Sciences*, **5**, 3139-3145.

[30] Rabieyan, Z. and Kashani, Z.F. (2012) The Effect of Irrigation and Biofertilizer on Seed and Yield Index on Two Chickpea (*Cicer arietinum* L.) Varittes. *Advances in Environmental Biology*, **6**, 1528-1533.

[31] Yagmur, M. and Kaydan, D. (2011) Plant Growth and Protein Ratio of Spring Sown Chickpea with Various Combinations of Rhizobium Inoculation, Nitrogen Fertilizer and Irrigation under Rain Fed Condition. *African Journal of Agricultural Research*, **6**, 2648-2654.

[32] Zaman, A. and Malik, S. (1988). Effect of Irrigations Levels and Mulches on Yield Attributes and Yield of Green Gram in Laterite Soil. *Environmental Ecology*, **6**, 437-440.

[33] Ali, A., Ali, Z., Iqbal, J., Nadeem, M.A., Akhtar, N., Akra, H.M. and Sattar, A. (2010) Impact of Nitrogen and Phosphorus on Seed Yield of Chickpea. *Journal of Agricultural Research*, **48**, 335-343.

[34] Mohamed, A.K. (1990) Effect of Planting Method, Irrigation and Nitrogen Fertilizer Application on Grain Yield and Yield Components of Chickpea (*Cicer arietinum*) in Shendi Area, Sudan. *Acta Agronomica Hungarica*, **39**, 393-399.

18

Greywater Treatment by High Rate Algal Pond under Sahelian Conditions for Reuse in Irrigation

Ynoussa Maiga[1*], Masahiro Takahashi[2], Thimotée Yirbour Kpangnane Somda[3], Amadou Hama Maiga[3]

[1]Laboratory of Microbiology and Microbial Biotechnology, University of Ouagadougou, Ouagadougou, Burkina Faso
[2]Environmental Engineering and Science, Hokkaido University, Sapporo-shi, Japan
[3]Laboratory of Water Decontamination, Ecosystems and Health, International Institute for Water and Environmental Engineering, Ouagadougou, Burkina Faso
Email: [*]ynoussa.maiga@gmail.com, m-takaha@eng.hokudai.ac.jp, thimotekpangnane@yahoo.fr, amadou.hama.maiga@2ie-edu.org

Abstract

High Rate Algal Pond (HRAP) was constructed and operated using a mixer device to investigate its capability in treating greywater for reuse in gardening. Physico-chemical and microbiological parameters were monitored. With a hydraulic retention time of 7.5 days and a solid retention time of 20 days, the average removal efficiencies (ARE) were 69% and 62% for BOD_5 and COD respectively. The ARE for NO_3^-, NH_4^+ and PO_4^{3-} were 23%, 52% and 43% respectively. The removal of suspended solids (SS) was unsatisfactory, which could be attributed to the low average algal settling efficiencies of 9.3% and 16.0% achieved after 30 and 60 minutes respectively. The ARE of fecal coliforms, *Escherichia coli* and enterococci were 2.65, 3.14 and 3.17 log units respectively. In view of the results, the HRAP technology could be adapted for greywater treatment in sahelian regions. However, further studies on the diversity of the algal species growing in the HRAP unit are necessary in order to increase the removal of SS. Hazards of a reuse of the effluents are discussed on the basis of the various qualitative parameters. The residual content of *E. coli* was varying from <1 to 1.77×10^4 CFU per 100 mL. Based on WHO guidelines for greywater reuse in irrigation, the effluents could be used for restricted irrigation (*E. coli* < 10^5 CFU per 100 mL). Furthermore, the reuse potential is discussed on the basis of FAO guidelines using SAR (3.03 to 4.11), electrical conductivity (482 to 4500 µS/cm) and pH values (6.45 to 8.6).

---

[*]Corresponding author.

Keywords

Greywater Treatment, High Rate Algal Pond, Irrigation Reuse, Sahelian Region

1. Introduction

Wastewater reuse in irrigation has been reported worldwide [1] including in low-income arid and semi-arid countries, where water shortage has promoted the use of alternative sources. In Burkina Faso, treated and untreated wastewater are used in gardening and horticulture [1] [2], in spite of the health and environmental risks. Wastewater recycling for reuse in irrigation can have multiple benefits especially for low-income arid and semi-arid regions, since it can contribute to reducing water related diseases with increased possibilities for food production and increased employment opportunities for poor population. Due to the low levels of microorganisms, greywater which constitutes 50% to 80% of the total household wastewater [3], is receiving more and more attention. However, many different kinds of pathogen of fecal origin have been found in greywater [4]. Besides, irrigation with untreated greywater has been demonstrated to contribute to increased soil hydrophobicity [5] and levels of fecal bacteria in the soil [6]. Treatment methods that reduce the number of pathogens are thus necessary if greywater is to be used for vegetables irrigation. High Rate Algal Pond (HRAP) is one of the promising wastewater treatment technologies: it provides cost-effective and efficient treatment with minimal energy consumption and has considerable potential to upgrade oxidation ponds [7]. In addition, the algal biomass harvested from this treatment system could be converted to biofuels, biogas and bioethanol [8] [9]. Previous studies have reported wastewater treatment by HRAP system equipped with an air-lift [10] and a paddlewheel [11]. The present study deals with HRAP system operated with a mixer agitation system for greywater treatment under real sahelian conditions. Furthermore, to our knowledge, it is the first time to assess the operation of a HRAP system in Burkina Faso where more than 300 days per year can be expected to be sunny [12]. HRAP is characterized by their shallow depth and high algal productivity that can negatively impact the irrigated soil. Recycling gravity harvested algae could be a simple and effective operation strategy to maintain the dominance of readily settleable algal species, and enhance algal harvest by gravity sedimentation [11]. This consideration is particularly important when the treated water is intended for reuse in irrigation, since it minimizes the clogging of irrigated soil. Therefore, the experimental HRAP is equipped with an algal recycling system.

The purpose of this study was to evaluate the potential of a HRAP system operated with a mixer, to adequately treat greywater under sahelian climatic conditions for reuse in gardening. The specific objectives were to:

- Evaluate the efficiency of the HRAP in terms of greywater chemical pollutants removal;
- Evaluate the efficiency of the HRAP related to greywater microbial removal;
- Discuss the reuse potential of the treated greywater.

2. Material and Methods

2.1. Experimental HRAP: Characteristics, Operation and Greywater Source

Experiments were carried out using a pilot-scale single-loop race truck configuration HRAP treating greywater at the International Institute for Water and Environmental Engineering (2iE) campus of "Kamboinsé", Ouagadougou, Burkina Faso (12.46N, 1.55W). The HRAP had a surface area of 84.4 m^2, a depth of 0.3 m and a total volume of 21.09 m^3. The design characteristics are presented in **Table 1**.

The pond water was continuously mixed (from 6:00 am to 6:00 pm) by a mixer (Satake Model A640 SATAKE chemical Equipment) allowing a speed of water at the surface of 15 m/s. A top view of the pilot-scale HRAP is shown in **Figure 1**. Greywater was collected from a dormitory of 40 students at the 2iE campus of "Kamboinsé". Shower, laundry and washbasin greywater are discharged into a single outlet pipe from which, it flowed by gravity to the water receiving pond (RP) of the treatment unit (**Figure 1**). The greywater is pumped to the Imhoff tank (IT) using a peristaltic pump (Master flex Model 07591-55) at a flow rate of 2.8 m^3/day, the remaining greywater being discharged to the infiltration pond (IP) for infiltration. From IT, the greywater

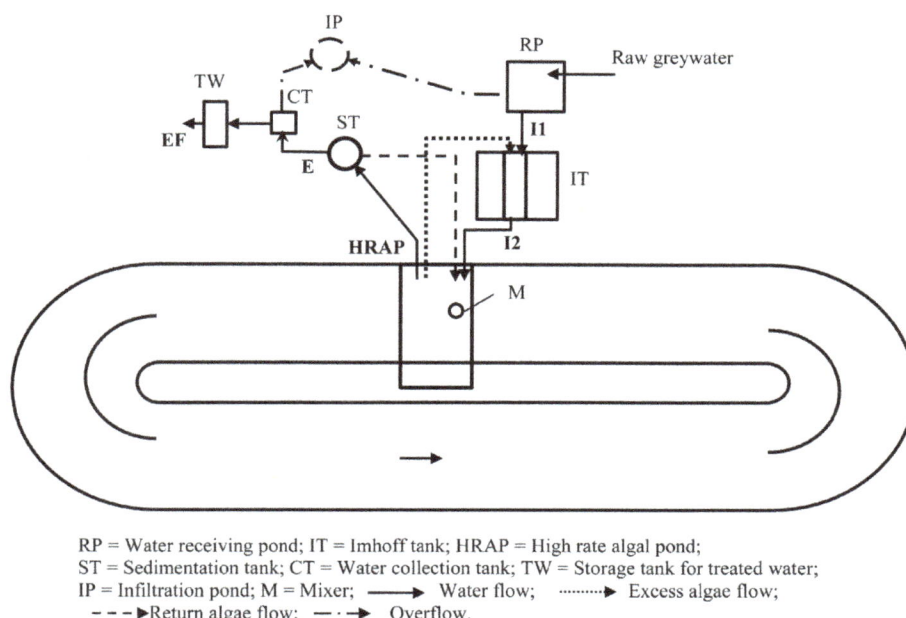

RP = Water receiving pond; IT = Imhoff tank; HRAP = High rate algal pond;
ST = Sedimentation tank; CT = Water collection tank; TW = Storage tank for treated water;
IP = Infiltration pond; M = Mixer; ——▶ Water flow; ·········▶ Excess algae flow;
– – –▶Return algae flow; —·–▶ Overflow.

Figure 1. Top view of the greywater treatment unit showing the high rate algal pond, the associated ponds, the water flow directions and the sampling points.

Table 1. Characteristics and operational conditions of the HRAP.

Characteristics		Size
Water depth		0.3 m
Length		23.96 m
Width		6.8 m
Length of linear channels		17.16 m
Useful channel width (from top)		1.8 m
Useful channel width (from bottom)		1.2 m
Effective diameter of the semi-circular channels	External edge	5.8 m
	Internal edge	2.2 m
Surface area		84.4 m^2
Effective volume		21.09 m^3
Influent flow rate		2.8 m^3/day
Excess algae flow rate		1.05 m^3/day
Return algae flow rate		2.8 m^3/day
Hydraulic retention time		7.5 days
Solid retention time		20 days

entered the HRAP by gravity. Greywater circulation and homogenization in the HRAP was obtained using a mixer (M) (Satake Model A640 SATAKE chemical Equipment) rotating at a speed of 150 rpm. The hydraulic retention time was about 7.5 days. From the HRAP, greywater flowed to the Sedimentation tank (ST) by gravity. The algal biomass settled down in ST and the supernatant flowed by gravity into the water collection tank (CT). From CT, treated greywater was pumped to the Storage tank for treated water (TW) using an automatic pump (Master flex Model 07591-55). In order to select settleable algae and allow their growth in the HRAP, the algal biomass collected at the bottom of ST was removed using a peristaltic pump (Master flex Model 07591-55) at a flow of 2.8 m³/day and recycled back to the HRAP (return algae). To avoid an overproduction of algae in the HRAP, greywater was pumped at mid depth from the HRAP and collected in the Imhoff tank using a peristaltic pump (Master flex Model 07591-55) at a flow of 1.05 m³/day (excess algae). The solid retention time was estimated at 20 days. The experimental system was operated from the start-up in March 2013. Preliminary assays were conducted during the first 5 months in order to test the robustness of the system (clogging of pumps, optimization of hydraulic retention time, and appropriate rotation speed of the mixer). The present study presents the results obtained from October 2013 to April 2014 (7 months).

2.2. Monitoring

In order to assess the efficiency of the HRAP system, field measurements and water samples was taken once a week (9:00-10:00 am) to analyze fecal indicators, physico-chemical and organic parameters using influent of IT (I1), influent of HRAP (I2), the HRAP water (HRAP), effluent of ST (E) and the effluent from storage tank (EF). Temperature, pH, dissolved oxygen (DO) and electrical conductivity (EC) were measured *in situ* using a portable electronic probe WTW multi 340i (WTW, GmbH, Weilheim, Germany). Chemical Oxygen Demand (COD), 5-days Biochemical Oxygen Demand (BOD_5), Suspended Solids (SS) were measured from homogenized samples to assess the removal efficiency of organic parameters. SS were determined by a gravimetric method using glass microfiber filters Whatman (porosity 1.5 μm). Nitrate, ammonia and orthophosphate were measured as nutrients by spectrophotometry, using filtered samples. Calcium, sodium and magnesium were determined in the treated water using atomic spectrophotometer (Perkin Elmer analyst 200). All analyses were performed according to Standard Methods for the Examination of Water and Wastewater [13]. Sodium adsorption ratio (SAR) was evaluated using the results from calcium, sodium and magnesium measurements "Equation (1)" to determine the suitability of the treated greywater for irrigation.

$$SAR = \frac{Na^+}{\sqrt{\dfrac{Mg^{2+} + Ca^{2+}}{2}}} \tag{1}$$

where Na^+, Ca^{2+} and Mg^{2+} were expressed in milli-equivalents per litre (meq/L).

Escherichia coli, fecal coliforms and enterococci were monitored as indicator bacteria for microbiological pollution assessment. The spread plate method was used after an appropriate dilution of the samples in accordance with the procedure in Standard Methods [13]. Chromocult Coliform Agar (Merck KGaA 64271, Darmstadt, Germany) was used as the culture medium for both *E. coli* and fecal coliforms assessment whereas Slanetz and Bartley medium (Biokar Diagnostics, France) was used for enterococci assessment.

2.3. Measurement of Algal Settling Efficiency

Samples from HRAP water was collected once a week for the measurement of SS according to standard methods [13]. Algal settling efficiency (ASE) was measured once a week based on the method described by Park *et al.* [14]. A 1 liter imhoff cone was filled with HRAP water and left under laboratory conditions for sedimentation. To determine ASE (after 30 and 60 minutes), 50 ml of water samples were taken after 30 and 60 minutes respectively, using a syringe from the top of the imhoff cone. SS were assessed in each sample and compared with the initial *SS* to determine ASE_{30} and ASE_{60} according to "Equation (2)".

$$ASE_x = \left[\frac{(SS_i - SS_x)}{SS_i} \right] \times 100 \tag{2}$$

with SS_i = initial *SS*; SS_x = *SS* remaining in the supernatant after "*x*" (30 or 60) minutes.

3. Results and Discussion

3.1. Climatic Conditions of the Experimental Site

The climatic data at the treatment plant (12.46N, 1.55W) were obtained from the NASA Amospheric Science Data Center (https://eosweb.larc.nasa.gov). The daily average (22-years average) solar radiation (MJ/m²/day) and the minimum and maximum temperature for each month during the period of the study (October-April) are shown on **Figure 2**. December, January and February are marked by lower temperature and solar radiation. This period is the cold season in Burkina Faso which is marked by dusty weather that could have decreased the radiation reaching the earth. During the study period, the daily solar radiation was varying from 18.61 to 23.22 MJ/m^2/day. The lowest solar radiation was registered in January while the highest value was obtained in April. The lowest minimum temperature (16.3°C) was registered in January while the highest maximum temperature (52.2°C) was obtained in March.

3.2. Physico-Chemical Characteristics of the Raw Greywater

The raw greywater produced at the students' residence was slightly polluted with organic matter when compared to data obtained in rural area in Burkina Faso. The values were varying from 14 to 88 mg/L for *SS*, 65 to 170 mg/L for BOD$_5$ and 145 to 958 mg/L for COD (**Table 2**). In rural area, for shower, laundry and dishwashing greywater, mean values of 1093 to 3060 mg/L, 533 to 2743 mg/L and 1240 to 6497 mg/L have been reported for *SS*, BOD$_5$ and COD respectively [15]. This situation could find an explanation in the effect of dilution due to the differences in amounts of water used to perform the activities. Indeed, greywater production in a household is directly influenced by water consumption which is dependent on a number of factors including the existing water supply service and infrastructure, the number of household members, the age distribution, the life style characteristics, the typical water usage patterns [16]. The authors highlighted that, in areas with water scarcity and rudimentary forms of water supply, water consumption varies from 20 to 30 L/capita/day whereas a household member in a richer area with piped water may generate several hundred liters per day [16]. Mean

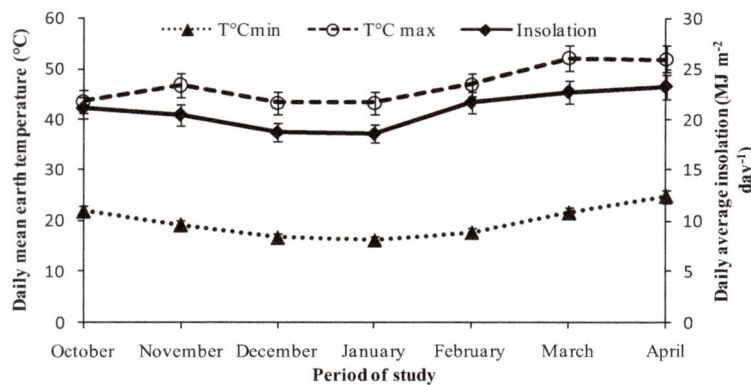

Figure 2. Mean monthly (22 years) temperature and solar radiation in the treatment plant Data obtained from NASA at: http://eosweb.larc.nasa.gov/cgi-bin/sse/grid.cgi. T°C min: minimum temperature; T°C max: maximum temperature.

Table 2. Characteristics of the raw greywater pumped from the water receiving pond.

Parameter	Range	Average ± SD
Temperature (°C)	23.3 - 30.8	27.13 ± 2.25
pH	6.29 - 7.54	6.95 ± 0.30
EC (µS/cm)	669 - 5620	3677.05 ± 1549.03
SS (mg/L)	14 - 88	40.95 ± 16.06
BOD$_5$ (mg/L)	65 - 170	109.25 ± 27.21
COD (mg/L)	146 - 958	464.4 ± 252
NH$_4^+$ (mg/L)	1.73 - 34.19	22.55 ± 11.63
NO$_3^-$ (mg/L)	6.2 - 48.29	21.52 ± 10.9
PO$_4^{3-}$ (mg/L)	0.82 - 6.6	3.36 ± 1.58

SD = standard deviation; EC = electrical conductivity; *SS* = suspended solids; BOD$_5$ = 5 days-Biochemical oxygen demand; COD = Chemical oxygen demand.

water consumption of 11 and 24 L/capita/day with a greywater production of 8 and 13 L/capita/day have been reported in rural area in Burkina Faso [15] whereas water consumption in France was estimated at 150 L/capita/day.

3.3. Evolution of the Physico-Chemical Parameters through the Pond System

Figure 3(a) shows the distribution of the pH in the influent (I1), the HRAP (HRAP) and the effluent (EF). The pH of the influent greywater was varying from 6.29 to 7.54. In the HRAP, the pH increased to reach values varying between 7.38 and 9.46. Aguirre *et al.* [17] reported values of 7 to 8.2 in pilot HRAP treating pretreated piggery wastewater while Santiago *et al.* [18] reported mean pH values of 7.7 ± 0.7 and 8.1 ± 1 in HRAP treating non-disinfected and disinfected effluents from an upflow anaerobic sludge blanket respectively. Diurnal variation of pH has been reported in HRAP [10] [19] with maximum values reached between 1:00 pm and 3:00 pm [19]. Thus, the maximum pH of 9.46 reached in our HRAP, could be higher than this value, due to the fact that the measurements were performed in the morning (9:00 am-10: 00 am). The pH of the treated greywater is following the same trend as that of the HRAP water, however at a lower level (6.45 to 8.6).

Figure 3(b) shows the variation of the mean values of pH, DO, temperature and EC through the treatment system. The pH and DO values are higher in the HRAP compared to that of the other ponds. This finding could be explained by the fact that pH and DO are variables associated with photosynthetic activity and that algal photosynthesis in HRAP can raise pH often exceeding pH > 11 [7]. Morning hours are marked by low DO values

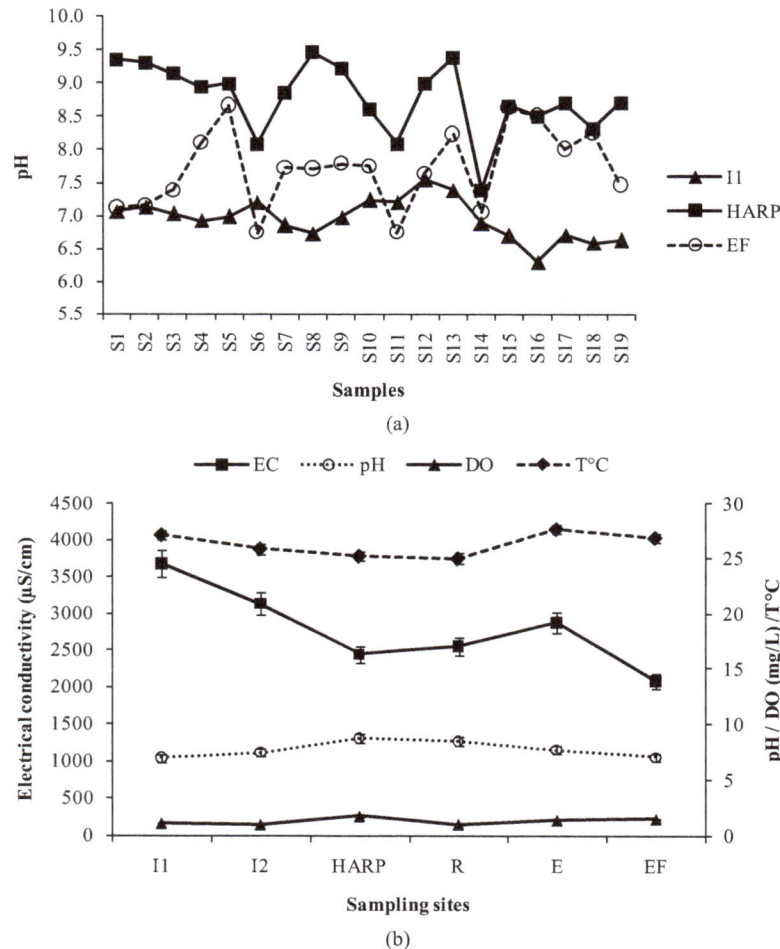

(a)

(b)

Figure 3. Variation of physico-chemical parameters. (a) Variation of pH in the influent (I1), the HRAP and the effluent (EF) during the study period; (b) Variation of mean values of pH, DO, temperature and EC through the pond system.

compared to that of the afternoon, with maximum values reached between 1:00 pm and 3:00 pm [19]. In addition, Narcir *et al.* [10] reported diurnal variation of DO in HRAP with minimum values of 0.5 mg/L. Therefore, the low values of DO in the HRAP compared to reported values is probably due to the fact that the sampling was conducted during the morning.

The EC values in the raw greywater were ranging from 669 to 5620 µS/cm for water temperature ranging from 23.3°C to 30.8°C (**Table 2**). As EC is varying with temperature, the trend of the later could explain the trend of EC in **Figure 3(b)**.

3.4. Distribution and Removal of Organic Compound

Figures 4(a)-(c) show respectively the variations of the concentrations of SS, BOD_5 and COD in the influent and the effluent of the pond system. In the influent, SS was varying from 14 to 88 mg/L, BOD_5 from 65 to 170 mg/L and COD from 145 to 958 mg/L (**Table 2**).

The corresponding concentrations in the effluent were varying from 1 to 128 mg/L for *SS*, 5 to 70 mg/L for BOD_5 and 54 to 366 mg/L for COD. The average removal efficiencies were 69% and 62% for BOD_5 and COD respectively (**Figure 4(d)**). Narcir *et al.* [10], using a HRAP equipped with an air-lift have reported a removal efficiency of 44% for BOD_5 while Chen *et al.* [20] have reported an annual removal efficiency of 50 % for COD. In addition, El Hamouri *et al.* [21] have reported values of 65% removal for BOD and COD under Moroccan climate. The shallow depth (30 cm) and the sunlight (**Figure 2**) contributed to enhance the algal productivity

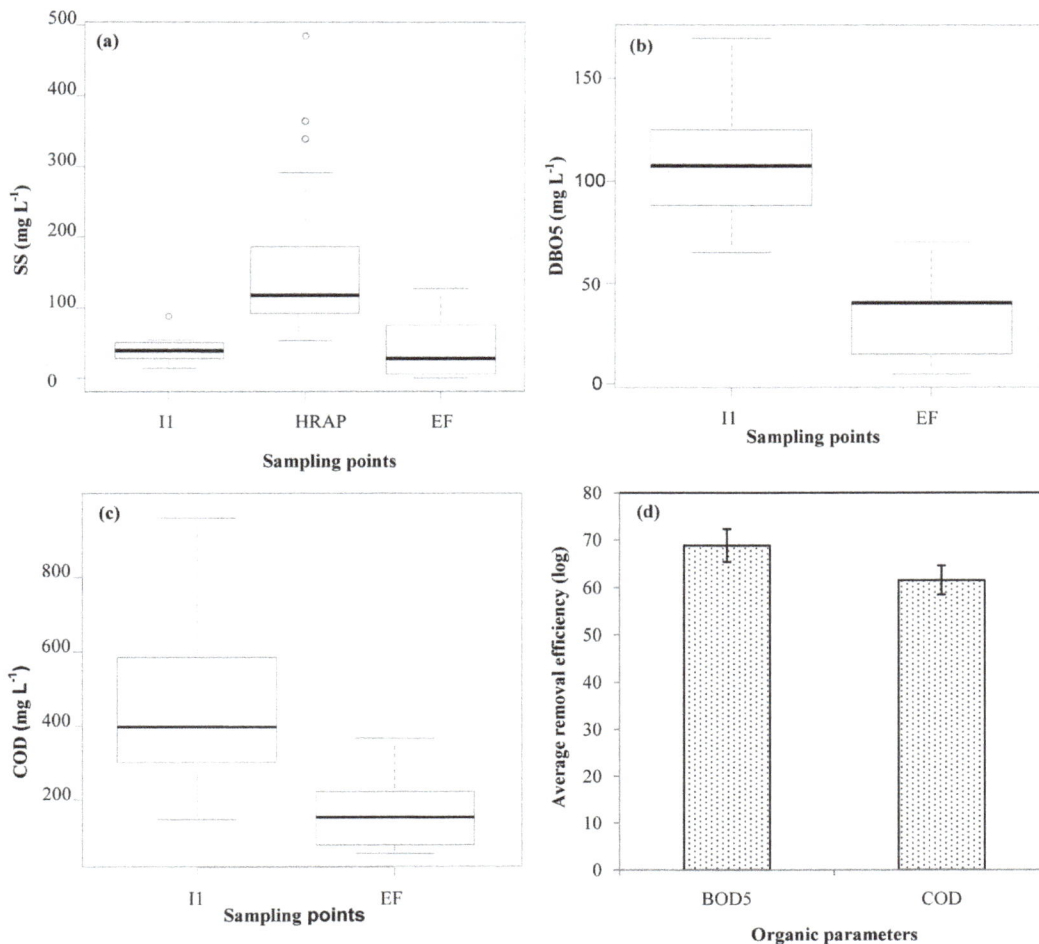

Figure 4. Variation of organic parameters *SS* (a), BOD_5 (b) and COD (c) through the treatment and the average removal efficiencies for BOD5 and COD (d) during the period of study. I1 = influent of imhoff tank (influent); EF = effluent of storage tank (treated greywater).

and then, to increase *SS* in the HRAP, with values varying from 53 to 484 mg/L (**Figure 4(a)**). Consequently, *SS* removal in the ST was unsatisfactory with values higher in the effluent compared to the influent in most of the cases. This finding is common in HRAP system. For instance, Chen *et al.* [20] reported that variables associated with photosynthetic activity such as DO, pH and *SS* are significantly higher in the effluent compared to the influent.

3.5. Nutrient Removal

Nitrogen and phosphorus are important parameters given their fertilizing value for plant, their relevance for natural treatment processes and their potential negative impact on aquatic environment [16]. **Figure 5(a)** and **Figure 5(b)** show the variations in the concentrations of different forms of nutrients (NO_3^-, NH_4^+ and PO_4^{3-}) in the influent and the effluent of the pond system and the corresponding removal efficiencies. The presence of phosphates and nitrates in greywater has been attributed to detergents and washing powders [16]. The nutrient content in the influent is in the range of reported values from Burkina Faso [15]. Nitrogen values are found within 6.3 and 48.29 mg/L for NO_3^- and within 1.73 and 34.19 for NH_4^+ (**Table 1**; **Figure 5**). In addition, these values are in the range of typical values of nitrogen in mixed household greywater from different countries which are found within 5 to 50 mg/L [16]. The concentrations of PO_4^{3-} in the raw greywater are relatively low (**Table 1**). The relative use of phosphorus containing detergents and the dilution potential due to high water consumption could explain this finding.

The average removal efficiencies for nitrate and ammonia are 23% and 52% respectively (**Figure 5(b)**). Generally, the removal efficiency of ammonia by HRAP system is high. Chen *et al.* [20] and Narcir *et al.* [10] have reported removal efficiencies of 87% and up to 90% respectively. Furthermore, Aguirre *et al.* [17] have reported removal efficiencies varying from 68% to 85% for ammonia in a HRAP system treating piggery wastewater. However, El Hamouri *et al.* [21] have reported a low removal efficiency of 48% for ammonia under arid conditions. More recently, Derabe *et al.* [22] using artificial greywater at lab. scale have reported removal efficiencies of 20.07% and 53.39% for NH_4-N in continuous and batch experiments respectively. The low removal efficiency of nitrate compared to ammonia in our study could be explained by the mechanisms involved in the removal process. Indeed, nitrate is mainly removed through algal uptake while ammonia is removed through stripping and algal uptake [23]. In addition, the same authors, dealing with nitrogen removal in HRAP systems, have reported that ammonia stripping was the most important mechanism for nitrogen removal followed by algal uptake and subsequent algal separation in the clarifiers. Furthermore, the low removal efficiency for nitrate despite the algae uptake could be attributed to the potential occurrence of nitrification during the process [20].

The average removal efficiency for orthophosphate is 43% (**Figure 5(b)**). Similar removal efficiencies of 40% [20] and 54% have been reported [21]. Orthophosphate is removed through algal uptake allowing algal growth

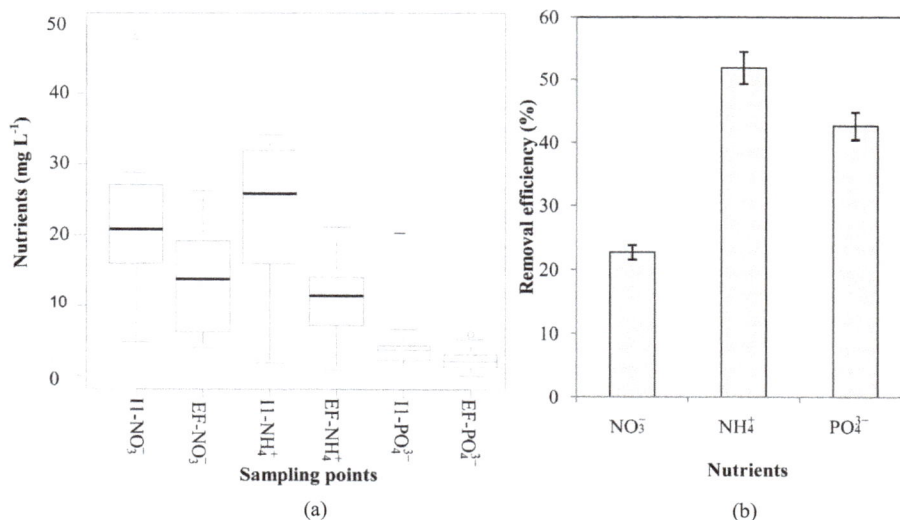

Figure 5. (a) Variation of concentration of nutrients (NO_3^-, NH_4^+, PO_4^{3-}) in the influent (I1) and the effluent (EF) and (b) the corresponding removal efficiencies.

that in terns raises the pH of the mixed liquor, resulting in orthophosphate precipitation. The low concentration of orthophosphate in the influent (**Table 1**) could explain its low removal efficiency since phosphorus removal is positively linked to the influent phosphate concentration and the retention time [20].

3.6. Distribution and Removal of Fecal Indicators

During the study period, fecal bacteria content in the raw greywater pumped from the water receiving pond, varied from 2.60×10^4 to 1.59×10^5 CFU per 100mL for fecal coliforms, 6.67×10^2 to 6.27×10^4 CFU per 100 mL for *E. coli* and from 2.33×10^3 to 1.73×10^5 CFU per 100 mL for enterococci (**Figure 6(a)**). The mean values were 6.73×10^4 CFU per 100 mL, 1.17×10^4 CFU per 100 mL and 5.71×10^4 CFU per 100 mL for fecal coliforms, *E. coli* and enterococci respectively. In terms of fecal bacteria content, the greywater produced in rural area in Burkina Faso is much more polluted than the greywater from the students' dormitory, used for the treatment. Indeed, values of up to 6×10^8 CFU per 100 mL for *E. coli* and up to 1.63×10^9 CFU per100 mL for fecal coliforms have been reported from a household in rural area [24]. The effect of the dilution due to the high amount of water used in urban area than rural area to perform the activities, could explain this difference.

The average removal efficiencies for the whole system were 2.65 log units for fecal coliforms, 3.14 log units for *E. coli* and 3.17 log units for enterococci. The efficiency of the system is in the same range of reported values from previous studies. El Hamouri *et al.* [21] have reported a removal efficiency of 2.44 log units for fecal coliforms from the whole HRAP system treating wastewater. The HRAP component of the system, which operated under paddle wheel mixing contributed to a removal efficiency of 1.59 log units. More recently, removal efficiencies of 2.48 log units for fecal coliforms and 2.62 log units for streptococci have been reported from a whole HRAP system treating wastewater in Morocco [10]. The contribution of the HRAP component operated by air-lift mixing was estimated to 1.43 and 1.47 log units for fecal coliforms and streptococci respectively.

Previous studies showed that sunlight, pH, protozoan grazing and reactive byproducts of oxygen such as superoxide anion radical, hydrogen peroxide and reactive hydroxyl radicals are factors involved in the inactivation of bacteria during the treatment [25]-[27]. Sunlight seems to be a major factor because of its high intensity ($18.61 - 23.22$ MJ/m^2/day) (**Figure 2**) and the low depth of the pond which allows its lower attenuation. Sunlight is detrimental to bacteria and beneficial to algal growth. UV-B (280 - 320 nm), UV-A (320 - 400 nm) and the Photosynthetically Available Radiation (PAR > 400 nm) of the solar spectrum are responsible of inactivating bacteria [28]. Algal growth increased the pH (highest value of 9.46 from morning sampling) which is detrimental to bacteria. In this connection, Benchokroun *et al.* [26] have reported that *E. coli* was inactivated more rapidly when the pH was elevated above 8.5 than at lower pH. In addition, algal growth promoted oxygen production

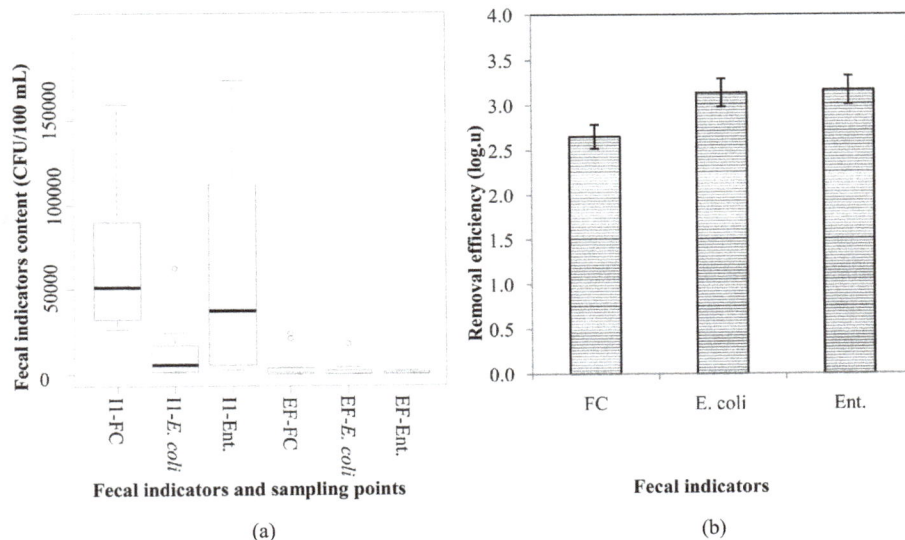

Figure 6. (a) Distribution of fecal indicators (FC, *E. coli* and Ent.) in the influent (I1) and the effluent (EF) and (b) the corresponding removal efficiencies. FC = fecal coliforms; Ent. = enterococci; log.u = log units.

which is detrimental to fecal bacteria since it has been reported that survival of fecal coliforms in sunlight was completely dependent on the presence of oxygen and decreased with increasing oxygen concentration [26]. Molecular oxygen promotes solar photoinactivation mediated by endogenous photosensitizers [28]. As previously reported [27] [29], the results show that enterococci (Gram positive) (3.17 log units removal) were more affected that fecal coliforms (Gram negative) (2.65 log units removal). The difference in the characteristics of the bacterial cell wall was previously used to explain this finding [30]. The Gram negative bacterial (fecal coliform) cell wall lipopolysaccharide coat offers some protection from the toxic effects of exogenous agents.

3.7. Algal Settling Efficiency

Average algal settling efficiencies of 9.3% ± 3.4% and 16.0% ± 7.9% were achieved after 30 and 60 minutes of settling respectively. These low values of ASE corroborate with the unsatisfactory removal of SS. Park *et al.* [14] have reported ASE values of 86% ± 9.1% and 93.6% ± 2.8% after 30 and 60 minutes when *Pediastrum* sp. was present at over 80% dominance in the HRAP; ASE reduced to 25.6% ± 10.2% and 35.2% ± 10.1% when *Pediastrum* sp. dominance declined to less than 40%, for respective settling periods. Colonial algae settle faster than unicellular. Therefore, the characteristics of the algal species growing in the HRAP could explain the low ASE values. For this reason, further studies on the diversity and the identity of the algal species growing in the HRAP are necessary in order to enhance the settling efficiency. In addition, a previous study has reported that algal settling was promoted by calcium and orthophosphate concentrations in alkaline conditions (pH: 10 - 11) [31]. Therefore, the low pH values recorded in the HRAP (maximum value of 9.46) and the low concentration of orthophosphate in the influent could explain the low ASE values.

3.8. Reuse Potential

In the final effluent, the range of the concentration of *E. coli* was varying from less than 1 to 1.77×10^4 CFU per 100 mL with a mean value of 1.96×10^3 CFU per 100 mL. The maximum values for fecal coliforms and enterococci were 2.37×10^4 and 1.67×10^3 CFU per 100 mL respectively; their mean values were 3.98×10^3 and 5.12×10^2 CFU per 100 mL respectively. Based on the WHO guidelines for greywater reuse in restricted (*E. coli* < 10^5 CFU per 100 mL) and unrestricted irrigation (*E. coli* < 10^3 CFU per 100 mL) [32], the final effluent could be used for restricted irrigation.

pH is also an important environmental parameter to consider when greywater is intended for reuse in irrigation. The treated greywater was characterized by pH values varying from 6.45 to 8.6 which can have beneficial effect on the bacteria of the irrigated soil, since most bacteria prefer neutral or slightly alkaline conditions, around 6.5 - 8.5 [33]. In turn, bacterial growth can promote irrigated vegetables development by increasing nutrients availability from organic matter.

The concentrations of nutrients found in the effluents (3.99 - 26.14 mg/L for nitrate, 0.71 - 21.03 mg/L for ammonia, 0.07 - 6 mg/L for orthophosphate) can be beneficial for irrigated vegetables, since nitrogen and phosphorus are essential plant nutrient. However, greywater rich in nitrate can have a negative impact as nitrate is highly soluble and can move easily in soils irrigated with wastewater [1].

The electrical conductivity (EC) values of the treated greywater varied from 482 and 4500 µS/cm. High EC in irrigated water can result in an increase in osmotic potential in the soil solution and interfere with extraction of water by plants [1]. Permissible EC for greywater reuse in irrigation are strongly dependent on soil characteristics and the suggested limits differ in the literature reviewed [16]. According to WHO [32] guidelines, the recommended maximum value for greywater reuse in irrigation is 3000 µS/cm. Grattan [34] reported that EC below 1300 µS/cm should normally not cause problems whereas irrigation with saline greywater (EC exceeding 1300 µS/cm) requires special precautions (use of salt-tolerant plants). According to FAO [35], water with EC varying from 700 to 3000 µS/cm and EC exceeding 3000 µS/cm are considered as "slight to moderate" and "severe degree of restriction on use" respectively. Therefore, the treated greywater is classified between "slight to moderate" and "severe degree of restriction on use" based on FAO guidelines.

In addition, it has been reported that the effect of sodium ions in irrigation water is dependent on the total salt concentration and the sodium ions concentration relative to the concentration of calcium and magnesium ions (as indicated by SAR) [35]. The SAR values of the treated greywater were varying between 3.03 and 4.11 (**Table 3**). Thus, for surface irrigation, when we consider specific sodium toxicity, the treated greywater is classified as "slight to moderate degree of restriction on use" according to FAO guidelines [35].

Table 3. Range of the concentrations of Ca^{2+}, Mg^{2+}, Na^+ and the corresponding SAR values of the treated greywater.

Parameters	Minimum	Maximum
Calcium (mg/L)	5.89	7.21
Magnesium (mg/L)	2.73	5.73
Sodium (mg/L)	22.74	24.99
SAR	3.03	4.11

SAR = Sodium Adsorption Ratio.

Furthermore, it has been reported that for a given SAR value, an increase in total salt concentration (EC) is likely to increase soil permeability and, for a given total salt concentration, an increase in SAR will decrease soil permeability. Irrigation water with SAR values of 3 to 6 and EC exceeding 1200 µS/cm is considered as "none degree of restriction on use" [35]. Therefore, the possible effect of the high EC values of the treated greywater can be moderated by the SAR values.

4. Conclusions

This study demonstrated that High Rate Algal Pond technique could be suitable for greywater treatment under climatic conditions of Ouagadougou (Burkina Faso). The mean removal efficiencies were 69% and 62% for BOD_5 and COD respectively. The average removal efficiencies for NO_3^-, NH_4^+ and PO_4^{3-} were 23%, 52% and 43% respectively. The relative low removal of nutrients compared to previous studies could be beneficial since the effluents are intended for reuse in gardening, the residual nutrients, being important for vegetables.

The whole treatment system allowed average removal efficiencies of 2.65 log units for fecal coliforms, 3.14 log units for *E. coli* and 3.17 log units for enterococci. The residual content of *E. coli* was varying from <1 to 1.77×10^4 CFU per 100 mL with a mean value of 1.96×10^3 CFU per 100 mL. Based on the WHO guidelines for greywater reuse in restricted (*E. coli* $< 10^5$ CFU per 100 mL) and unrestricted irrigation (*E. coli* $< 10^3$ CFU per 100 mL), the final effluent could be used for restricted irrigation. The pH values of the effluent were in compliance with the recommended values for irrigation. The EC values of the treated greywater were high, sometimes over the recommended values of 3000 µS/cm. However, the effect of these high values on the irrigated soil can be moderated by the SAR values.

Low average algal settling efficiencies of 9.3% ± 3.4% and 16.0% ± 7.9% were achieved after 30 and 60 minutes of settling respectively, which could explain the unsatisfactory removal of SS. To increase the settleability of algae, and then SS removal, further studies on the algal species involved are necessary.

Acknowledgements

The authors thank Japan International Cooperation Agency (JICA) for providing the funds.

References

[1] International Water Management Institute (IWMI) (2010) Wastewater Irrigation and Health: Assessing and Mitigating Risk in Low-Income Countries. Earthscan, London.

[2] Kêdowidé, C.M.G., Sedogo, M.P. and Cissé, G. (2010) Dynamique spatio temporelle de l'agriculture urbaine à Ouagadougou: Cas du Maraîchage comme une activité montante de stratégie de survie. *Vertigo—La Revue Electronique en Sciences de l'Environnement*, **10**, [En Ligne]. http://vertigo.revues.org/10312

[3] Friedle, R.E. and Hadari, M. (2006) Economic Feasibility of On-Site Greywater Reuse in Multi-Storey Building. *Desalination*, **190**, 221-234. http://gwri.technion.ac.il/pdf/gwri_abstracts/2006/57.pdf http://dx.doi.org/10.1016/j.desal.2005.10.007

[4] Winward, G.P., Avery, L.M., Frazer-Williams, R., Pidou, M., Jeffrey, P., Stephenson, T. and Jefferson, B. (2008) A Study of the Microbial Quality of Greywater and an Evaluation of Treatment Technologies for Reuse. *Ecological Engineering*, **32**, 187-197. http://www.sciencedirect.com/science/article/pii/S092585740700211X http://dx.doi.org/10.1016/j.ecoleng.2007.11.001

[5] Tarchitzky, J., Lerner, O., Shani, U., Arye, G., Lowengart-Aycicegi, A., Brener, A. and Chen, Y. (2007) Water Distri-

bution Pattern in Treated Wastewater Irrigated Soils: Hydrophobicity Effect. *European Journal of Soil Science*, **58**, 573-588. http://onlinelibrary.wiley.com/doi/10.1111/j.1365-2389.2006.00845.x/abstract
http://dx.doi.org/10.1111/j.1365-2389.2006.00845.x

[6]   Travis, M.J., Wiel-Shafran, A., Weisbrod, N., Adar, E. and Gross, A. (2010) Greywater Reuse for Irrigation: Effect on Soil Properties. *Science of the Total Environment*, **408**, 2501-2508.
http://www.sciencedirect.com/science/article/pii/S0048969710002524
http://dx.doi.org/10.1016/j.scitotenv.2010.03.005

[7]   Park, J.B.K. and Craggs, R.J. (2010) Wastewater Treatment and Algal Production in High Rate Algal Ponds with Carbon Dioxide Addition. *Water Science and Technology*, **63**, 633-639.
http://www.ncbi.nlm.nih.gov/pubmed/20150699
http://dx.doi.org/10.2166/wst.2010.951

[8]   Craggs, R.J., Heubeck, S., Lundquist, T.J. and Benemann, J.R. (2011) Algae Biofuel from Wastewater Treatment High Rate Algal Ponds. *Water Science and Technology*, **63**, 660-665. http://www.ncbi.nlm.nih.gov/pubmed/21330711
http://dx.doi.org/10.2166/wst.2011.100

[9]   Sukias, J.P.S. and Craggs, R.J. (2011) Digestion of Wastewater Pond Microalgae and Inhibition from Ammonium and Alum. *Water Science and Technology*, **63**, 835-840. http://www.ncbi.nlm.nih.gov/pubmed/21411930
http://dx.doi.org/10.2166/wst.2011.101

[10] Nacir, S., Ouazzani, N., Vasel, J.L., Jupsin, H. and Mandi, L. (2010) Traitement des eaux usées domestiques par un chénal algal à haut rendement (CAHR) agité par air lift sous climat semi-aride. *Revue des sciences de l'eau/Journal of Water Science*, **23**, 57-72. http://www.erudit.org/revue/rseau/2010/v23/n1/038925ar.html?vue=resume

[11] Park, J.B.K., Craggs, R.J. and Shilton, A.N. (2011) Wastewater Treatment High Rate Algal Ponds for Biofuel Production. *Bioresource Technology*, **102**, 35-42. http://www.sciencedirect.com/science/article/pii/S0960852410011636
http://dx.doi.org/10.1016/j.biortech.2010.06.158

[12] Kenfack, S. (2006) Helio-Photocatalytic Enhancement of the Biodegradation of Biorecalcitrant Pollutants in Water: Physicochemical and Technical Aspects. PhD Thesis, EPFL, Lausanne.

[13] American Public Health Association (APHA), American Water Works Association (AWWA) and Water Environment Federation (WEF) (1998) Standard Methods for the Examination of Water and Wastewater. 20th Edition, APHA/AWWA/WEF, Washington DC.

[14] Park, J.B.K., Craggs, R.J. and Shilton, A.N. (2011) Recycling Algae to Improve Species Control and Harvest Efficiency from a High Rate Algal Pond. *Water Research*, **45**, 6637-6648.
http://www.sciencedirect.com/science/article/pii/S0043135411005720

[15] Maiga, Y., Moyenga, D., Ushijima, K., Sou, M. and Maiga, A.H. (2014) Greywater Characteristics in Rural Areas of Sahelian Region for Reuse Purposes: The Case of Burkina Faso. *Revue des sciences de l'eau/Journal of Water Science*, **27**, 39-54. www.erudit.org/revue/rseau/2014/v27/n1/1021981ar.html?vue=resume&modes=restriction.

[16] Morel, A. and Diener, S. (2006) Greywater Management in Low and Middle-Income Countries, Review of Different Treatment Systems for Households and Neighborhood. Swiss Federal Institute of Aquatic Science and Technology (Eawag), SANDEC, Dübendorf.

[17] Aguirre, P., Alvarez, E., Ferrer, I. and Garcia, J. (2011) Treatment of Piggery Wastewater in Experimental High Rate Algal Ponds. *Revista Latinoamericana de Biotecnologia Ambiental y Algal*, **2**, 57-66.
http://uniciencia.ambientalex.info/revistas/vol2n21.pdf

[18] Santiago, A.N., Calijuri, M.L., Assemany, P.P., Calijuri, M.C. and dos Reis, A.J.D. (2013) Algal Biomass Production and Wastewater Treatment in High Rate Algal Ponds Receiving Disinfected Effluent. *Environmental Technology*, **34**, 1877-1885. http://dx.doi.org/10.1080/09593330.2013.812670

[19] Picot, B., Moersidik, S., Casellas, C. and Bontoux, J. (1993) Using Diurnal Variations in a High Rate Algal Pond for Management Pattern. *Water Science and Technology*, **28**, 169-175.
http://www.iwaponline.com/wst/02810/wst028100169.htm

[20] Chen, P., Zhou, Q., Paing, J., Le, H. and Picot, B. (2003) Nutrient Removal by the Integrated Use of High Rate Algal Ponds and Macrophyte Systems in China. *Water Science and Technology*, **48**, 251-257.
http://www.ncbi.nlm.nih.gov/pubmed/14510218

[21] El Hamouri, B., Rami, A. and Vasel, J.L. (2003) The Reasons behind the Performance Superiority of a High Rate Algal Pond over Three Facultative Ponds in Series. *Water Science and Technology*, **48**, 269-276.
http://www.ncbi.nlm.nih.gov/pubmed/14510220.

[22] Derabe, H.M., Onodera, M., Takahashi, M., Satoh, H. and Fukazawa, T. (2014) Control of Algal Production in a High Rate Algal Pond: Investigation through Batch and Continuous Experiments. *Water Science and Technology*, **69**, 2519-2525. http://www.ncbi.nlm.nih.gov/pubmed/24960016

http://dx.doi.org/10.2166/wst.2014.174

[23] García, J., Mujeriego, R. and Hernández-Mariné, M. (2000) High Rate Algal Ponds Operating Strategies for Urban Wastewater Nitrogen Removal. *Journal of Applied Phycology*, **12**, 331-339. http://link.springer.com/article/10.1023%2FA%3A1008146421368 http://dx.doi.org/10.1023/A:1008146421368

[24] Maiga, Y., Moyenga, D., Nikiema, B.C., Ushijima, K., Maiga, A.H. and Funamizu, N. (2014) Designing Slanted Soil System for Greywater Treatment for Irrigation Purposes in Rural Area of Arid Regions. *Environmental Technology*, **35**, 3020-3027. http://dx.doi.org/10.1080/09593330.2014.929180 http://dx.doi.org/10.1080/09593330.2014.929180

[25] Arana, I., Irizar, A., Seco, C., Muela, A., Fernández-Astorga, A. and Barcina, I. (2003) gfp-Tagged Cells as a Useful Tool to Study the Survival of *Escherichia coli* in the Presence of the River Microbial Community. *Microbial Ecology*, **45**, 29-38. http://www.ncbi.nlm.nih.gov/pubmed/12447583 http://dx.doi.org/10.1007/s00248-002-1029-9

[26] Benchokroun, S., Imziln, B. and Hassani, L. (2003) Solar Inactivation of Mesophilic *Aeromonas* by Exogenous Photooxidation in High-Rate Algal Pond Treating Wastewater. *Journal of Applied Microbiology*, **94**, 531-538. http://www.ncbi.nlm.nih.gov/pubmed/12588563 http://dx.doi.org/10.1046/j.1365-2672.2003.01867.x

[27] Maiga, Y., Wethe, J., Denyigba, K. and Ouattara, A.S. (2009) The Impact of Pond Depth and Environmental Conditions on Sunlight Inactivation of *E. coli* and Enterococci in Wastewater in a Warm Climate. *Canadian Journal of Microbiology*, **55**, 1364-1374. http://www.ncbi.nlm.nih.gov/pubmed/20029528 http://dx.doi.org/10.1139/W09-104

[28] Muela, A., Garcia-Bringas, J.M., Seco, C., Arana, I. and Barcina, I. (2002) Participation of Oxygen and Role of Exogenous and Endogenous Sensitizers in the Photoinactivation of *Escherichia coli* by Photosynthetically Active Radiation, UV-A and UV-B. *Microbial Ecology*, **44**, 354-364. http://www.ncbi.nlm.nih.gov/pubmed/12375094 http://dx.doi.org/10.1007/s00248-002-1027-y

[29] Anderson, K.L., Whitlock, J.E. and Harwood, V.J. (2005) Persistence and Differential Survival of Fecal Bacteria in Subtropical Waters and Sediments. *Applied and Environmental Microbiology*, **71**, 3041-3048. http://www.ncbi.nlm.nih.gov/pmc/articles/PMC1151827/pdf/1861-04.pdf http://dx.doi.org/10.1128/AEM.71.6.3041-3048.2005

[30] Jori, G. and Brown, S.B. (2004) Photosensitized Inactivation of Microorganisms. *Photochemical & Photobiological Sciences*, **3**, 403-405. http://www.ncbi.nlm.nih.gov/pubmed/15122355 http://dx.doi.org/10.1039/b311904c

[31] Baya, D.G.S.T. (2012) Etude de l'autofloculation dans un chénal algal à haut rendement. Thèse de Doctorat, Université de Liège, Liège. (In French)

[32] WHO (2006) Guidelines for the Safe Use of Wastewater, Excreta and Greywater, Volume 4: Excreta and Greywater Use in Agriculture. WHO Press, Geneva.

[33] Mara, D. (2004) Domestic Wastewater Treatment in Developing Countries. Earthscan, London.

[34] Grattan, S.R. (2002) Irrigation Water Salinity and Crop Production. Publication 8066, University of California, Oakland. http://vric.ucdavis.edu/pdf/Irrigation/IrrigationWaterSalinityandCropProduction.pdf

[35] Food and Agriculture Organization of the United Nations (1985) Water quality for agriculture. FAO, Rome. http://www.fao.org/DOCReP/003/T0234e/T0234e00.htm

19

Global Warming Effects on Irrigation Development and Crop Production: A World-Wide View

Daniele De Wrachien[1], Mudlagiri B. Goli[2]

[1]Department of Agricultural Engineering, State University of Milano, Milano, Italy
[2]Mississippi Valley State University, Itta Bena, USA
Email: daniele.dewrachien@unimi.it

Abstract

Despite the enormous advances in our ability to understand, interpret and ultimately manage the natural world, we have reached the 21st century in awesome ignorance of what is likely to unfold in terms of both the natural changes and the human activities that affect the environment and the responses of the Earth to those stimuli. One certain fact is that the planet will be subjected to pressures hitherto unprecedented in its recent evolutionary history. The "tomorrow's world" will not simply be an inflated version of the "today's world", with more people, more energy consumption and more industry, rather it will be qualitatively different from today in at least three important respects. First, new technology will transform the relationship between man and the natural world. An example is the gradual transition from agriculture that is heavily dependent on chemicals to one that is essentially biologically intensive through the application of bio-technologies. Consequently, the release of bio-engineered organisms is likely to pose new kinds of risks if the development and use of such organisms are not carefully controlled. Second, society will be moving beyond the era of localized environmental problems. What were once local incidents of natural resource impairment shared throughout a common watershed or basin, now involve many neighboring countries. What were once acute, short-lived episodes of reversible damage now affect many generations. What were once straightforward questions of conservation versus development now reflect more complex linkages. The third major change refers to climate variations. It is nowadays widely accepted that the increasing concentration of the so-called greenhouse gases in the atmosphere is altering the Earth's radiation balance and causing the temperature to rise. This process in turn provides the context for a chain of events which leads to changes in the different components of the hydrological cycle, such as evapotranspiration rate, intensity and frequency of precipitation, river flows, soil moisture and groundwater recharge. Mankind is expected to respond to these effects by taking adaptive measures including changing patterns of land use, adopting new strategies for soil and water management and looking for non-conventional water resources (e.g. saline/brackish waters, desalinated water, and treated wastewater). All these problems will become more pronounced in the years to come, as society enters an era of increasingly complex paths towards the global economy. In this context, engineers and decision-makers need to systematically review planning principles, design criteria, operating rules, contingency plans and management policies for new infra-structures. In relation to these issues and based on available information, this report gives an overview of current and future (time horizon 2025) irrigation and food production development around the world. Moreover, the paper analyses the results of the most recent and advanced General Circulation Models for assessing the hydrological impacts of climate variability on crop requirements, water availability and the planning and design process of irrigation systems. Finally, a five-step planning and design procedure is proposed that is able to integrate, within the development process, the hydrological consequences of climate change. For researchers interested in irrigation and drainage and in crop production under changing climate conditions, references have been included, under developments in irrigation section on Page 3. Many climate action plans developed by few cities, states and various countries are cited for policy makers to follow or to make a note off. Few citations are also included in the end to educate every one of us, who are not familiar with the scientific work of our colleagues, related to global warming. The colleagues are from different areas, physics, mathematics, agricultural engineering, crop scientists and policy makers in United Nations. Most of the citation links do open, when you click on them. If it does not, copy and paste the link on any web browsers.

Keywords

Global Warming Prediction Models, Irrigation, Food, Land and Water Shortage and Few Sample Development Plans in Operation for Global Warming

1. Introduction

In recent years, climate change issues have become the focus of the world opinion. Early in the 1970s, scientists had put forward climate warming as a global environmental issue. In 1988 the World Meteorological Organization and the United Nations Environmental Program established "the Intergovernmental Panel on Climate Change (IPCC)" [1] [2]. In 1992, the Rio Conference on Environment and Development passed "the United Nations Framework Convention on Climate Change "(the so-called Convention on Climate Change)" which prescribed that developed countries should combat climate change and its adverse effects [3]. Moreover, the Convention declared that "responses to climate change should be coordinated with social and economic development in an integrated manner for the achievement of sustained economic growth and the eradication of poverty". The Convention recognized that all countries, especially developing countries, needed access to resources required to achieve sustainable social and economic development (Yang, 2012) [4]. In this context, the role of Agriculture is to meet the future challenges posed by food security by increasing production while conserving natural resources.

In the past, the increased demand for food has been satisfied by the expansion of agricultural land. Today, the prospects of increasing the gross cultivated area, in both the developed and developing countries are limited by the dwindling number of economically attractive sites for new large-scale irrigation and drainage projects. Therefore, any increase in agricultural production will necessarily rely largely on a more accurate estimation of crop water requirements on the one hand, and on major improvements in the operation, management and performance of existing irrigation and drainage systems, on the other hand.

The failings of present systems and the inability to sustainably exploit surface and ground water resources can be attributed essentially to poor planning, design, systems management and development. With a population that is expected to grow from 6 billion today to at least 8 billion by the year 2025, bold measures are essential if the problems of irrigation systems and shortage of food are to be avoided.

Concerning agricultural development, most of the world's 270 million ha of irrigated lands and 130 million ha of rain-fed lands with drainage facilities were developed on a step-by-step basis over the centuries. Many of the systems structures have aged or are deteriorating. Added to this, the systems have to withstand the pressures of changing needs, demands and social and economic evolution. Consequently, the infrastructures in most irrigated and drained areas need to be renewed or even replaced and thus redesigned and rebuilt, in order to achieve improved sustainable production. This process depends on a number of common and well-coordinated factors, such as advanced technology, environmental protection, institutional strengthening, economic and financial assessment and human resource development. Most of these factors are well known and linked to uncertainties associated with climate change, world market prices and international trade. All the above factors and constraints compel decision-makers to review the strengths and weaknesses of current trends in irrigation and drainage and rethink technology, institutional and financial patterns, research thrust and manpower policy, so that service levels and system efficiency can be improved in a sustainable manner.

2. Irrigation Development and the Global Food Challenge

To solve the above problems massive investments have been made over the last few decades by governments and individuals and a concerted effort by the International Community. The challenge was to provide enough food for 2 billion more people, while increasing domestic and industrial water demand. Different scenarios have been developed to explore a number of issues, such as the expansion of irrigated agriculture, massive increases in food production from rain-fed lands, water productivity trends and public acceptance of genetically modified crops. Opinions differ among the experts as to some of the above issues. However, there is broad consensus that irrigation can contribute substantially to increasing food production.

Today, the world's food production comes from a cultivated area of about 1.5 billion ha, representing 12% of the total land area (Schultz and De Wrachien, 2002) [5]. About 1.1 billion ha of cultivated land have no water management systems, though this area supplies 45% of food production. At present irrigation covers 270 million ha, *i.e.* 18% of the world's arable land. Overall, irrigated land contributes to 40% of agricultural output and employs about 30% of population in rural areas. It uses about 70% of water withdrawn from global river systems. About 60% of this water is used consumptively, the rest returning to the river systems. Drainage of rain-fed crops covers about 130 million ha, *i.e.* 9% of the world's arable land. In about 60 million ha of the irrigated lands there is a drainage system, as well. The 130 million or so hectares of drained rainfed land produces around 15% of crop output.

2.1. Developments in Irrigation

Over the last forty years, the irrigation has been a major contributor to the growth of food and fiber supply for a global population that has more than doubled, from 3 to over 6 billion people. Global irrigated area increased by around 2% a year in the 1960s and 1970s, slowing down to around 1% in the 1980s, and lower still in the 1990s. Between 1965 and 1995 the world's irrigated land grew from 150 to 260 million ha. Nowadays it is increasing at a very slow rate because of the significant slowdown in new investments, combined with the loss of irrigated areas due to salination and urban encroachment.

Notwithstanding these achievements, today the majority of agricultural land (1.1 billion ha) still has no water management system. In this context it is expected that 90% of the increase in food production will have to come from existing cultivated land and only 10% from conversion from other uses. In the rainfed areas with no water management systems some improvements can be achieved with water harvesting and watershed management. However, in no way can the cultivated area with no water management contribute significantly to the required increase in food production. For this reason, the share of irrigated and drained areas in food production will have to increase. This can be achieved either by installing irrigation or drainage facilities in the areas without a system or by improving and modernizing existing systems. The International Commission on Irrigation and Drainage (ICID) estimates that within the next 25 years, this process may result in a shift of the contribution to the total food production to around 30% for the areas with no water management system, 50% for the areas with an irrigation system and 20% for the rainfed areas with a drainage system (Schultz, 2002) [6] [7]. Researchers/ readers interested in in irrigation and drainage and in crop production under changing climate conditions   may like to read articles quoted in [35] [37] [40] [47].

2.2. The Global Food Challenge

As the world population continues to grow so too does the need to constantly increase the food production. Several actions are required to cope with this increasing demand. Globally, the core challenge must be to improve water productivity. Where land is limiting, yields per unit area must also be enhanced. These measures lead to two basic development directions [8]:

♦ increasing the yield frontier in those areas where present levels of production are close to their potential;
♦ closing the yield gap where considerable production gains can be achieved with current technology.

Based on the above assumptions, three models of food and irrigation water demand have been developed by non-governmental organizations for the time horizon 2025 (Plusquellec, 2002) [9]. These three models predict that present irrigated agriculture would have to increase by 15% - 22%. Moreover, water withdrawals for irrigation are also expected to increase at unprecedented rates, a major challenge considering that environmentalists argue that irrigation withdrawals should be reduced, as they have great expectations in the potential of biotechnology in agriculture.

Although the scenarios differ considerably, it is generally agreed that the world is entering the twenty-first century on the brink of a new food crisis, as ominous, but far more complex, that the famine it faced in the 1960s. Some analysts believe that what is needed is a new and "greener revolution" to increase productivity again and boost production. But the challenges are far more complex than simply producing more food, because global conditions have changed since the green revolution years.

3. Climate and Climatic Change

3.1. The Greenhouse Effect

Over the past centuries, the Earth's climate has been changing due to a number of natural processes, such as gradual variation in solar radiation, meteorite impacts and, more importantly, sudden volcanic eruptions in which solid matter, aerosols and gases are ejected into the atmosphere. Ecosystems have adapted continuously to these natural changes in climate, and flora and fauna have evolved in response to the gradual modifications to their physical surroundings, or have become extinct.

Human beings have also been affected by and have adapted to changes in local climate, which, in general terms, have occurred very slowly. Over the past century, however, human activities have begun to affect the global climate. These effects are due not only to population growth, but also to the introduction of technologies developed to improve the standard of living. Human-induced changes have taken place much more rapidly than natural changes. The scale of current climate forcing is unprecedented and can be attributed to greenhouse gas emissions, deforestation, urbanization, and changing land use and agricultural practices. The increase in greenhouse gas emissions into the atmosphere is responsible for the increased air temperature, and this, in turn, induces changes in the different components making up the hydrological cycle such as evapotranspiration rate, intensity and frequency of precipitation, river flows, soil moisture and groundwater recharge. Mankind will certainly respond to these changing conditions by taking adaptive measures such as changing patterns in land use. However, it is difficult to predict what adaptive measures will be chosen, and their socio-economic consequences [10]-[15].

Concern global patterns the following considerations can be drawn from analysis of the hydrologic and meteorological time series available:

- Average global temperature rose by 0.6°C during the 20th century [16].
- 1990's was the warmest decade and 1998 the warmest year since 1861 [17].
- The extent of snow cover has decreased by 10% since the late 1960s [18].
- Average global sea level rose between 0.1 and 0.2 meters during the 20th century [19] [20].
- Precipitation increased by 0.5% to 1% per decade in the 20th century over the mid and high latitudes of the northern hemisphere and by between 0.2% and 0.3% per decade over the tropics (10°N to 10°S) [21].
- Precipitation decreased over much of the northern sub-tropical (10°N to 30°N) land areas during the 20th century by about 0.3% per decade [22].
- The frequency of heavy rain events increased by 2% to 4% in the mid and high latitudes of the northern hemisphere in the second half of the 20th century. This could be the result of changes in atmospheric moisture, thunderstorm activity, large-scale storm activity, etc. [23]
- Over the 20th century land areas experiencing severe drought and wetness have increased [23].
- Some regions of Africa and Asia recorded an increase in the frequency and intensity of drought in the last decade [24].

- CO_2 concentration has increased by 31% since 1750 [25].
- 75% of CO_2 emissions is produced by fossil fuel burning, the remaining 25% by land use change especially deforestation [25].
- Methane CH_4 has increased by 151% since 1750 and continues to increase. Fossil fuel burning, livestock, rice cultivation and landfills are responsible for emissions [26].
- Nitrous Oxide (N_2O) has increased by 17% since 1750 and continues to increase. This gas is produced by agriculture, soils, cattle feed lots and the chemical industry [27].
- Stratospheric Ozone (O_3) layer has been depleting since 1979 [28].

3.2. Climate Change Scenarios

Current scientific research is focused on the enhanced greenhouse effect as the most likely cause of climate change in the short-term. Until recently, forecasts of anthropogenic climate change have been unreliable, so that scenarios of future climatic conditions have been developed to provide quantitative assessments of the hydrologic consequences in some regions and/or river basins. Scenarios are "internally-consistent pictures of a plausible future climate" (Wigley et al., 1986) [29]. These scenarios can be classified into three groups:

- hypothetical scenarios;
- climate scenarios based on General Circulation Models (GCMs) [30];
- scenarios based on reconstruction of warm periods in the past (paleoclimatic reconstruction).

The plethora of literature on this topic has been recently summarized by the Intergovernmental Panel on Climate Change [31].

The scenarios of the second group have been widely utilized to reconstruct seasonal conditions of the change in temperature, precipitation and potential evapotraspiration at basin scale over the next century. GCMs are complex three-dimensional computer-based models of the atmospheric circulation, which provide details of changes in regional climates for any part of the Earth. Until recently, the standard approach has been to run the model with a nominal "pre-industrial" atmospheric carbon dioxide (CO_2) concentration (the control run) and then to rerun the model with doubled (or sometimes quadrupled) CO_2 (the perturbed run). This approach is known as "the equilibrium response prediction". The more recent and advanced GCMs are, nowadays, able to take into account the gradual increase in the CO_2 concentration through the perturbed run. However, current results are not sufficiently reliable.

4. Climate Change and Irrigation Requirements

Agriculture is a human activity that is intimately associated with climate. It is well known that the broad patterns of agricultural growth over long time scales can be explained by a combination of climatic, ecological and economic factors. Modern agriculture has progressed by weakening the downside risk of these factors through irrigation, the use of pesticides and fertilizers, the substitution of human labor with energy intensive devices, and the manipulation of genetic resources. A major concern in the understanding of the impacts of climate change is the extent to which world agriculture will be affected. Thus, in the long term, climate change is an additional problem that agriculture has to face in meeting global and national food requirements. This recognition has prompted recent advances in the coupling of global vegetation and climate models.

In the last decade, global vegetation models have been developed that include parameterizations of physiological processes such as photosynthesis, respiration, transpiration and soil water in-take (Bergengren et al.) [32]. These tools have been coupled with General Circulation Models (GCMs) and applied to both paleoclimatic and future scenarios (Doherty et al. and Levis et al. [33] [34]. The use of physiological parameterizations allows these models to include the direct effects of changing CO_2 levels on primary productivity and competition, along with the crop water requirements. In the next step the estimated crop water demands could serve as input to agro-economic models which compute the irrigation water requirements (IR), defined as the amount of water that must be applied to the crop by irrigation in order to achieve optimal crop growth.

On the global scale, scenarios of future irrigation water use have been developed by Seckler et al. [35] and Alcamo et al. (2000). Alcamo et al. employed the raster-based Global Irrigation Model (GIM) [37] of Döll and Siebert (2002) [36], with a spatial resolution of 0.50 by 0.50. This model represents one of the most advanced tools today available for exploring the impact of climate change on IR at worldwide level.

More recently, the GIM has been applied to explore the impact of climate change on the irrigation water requirements of those areas of the globe equipped for irrigation in 1995 (Döll, 2002) [36]. Estimates of long-term

average climate change have been taken from two different GCMs:
♦ the Max Planck Institute for Meteorology (MPI-ECHAM4), Germany;
♦ the Hadley Centre for Climate Prediction and Research (HCCPR-CM3), UK.

The following climatic conditions have been computed:
♦ present-day long-term average climatic conditions, *i.e.* the climate normal 1961-1990 (baseline climate) [38];
♦ future long-term average climatic conditions of the 2020s and 2070s (climatic change) [39].

For the above climatic conditions, the GIM computed both the net and gross irrigation water requirements in all 0.50 by 0.50 raster cells with irrigated areas. "Gross irrigation requirement" is the total amount of water that must be applied such that evapotraspiration may occur at the potential rate and optimum crop productivity may be achieved. Only part of the irrigated water is actually used by the plant and evapotranspirated. This amount, *i.e.* the difference between the potential evapotranspiration and the evapotranspiration that would occur without irrigation, represents the "net irrigation requirement" (IRnet. [40]).

The simulations show that irrigation requirements increase in most irrigated areas north of 40°N, by up to 30%, which is mainly due to decreased precipitation, in particular during the summer. South of this latitude, the pattern becomes complex. For most of the irrigated areas of the arid northern part of Africa and the Middle East, IRnet diminishes. In Egypt, a decrease of about 50% in the southern part is accompanied by an increase of about 50% in the central part [41]. In central India, baseline IRnet values of 250 - 350 mm are expected to more than double by the 2020s [42]. In large parts of China the impact of climate change is negligible (less than 5%), with decreases in northern China, as precipitation is assumed to increase [43]. When the cell-specific net irrigation requirements are summed up over the world regions, increases and decreases of the cell values caused by climate change almost average out, increasing by 3.3% in the 2020s and by 5.5% in the 2070s [5] [40] (**Table 1**). Climate Change and Water Availability.

Table 1. The simulations also show that in areas equipped for irrigation in 1995 IRnet is likely to increase in 66% of these areas by the 2020s and in 62% by the 2070s.

	Irrigated Cropping Long-Term Average IRnet, km^3/yr					
	Area 1995, 1000 km^2	Intensity	Baseline	2020s ECHAM4	HadCM3	2070s ECHAM4	HadCM3
Canada	7.1	1.0	2.4	2.9	2.7	3.3	2.9
USA	235.6	1.0	112.0	120.6	117.9	123.0	117.9
Central America	80.2	L.O	17.5	17.0	17.6	18.1	19.7
South America	98.3	LO	26.6	27.1	27.S	28.2	29.1
Northern Africa	59.4	LS	66.4	62.7	65.3	56.0	57.7
Western Africa	8.3	1.0	2.5	2.2	2.4	2.4	2.6
Eastern Africa	35.8	1.O	12.3	13.1	12.2	14.5	14.3
Southern Africa	18.6	1.0	7.1	7.0	7.4	6.4	7.2
OECD Europe	118.0	1.0	52.4	55.8	55.2	56.5	57.8
Eastern Europe	49.4	LO	16.7	18.4	19.0	19.7	22.1
Former U.S.S.R.	218.7	0.8	104.6	106.6	112.1	104.4	108.7
Middle East	185.3	1.O	144.7	138.7	142.4	126.5	137.8
South Asia	734.6	L3	366.4	389.8	400.4	410.7	422.0
East Asia	492.5	1.5	123.8	126.0	126.6	131.3	127.1
Southeast Asia	154.4	1.2	17.1	20.3	18.8	30.4	28.6
Oceania	26.1	L5	17.7	17.8	17.6	18.2	19.7
Japan	27.0	L5	1.3	1.3	L8	1.4	1.5
World	2549.1		1091.5	1127.5	1147.0	1151.0	1176.8

Irrigated areas of 1995, under 1961-1990 average observed climate ("baseline"), and scaled with MPI-ECHAM4 or HCCPR-CM3 climate change scenarios for 2020-2029 ("2020s") and 2070-2090 ("2070s").

In order to assess the problem of water scarcity, the appropriate averaging units are not world regions but river basins.

Climate predictions from four state-of-the-art General Circulation Models were used to assess the hydrologic sensitivity to climate change of nine large, continental river basins (Nijssen *et al.*, 2001). The river basins were selected on the basis of the desire to represent a range of geographic and climatic conditions. Four models have been used:

♦ the Hadley Centre for Climate Prediction and Research (HCCPR-CM2), UK;
♦ the Hadley Centre for Climate Prediction and Research (HCCPR-CM3), UK;
♦ the Max Planck Institute for Meteorology (MPI-ECHAM4), Germany;
♦ the Department of Energy (DOE-PCM3), USA.

All predicted transient climate response to changing greenhouse gas concentrations and incorporated modern land surface parameterizations. The transient emission scenarios differ slightly from one model to another, partly because they represent greenhouse gas chemistry differently.

Changes in basin-wide, mean annual temperature and precipitation were computed for three decades in the transient climate model runs (2025, 2045 and 2095) and hydrologic model simulations were performed for decades centered on 2025 and 2045 [43].

The main conclusions are summarized below.

♦ All models predict a warming for all nine basins, but the amount of warming varies widely between the models, especially for the longer time horizon. The greatest warming is predicted to occur during the winter months in the highest latitudes [44]. Precipitation generally increases for the northern basins, but the signal is mixed for basins in the mid-latitudes and tropics, although on average slight precipitation increases are predicted [45].

♦ The largest changes in hydrological cycle are predicted for the snow-dominated basins of mid to higher latitudes, as a result of the greater amount of warming that is predicted for these regions. The presence or absence of snow fundamentally changes the water balance, due to the fact that water stored as snow during the winter does not become available for runoff or evapotranspiration until the following spring's melt period [45].

♦ Globally, the hydrological response predicted for most of the basins in response to the GCMs predictions is a reduction in annual stream flow in the tropical and mid-latitudes. In contrast, high-latitude basins tend to show an increase in annual runoff, because most of the predicted increase in precipitation occurs during the winter, when the available energy is insufficient for an increase in evaporation. Instead, water is stored as snow and contributes to stream flow during the subsequent melt period [45].

5. Planning and Design of Irrigation Systems under Climate Change

Uncertainties as to how the climate will change and how irrigation systems will have to adapt to these changes, are challenges that planners and designers will have to cope with. In view of these uncertainties, planners and designers need guidance as to when the prospect of climate change should be embodied and factored into the planning and design process (De Wrachien and Feddes, 2004) [46]. An initial question is whether, based on GCM results or other analyses, there is reason to expect that a region's climate is likely to change significantly during the life of a system. If significant climate change is thought to be likely, the next question is whether there is a basis for forming an expectation about the likelihood and nature of the change and its impacts on the infrastructures [47].

The suitability and robustness of an infrastructure can be assessed either by running "what if scenarios" that incorporate alternative climates or through synthetic hydrology by translating apparent trends into enhanced persistence [46].

When there are grounds for formulating reasonable expectations about the likelihood of climate changes, the relevance of these changes will depend on the nature of the project under consideration. Climate changes that are likely to occur several decades from now will have little relevance for decisions involving infrastructure development or incremental expansion of existing facilities' capacity. Under these circumstances planners and designers should evaluate the options under one or more climate change scenario to determine the impacts on the project's net benefits. If the climate significantly alters the net benefits, the costs of proceeding with a decision assuming no change can be estimated. If these costs are significant, a decision tree can be constructed for evaluating the alternatives under two or more climate scenarios (Hobbs *et al.*, 1997) [48].

Delaying an expensive and irreversible project may be a competitive option, especially in view of the prospect that the delay will result in a better understanding as to how the climate is likely to change and impact the effectiveness and performance of the infrastructure [46].

Aside from the climate change issue, the high costs of and limited opportunities for developing new large scale projects, have led to a shift away from the traditional, fairly inflexible planning principles and design criteria for meeting changing water needs and coping with hydrological variability and uncertainty. Efficient, flexible works designed for current climatic trends would be expected to perform efficiently under different environmental conditions. Thus, institutional flexibility that might complement or substitute infrastructure investments is likely to play an important role in irrigation development under the prospect of global climatic change. Frederick *et al.* (1997) proposed a five-step planning and design process for water resource systems, for coping with uncertain climate and hydrologic events, and potentially suitable for the development of large irrigation schemes [49].

If climate change is recognized as a major planning issue (first step), the second step in the process would consist of predicting the impacts of climate change on the region's irrigated area. The third step involves the formulation of alternative plans, consisting of a system of structural and/or non-structural measures and hedging strategies that address, among other concerns, the projected consequences of climate change. Non-structural measures that might be considered include modification of management practices, regulatory and pricing policies. Evaluation of the alternatives, in the fourth step, would be based on the most likely conditions expected to exist in the future with and without the plan [50]. The final step in the process involves comparing the alternatives and selecting a recommended development plan. Here in the authors have cited some sample plans by various governemnts [47]-[99]. We have listed a reference [79], to find the proper names for various countries. If the reader likes to find a climate action plan for his country of interest, just go to google.com and type in or search for *climate action* for country of your interest. We have cited climate action plans proposed by few countries like USA, India, China, and Europen contintent.

The planning and design process needs to be sufficiently flexible to incorporate consideration of and responses to many possible climate impacts. Introducing the potential impacts of and appropriate responses to climate change in planning and design of irrigation systems can be both expensive and time consuming. The main factors that might influence the worth of incorporating climate change into the analysis are the level of planning (local, national, international), the reliability of GCMs, the hydrologic conditions, the time horizon of the plan or life of the project [94] [95].

6. Concluding Remarks

♦ Agriculture will have to meet the future challenges posed by food security by increasing production while conserving natural resources.

♦ With a population that is expected to grow from 6 billion today to at least 8 billion by the year 2025 [100] [101], bold measures are essential if the problems of irrigation systems and shortage of food are to be avoided.

♦ Different scenarios have been developed to explore a number of issues, such as the expansion of irrigated agriculture, massive increases in food production from rainfed lands and water productivity trends. Opinions differ among experts as to some of the above issues. However, there is broad consensus that irrigation can contribute substantially to increasing food production in the years to come.

♦ Most of the world's irrigation systems were developed on a step-by-step basis over the centuries and were designed for a long life (50 years or more), on the assumption that climatic conditions would not change. This will not be so in the future, due to global warming and the greenhouse effect. Therefore, engineers and decision-makers need to systematically review planning principles, design criteria, operating rules, contingency plans and water management policies.

♦ Uncertainties as to how the climate will change and how irrigation systems will have to adapt to these changes are issues that water authorities are compelled to address. The challenge is to identify short-term strategies to cope with long-term uncertainties. The question is not what the best course for a project is over the next fifty years or more, but rather, what is the best direction for the next few years, knowing that a prudent hedging strategy will allow time to learn and change course.

♦ The planning and design process needs to be sufficiently flexible to incorporate consideration of and responses to many possible climate impacts. The main factors that will influence the worth of incorporating climate change into the process are the level of planning, the reliability of the forecasting models, the hydrological conditions and the time horizon of the plan or the life of the project.

The development of a comprehensive approach that integrates all these factors into irrigation project selection requires further research on the processes governing climate changes, the impacts of increased atmospheric carbon dioxide on vegetation and runoff, the effect of climate variables on crop water requirements and the impacts of climate on infrastructure performance.

Acknowledgements

Heartfelt acknowlegements to all the authors, cited in this global warming review article.   Because of the enormous challenge that our human race is going through in this 21st century, we the authors of this article, took liberty to share the knowledge and findings and plans that many of you have posted on the web sites. I hope we all scientists of diffrent countries share our know how in advance before we face the worst scenerios of human, plants and nature suffering due to our own negligence or unawareness.

References

[1]   Intergovernmental Climate Change (1988) https://www.ipcc.ch/organization/organization.shtml

[2]   Intergovernmental Climate Change Resolutions https://www.ipcc.ch/docs/WMO_resolution4_on_IPCC_1988.pdf

[3]   Climate Change 2001: Synthesis Report. http://www.ipcc.ch/ipccreports/tar/vol4/index.php?idp=204

[4]   Yang, Z.W. (2012) The Right to Carbon Emission: A New Right to Development. *American Journal of Climate Change*, 1, 108-116. http://dx.doi.org/10.4236/ajcc.2012.12009

[5]   De Wrachien, D. and Feddes, R. (2003) Drainage Development in a Changing Environment: Overview and Challenge. *9th International Drainage Workshop Drainage for a Secure Environment and Food Supply*, Utrecht, 10-13 September 2003, 1-31.
http://www.researchgate.net/publication/275977046_Drainage_Development_in_a_Changing_Environment__Overview_and_Challenges

[6]   Schultz, B. (2002) Opening Address. *Proceedings of the 18th Congress on Irrigation and Drainage (ICID)*. Montreal, 21-28 July 2002, 7-23. http://afeid.montpellier.cemagref.fr/old/Montreal2002final.pdf

[7]   Schultz, B. and De Wrachien, D. (2002) Irrigation and Drainage Systems. Research and Development in the 21st Century. *Irrigation and Drainage*, **51**, 311-327. http://dx.doi.org/10.1002/ird.67

[8]   Molden, D., *et al.* (2007) Trends in Water and Agricultural Development. IWMI Part 2 Ch2-3 final. indd 57.
http://www.iwmi.cgiar.org/assessment/Water%20for%20Food%20Water%20for%20Life/Chapters/Chapter%202%20Trends.pdf

[9]   Plusquellec, H. (2002) Is the Daunting Challenge of Irrigation Achievable? *Irrigation and Drainage*, **51**, 185-198. http://dx.doi.org/10.1002/ird.51

[10]   Stoddard, J.L., *et al.* (1999) Letters to Nature. Regional Trends in Aquatic Recovery from Acidification in North America and Europe. *Nature*, **401**, 575-578. http://www.nature.com/nature/journal/v401/n6753/full/401575a0.html

[11]   Schils, R., Kuikman, P., Liski, J., Van Oijen, M., Smith, P., Webb, J., *et al.* (2008) Review of Existing Information on the Interrelations between Soil and Climate Change. (ClimSoil). Final Report. Center for Ecology and Hydrology. Natural Environmental Research Council. European Commission, Brussels, 1-208.
http://nora.nerc.ac.uk/6452/1/climsoil_report_dec_2008.pdf
http://ec.europa.eu/environment/soil/review_en.htm

[12]   Walther, G.-R., Post, E., Convey, P., Menzel, A., Parmesan, C., Beebee, T.J.C., *et al.* (2002) Ecological Responses to Recent Climate Change. *Nature*, **416**, 389-395. http://www.nature.com/nature/journal/v416/n6879/full/416389a.html
http://dx.doi.org/10.1038/416389a

[13]   Seinfeld, J.H. and Pandis, S.N. (2012) Atmospheric Chemistry and Physics: From Air Pollution to Climate Change. John Wiley & Sons, 1-1232. ISBN: 978-0-471-72018-8.

[14]   US Department of Interior and US Geological Survey (2013) Artificial Groundwater Recharge.
http://water.usgs.gov/ogw/artificial_recharge.html

[15]   De Wrachien, D., Rag, A.R. and Giordano, A. (2003) Climate Change Land Degradation and Desertification in the Mediterranean Environment. NATO-CCMS Workshop. https://air.unimi.it/bitstream/2434/28644/1/Deser5rel.doc

http://www.epa.gov/esd/land-sci/desert/images/book-of-abstracts-nato-workshop.pdf

[16] World Nuclear Association (2014) Climate Change—The Science. Greenhouse Effect.
http://www.world-nuclear.org/info/Energy-and-Environment/Climate-Change---The-Science/

[17] Inter-Governmental Panel on Climate Change (IPCC) (2001) Climate Change 2001: The Scientific Basis. The Observed Changes in the Climate System Is the Earth's Climate Changing?
http://www.ipcc.ch/ipccreports/tar/vol4/082.htm
http://www.ipcc.ch/publications_and_data/publications_and_data_reports.shtml

[18] Inter-Governmental Panel on Climate Change (IPCC) (2001) Working Group I: The Scientific Basis. Observed Changes in Precipitation and Atmospheric Moisture. http://www.ipcc.ch/ipccreports/tar/wg1/013.htm

[19] Gammon, R. (Guest Lecturer) (2001) Climate Models over the Next 100 Years.
http://www.atmos.washington.edu/2001Q4/211/notes_climatechange.html

[20] USGCRP (2014) Our Changing Planet, Indicators of Change, Adapting to Change. USGCRP Released the Third National Climate Assessment, the Authoritative and Comprehensive Report on Climate Change and Its Impacts in the United States. http://www.globalchange.gov/

[21] Cline, W.R. (2004) Meeting the Challenge of Global Warming. Copenhagen Consensus 2004 Project.
http://www.copenhagenconsensus.com/sites/default/files/CP%2B-%2BGlobal%2BWarming%2BFINISHED.pdf

[22] Albritton, D.L., Allen, M.R., Baede, A.P.M., Church, J.A., Cubasch, U., Dai, X., et al. (2001) Summary for Policymakers. A Report of Working Group I of the Intergovernmental Panel on Climate Change (IPCC).
http://www.atmos.washington.edu/2003Q4/211/articles_required/IPCC_2001_SPM.pdf

[23] Inter-Governmental Panel on Climate Change (IPCC) (2001) Working Group I: The Scientific Basis. Droughts and Wet Spells. http://ipcc.ch/ipccreports/tar/wg1/092.htm

[24] Speranskaya, O., Eco-Accord and Olesen, G.B. (2001) Climate Change and Energy. The Climate Change Has Been Proven by Soli Scientific Facts. Published by ECO-Accord, Moscow, and Forum for Energy and Development, Denmark. http://www.ecoaccord.org/climate/2001-engl/1.htm

[25] Environmental Protection Agency (EPA) (2013) Overview of Greenhouse Gases. Carbon Dioxide Emissions.
http://www.epa.gov/climatechange/ghgemissions/gases/co2.html

[26] De Wrachien, D., Ragab, R. and Giordano, A. (2006) Climate Change, Land Degradation, and Desertification in the Mediterranean Environment. A Security Issue. NATO Security through Science Series, Vol. 3, 353-371.
http://link.springer.com/chapter/10.1007%2F1-4020-3760-0_16#page-1

[27] IPCC Assessment Report on Climate Change (2013) What Causes This Climate Change? Intergovernmental Panel on Climate Change (IPCC). http://www.greenfacts.org/en/climate-change-ar3/l-3/climate-change-2.htm

[28] Earth System Research Laboratory Stratospheric Ozone Layer Depletion and Recovery. U.S. Department of Commerce, National Oceanic and Atmospheric Administration, Earth System Research Laboratory.
http://www.esrl.noaa.gov/research/themes/o3/

[29] Wigley, T.M.L., Jones, P.D. and Kelly, P.M. (1986) Empirical Climate Studies. Warm World Scenarios and the Detection of Climate Change Induced by Radioactively Active Fases. In: Bolin, B., Doos, B.R., Jager, J. and Warrich, R.A., Eds., The Greenhouse Effect, Climatic Change and Ecosystems, SCOPE 26, Wiley Publisher, New York, 271-322.

[30] Numerous Scientists (2015) The Discovery of Global Warming. General Circulation Models of Climate.
http://www.aip.org/history/climate/GCM.htm

[31] Scott, D.B. and Collins, E.S. (1996) Late Mid-Holocene Sea-Level Oscillation: A Possible Cause. Quaternary Science Reviews, 15, 851-856. http://www.sciencedirect.com/science/article/pii/S0277379196000637
http://dx.doi.org/10.1016/s0277-3791(96)00063-7

[32] Bergengren, J.C., Thompson, S.L., Pollard, D. and Deconto, R.M. (2001) Modeling Global Climate-Vegetation Interactions in a Doubled CO_2 World. Climatic Change, 50, 31-75.
http://link.springer.com/article/10.1023/A:1010609620103#page-1
http://dx.doi.org/10.1023/A:1010609620103

[33] Doherty, R., Kutzbach, J., Foley, I. and Pollard, D. (2000) Fully-Coupled Climate/Dynamical Vegetation Model Simulation over Northern Africa during Mid-Holocene. Climate Dynamics, 16, 561-573.
http://www.research.ed.ac.uk/portal/en/publications/fully-coupled-climatedynamical-vegetation-model-simulations-over-northern-africa-during-the-midholocene%28e07a8ccb-e5bf-406e-9a45-baec5701ced5%29/export.html
http://dx.doi.org/10.1007/s003820000065

[34] Levis, S., Foley, J.A. and Pollard, D. (2000) Large-Scale Vegetation Feedbacks on a Double CO_2 Climate. Journal of Climate, 13, 1313-1325.
http://journals.ametsoc.org/doi/abs/10.1175/1520-0442%282000%29013%3C2217%3ARTCSTC%3E2.0.CO%3B2

[35] Seckler, D., Amarasinghe, V., Molden, D., de Silva, R. and Barker, R. (1997) World Water Demand and Supply, 1990 to 2025: Scenarios and Issues. Research Report 19, IWMI, Colombo, Sri Lanka.
http://dlc.dlib.indiana.edu/dlc/bitstream/handle/10535/4351/REPORT19.pdf?sequence=1

[36] Döll, P. (2002) Impact of Climate Change and Variability on Irrigation Requirements: A Global Perspective. *Climatic Change*, **54**, 269-293. http://link.springer.com/article/10.1023/A%3A1016124032231
http://dx.doi.org/10.1023/A:1016124032231

[37] Alcamo, J., Henrish, T. and Rösch, T. (2000) World Water in 2025. Global Modeling and Scenario Analysis for the World Commission on Water for the 21st Century. Kassel World Water Series Report 2. Centre for Environmental Systems Research, University of Kassel, Germany.
http://www.usf.uni-kassel.de/usf/archiv/dokumente/kwws/kwws.2.pdf

[38] The Climatic Research Unit Global Climate Dataset The Mean 1961-90 Climatology. Intergovernmental Climate Change (IPCC). http://www.ipcc-data.org/observ/clim/cru_climatologies.html

[39] Charabi, Y. and Al-Khoudh, S. (2013) Projection of Future Changes in Rainfall and Temperature Patterns in Oman. *Journal of Earth Science & Climatic Change*, **4**, 154.
http://omicsonline.org/projection-of-future-changes-in-rainfall-and-temperature-patterns-in-oman-2157-7617.1000154.pdf

[40] Döll, P. (2002) Impact of Climate Change and Variability on Irrigation Requirements: A Global Perspective. *Climatic Change*, **54**, 269-293. https://www.uni-frankfurt.de/45217733/doell_ClimaticChange2002_irrigation.pdf
http://dx.doi.org/10.1023/A:1016124032231

[41] El-Nahrawy, M.A. (2011) Egypt. Country Pasture/Forage Resource Profile.
http://www.fao.org/ag/AGP/AGPC/doc/Counprof/Egypt/Egypt.html

[42] De Wrachien, D. (2003) Global Warming and Irrigation Development. A World-Wide View. International Scientific Conference on Agricultural Water Management and Mechanization Factors for Sustainable Agriculture, Sofia, 8-10 October 2003. https://air.unimi.it/bitstream/2434/28642/1/Sofiarel.doc

[43] IPCC (2007) IPCC Fourth Assessment Report: Climate Change. Working Group II: Impacts, Adaptation and Vulnerability. How Will Climate Change Affect the Balance of Water Demand and Water Availability.
https://www.ipcc.ch/publications_and_data/ar4/wg2/en/ch3s3-5-1.html

[44] Nijssen, B., O'Donnell, G.M., Hamlet, A.F. and Lettenmaier, D.P. (2001) Hydrologic Sensitivity of Global Rivers to Climate Change. *Climatic Change*, **50**, 143-175.
http://climate.org/resources/climate-impacts/hydro/abstracts/index.html

[45] De Wrachien, D., Ragab, R., Hamdyand, A. and Trisorio Liuzz, G. (2004) Conference Paper. Conference: Food Security under Water Scarcity in the Middle East: Problems and Solutions.
http://www.researchgate.net/profile/Daniele_Wrachien/publication/275980991_Global_Warming_Water_Scarcity_and_Food_Security_in_the_Mediterranean_Environment/links/554dc38608ae93634ec59c8c

[46] Frederick, K.D. (2011) Principles and Concepts for Water Resources Planning under Climate Uncertainty.
http://opensiuc.lib.siu.edu/cgi/viewcontent.cgi?article=1238&context=jcwre

[47] Hobbs, B.F., Chao, P.T. and Venkatesh, B.M. (1997) Using Decision Analysis to Include Climate Change in Water Resources in Decision Making. *Climatic Change*, **37**, 177-202. http://dx.doi.org/10.1023/A:1005376622183

[48] Frederick, K.D., Major, D.C. and Stakhiv, E.Z. (1997) Water Resources Planning Principles and Evaluation Criteria for Climate Change. *Climatic Change*, **37**, 291-313. http://link.springer.com/chapter/10.1007/978-94-017-1051-0_16
http://dx.doi.org/10.1023/A:1005332807162

[49] Wrachien Daniele D. Wrachien, Feddes, R., Ragab, R. and Ba (2004) Agricultural Development and Food Security under Climate Uncertainty. New Medit N. 3, 12-19.
http://www.iamb.it/share/img_new_medit_articoli/140_12wrachien.pdf

[50] Reed, M.S., Fraser, E.D.G. and Dougill, A.J. (2006) An Adaptive Learning Process for Developing and Applying Sustainability Indicators with Local Communities. *Ecological Economics*, **59**, 406-418.
http://www.sciencedirect.com/science/article/pii/S0921800905005161
http://dx.doi.org/10.1016/j.ecolecon.2005.11.008

[51] Jankowsk, P. (1995) Integrating Geographical Information Systems and Multiple Criteria Decision-Making Methods. *International Journal of Geographical Information Systems*, **9**, 251-273.
http://www.tandfonline.com/doi/abs/10.1080/02693799508902036

[52] Rasmussen, B. and Lyons, W.M. (2008) Statewide and Regional Transportation Planning. 2-Puget Sound Regional Council Case Study. Integration of Climate Change Considerations into the Seattle Metropolitan Area Transportation Planning Process. http://climate.dot.gov/state-local/integration/chapter_02.html

[53] California (2010) Adaptation Strategies A Guide Book for Global Warming from California. 1-121.

http://www.climatechange.ca.gov/ecrcf/docs/CCSAdaptationGuidebook2011.pdf

[54] The Center for Climate Strategies (2015) Adaptation Guidebook Comprehensive Climate Action.
http://www.climatestrategies.us/

[55] EPA (United States Environmental Protection Agency) (2014) Being Prepared for Climate Change. A Workbook for Developing Risk-Based Adaptation Plans.
http://www2.epa.gov/sites/production/files/2014-09/documents/being_prepared_workbook_508.pdf

[56] Global Policy Form (2013) Climate Change and It's Impact on Poor Famers.
https://www.globalpolicy.org/social-and-economic-policy/the-environment/climate-change.html

[57] European Climate Adaption Form (2015) About Climate Change Adaptation in Europe.
http://climate-adapt.eea.europa.eu/home

[58] Sala, S., ICT for Development and Environment Specialist (2011) The Role of Information and Communication Technologies for Community-Based Adaptation to Climate Change. Food and Agriculture Organization of the United Nations, Rome, 2010. http://www.fao.org/uploads/media/ap606e_2.pdf

[59] Andrea Egan, N. (2011) Preparing Low-Emission Climate-Resilient Development Strategies. A UNDP Guidebook—Version 1. United Nations Development Programme, Anvil Creative, 1-24.
http://www.undp.org/content/dam/undp/library/Environment%20and%20Energy/Climate%20Strategies/UNDP-LECRDS-Guidebook-v17-web.pdf
http://www.undp.org/content/undp/en/home/librarypage.html

[60] US Army of Corps of Engineers (2009) Shared Vision Planning. Collaborative Planning Toolkit—A Web-Based Resource Guide with Extensive Hyperlinks to Existing Literature.
http://www.sharedvisionplanning.us/CPToolkit/PrintAllContent.asp

[61] Legrand Energy Efficiency. Social Governance (2015) Sustainable Development.
http://www.legrand.com/EN/sustainable-development-description_12847.html

[62] OREGON Scenario Planning Guidelines Resources for Developing and Evaluating Alternative Land Use and Transportation Scenarios.
http://www.oregon.gov/ODOT/TD/OSTI/docs/Scenario%20Planning%20Guidelines/ODOT-Guidelines-April2013-red.pdf

[63] Morrisette, P.M. (1989) The Evolution of Policy Responses to Stratospheric Ozone Depletion. *Natural Resources Journal*, **29**, 793-820. http://www.ciesin.org/docs/003-006/003-006.html

[64] Sathaye, J., Najam, A., Cocklin, C., Heller, T., Lecocq, F., Llanes-Regueiro, J., *et al.* (2007) Sustainable Development and Mitigation. In Climate Change 2007: Mitigation. Contribution of Working Group III to the Fourth Assessment Report of the Intergovernmental Panel on Climate Change. In: Metz, B., Davidson, O.R., Bosch, P.R., Dave, R. and Meyer, L.A., Eds., *Sustainable Development and Mitigation*, Chap. 12, Cambridge University Press, Cambridge, United Kingdom and New York, 693-743. https://www.ipcc.ch/pdf/assessment-report/ar4/wg3/ar4-wg3-chapter12.pdf

[65] Kelly, H.L. and His Group (2012) Urban Watershed Framework. A New Approach to Stormwater Management in San Francisco. 1-35. http://sfwater.org/Modules/ShowDocument.aspx?documentid=2552

[66] US Global Research Program. Introduction to Our Changing Climate.
http://nca2014.globalchange.gov/report/our-changing-climate/introduction

[67] US Global Research Program. Deision Support: Connecting Science, Risk Perception, and Decisions.
http://nca2014.globalchange.gov/report/response-strategies/decision-support

[68] International Encyclopedia of the Social Sciences (2008) Global Warming.
http://www.encyclopedia.com/topic/global_warming.aspx

[69] Pawlenty, T. (2007) Next Generation Energy Initiatives to Reduce Greenhouse Gases.
http://www.pca.state.mn.us/index.php/view-document.html?gid=20235

[70] Asian Develoment Bank (ADB) (2012) Environment Safeguards. A Good Practice Sourcebook Draft Working Document.
http://www.adb.org/sites/default/files/institutional-document/33739/files/environment-safeguards-good-practices-sourcebook-draft.pdf

[71] McLuhan, M. and Fiore, Q., and His Group (2012) SSHRC—Imagining Canada's Future—Readings. In: Schwarz-Herion, O. and Omran, A. Eds., *Strategies towards the New Sustainability Paradigm: Managing the Great Transition to Sustainable Global Democracy*, (e-Book) Springer. http://slab.ocadu.ca/sshrc-imagining-canadas-future-readings

[72] Leach, M., Rockström, J., Raskin, P., Scoones, I., Stirling, A.C., Smith, A., Thompson, J., Millstone, E., Ely, A., Arond, E., Folke, C. and Olsson, P. (2012) Transforming Innovation for Sustainability Ecology and Society. *Ecology and Society*, **17**, 11.

[73] Bodansky, D. (2009) Climate Change: Top 10 Precepts for U.S. Foreign Policy.
      http://www.rff.org/RFF/Documents/IB%2009-01.pdf

[74] Capellán-Pérez, I., Mediavill, M., de Castro, C., Carpintero, Ó. and Miguel, L.J. (2015) More Growth? An Unfeasible
      Option to Overcome Critical Energy Constraints and Climate Change.
      http://www.eis.uva.es/energiasostenible/wp-content/uploads/2015/05/AAM_More_growth_SS_2015.pdf
      http://link.springer.com/article/10.1007%2Fs11625-015-0299-3

[75] Worcester Polytechnique Institute WPI. Social Science & Policy Studies. Domestic Economic Problems and Policies
      such as Development Planning. http://www.wpi.edu/academics/ssps/ugrad-courses.html

[76] Your Internet Guide to Understand Policy Issues (2009) Environmmental Policy. What Are the Most Recent Develop-
      ments Regarding Environmental Matters? http://www.newsbatch.com/environment.htm

[77] Alaska (2011) The Final and Draft Reports of the Climate Change Advisory Groups.
      http://www.climatechange.alaska.gov/

[78] Philippines: M & E System for the National Climate Change (2011) National Climate Change Action Plan 2011-2028.
      http://adaptationmarketplace.org/data/library-documents/NCCAP_TechDoc.pdf

[79] Countries by Alphabetical Order. http://www.internetworldstats.com/list2.htm
      For a plan of a specific country type in google search "Climate Action Plan for the country of interest".

[80] The Whitehouse Offfice (USA) (2014) Climate Action Plan—Strategy to Cut Methane Emissions.
      https://www.whitehouse.gov/the-press-office/2014/03/28/fact-sheet-climate-action-plan-strategy-cut-methane-emission
      s

[81] India (2008) India's Natiuonal Action Plan on Climatre Change.
      http://www.c2es.org/international/key-country-policies/india/climate-plan-summary

[82] National Development and Reform Commission of People's Republic of China (2007) China's National Climate
      Change Programme. http://en.ndrc.gov.cn/newsrelease/200706/P020070604561191006823.pdf

[83] European Commissiion (2014) 2030 Framework for Climate and Energy Policies.
      http://ec.europa.eu/clima/policies/2030/index_en.htm

[84] Grogan, D.S., Zhang, F., Prusevich, A., Lammers, R.B., Wisser, D., Glidden, S., Li, C. and Frolking, S. (2015) Quan-
      tifying the Link between Crop Production and Mined Groundwater Irrigation in China. Science of the Total Environ-
      ment, 511, 161-175. http://www.ncbi.nlm.nih.gov/pubmed/25544335
      https://web.stanford.edu/group/emf-research/docs/CCIIA/2014/Frolking_additional.pdf

[85] Tai, A.P.K., Val Martin, M. and Heald, C.L. (2014) Threat to Future Global Food Security from Climate Change and
      Ozone Air Pollution. Nature Climate Change, 4, 817-821.
      http://www.nature.com/nclimate/journal/v4/n9/full/nclimate2317.html

[86] Pakistan Meteriology Division. Bibliography of Research Publications. Climate Dynamics.
      http://www.pmd.gov.pk/rnd/rndweb/rnd_new/publication.php

[87] List of Publications to Study Crop Yields/Loss in China to under Climate Change Scenerios (2014)
      https://www.apsim.info/Products/Publications.aspx

[88] Agricutural Meteriology Division (India). http://www.imdagrimet.gov.in/

[89] United States Agency for International Development (USAID) (2015) Proceedings of Workshop on Improving of Cli-
      mate Services for Farmers in Africa & South Asia (ICSFAFA).
      http://www.imdagrimet.gov.in/sites/default/files/Final%20%20Proceeding%20of%20Workshop%202-3%20February
      %202015_0.pdf

[90] National Center for Environmental Information (NOAA) (2015) U.S. Climate Divisions. Climate at Glance.
      http://www.ncdc.noaa.gov/monitoring-references/maps/us-climate-divisions.php

[91] Kan, Y., Ma, X. and Khan, S. (2014) Predicting Climate Change Impacts on Maize Crop Productivity in the Loess
      Plateau. Irrigation and Drainage, 63, 394-404. http://onlinelibrary.wiley.com/doi/10.1002/ird.1799/abstract
      http://dx.doi.org/10.1002/ird.1799

[92] Tyagy, A.C. (2015) Aligning the Role of ICID in the Changing Development Paradigm. Irrigation and Drainage, 64,
      152-153. http://onlinelibrary.wiley.com/doi/10.1002/ird.1923/abstract
      http://dx.doi.org/10.1002/ird.1923

[93] Yang, Z.W. (2012) The Right to Carbon Emission: A New Right to Development. American Journal of Climate
      Change, 1, 108-116. http://dx.doi.org/10.4236/ajcc.2012.12009

[94] De Wrachien, D. (2003) Paddy and Water Environment: Facilitation. Information Exchange and Identifying Future R
      & D Needs. Paddy and Water Environment, 1, 3-5.

[95] Fears, D. (2015) A "Megadrought" Will Grip U.S. in the Coming Decades, NASA Researchers Say. Healh & Science.

http://www.washingtonpost.com/national/health-science/todays-drought-in-the-west-is-nothing-compared-to-What-may-be-coming/2015/02/12/0041646a-b2d9-11e4-854b-a38d13486ba1_story.html

[96] Cook, B. (2015) A "Megadrought" Will Grip U.S. in the Coming Decades, NASA Researchers Say. https://www.youtube.com/watch?v=8p9oJWg2FSI

[97] Francis, P. (2015) In Sweeping Encyclical, Calls for Swift Action on Climate Change. The New York Times, 18 June 2015. http://www.nytimes.com/2015/06/19/world/europe/pope-francis-in-sweeping-encyclical-calls-for-swift-action-on-climate-change.html?_r=0

[98] Redford, R. (2015) Interview: "Climate Change Is in Everybody's Backyard". United Nations Headquarters, New York.

[99] Mann, M. (2014) The Hockey Stick and the Climate Wars. https://www.youtube.com/watch?v=OZ7dPyxr98M

[100] UN News Center (2013) World Population Projected to Reach 9.6 Billion by 2050—UN Report. http://www.un.org/apps/news/story.asp?NewsID=45165#.VaevevlViko

[101] UN News Center (2015) ADDIS: "Historic" Agreement Reached on Financing for New UN Sustainable Development Agenda. http://www.un.org/apps/news/story.asp?NewsID=51433#.VaenuPlViko

20

Sharp Expansion of Intensive Groundwater Irrigation, Semi-Arid Environment at the Northern Bekaa Valley Lebanon

Ihab Jomaa[1], Myriam Saadė Sbeih[2], Ronald Jaubert[2]

[1]Department of Irrigation and Agrometeorology, Lebanese Agriculture Research Institute, Zahle, Lebanon
[2]Graduate Institute of Geneva, Geneva, Switzerland
Email: ijomaa@lari.gov.lb

Abstract

This research focuses on the sharp expansion of groundwater irrigation in the Northern Beqaa, using Landsat satellite images and other auxiliary GIS relevant data sources. Topographic maps were used to assess the location and size of the irrigated area in the early 1960s as the initial years of agriculture expansion analysis. The first available Landsat image of the area was of the year 1972 followed with a series of accessible Landsat images until 2009. In the 1960s, agricultural practices were only limited to areas of surface water resources and open channels next to urban settlements. In the Early 1980s, farmers discovered the agricultural potential of the area. Only 3% of the area was cultivated before the 1970s. The cultivated area reached about the 20% in late 1990s. Weather conditions, shallowness of groundwater tables, low fuel costs and market opportunities have led to an agricultural boom in the area considered as prone to desertification and of low productivity by national authorities. The area is however poorly understood from its hydro geological characteristics and exposed to intensive and unsustainable use of its natural resources.

Keywords

Northern Bekaa, Agriculture Expansion, Landsat, NDVI, Surface Irrigation, Surface Water Sources, Groundwater

1. Introduction

In Lebanon, agriculture is trapped between four chief facts: a geographically small country, diversity in climate and terrains, soil types and national and regional political instability. The high Mount-Lebanon chain blocks the

Mediterranean climate from reaching inland areas. Relatively separated from the direct sea influence, the Bekaa Valley is by itself widely diverse in climate and land characteristics. Precipitation starts with about 800 mm to the south of the valley and it reaches less than 200 mm throughout its north. At the northern part of the Bekaa Valley, the Orontes River Watershed is, the zone of interest in this research, characterized by a low rainfall rate ranging between 300 mm to the south and 100 mm to the north.

The northern Bekaa Valley was always seen suitable for semi-arid climate crops [1]. Intensive agriculture in the semi-arid zone of the Bekaa was rarely investigated and it was thought to face salinity problems [2]. The National Action Plan to combat desertification classifies the Northern part of the Bekaa Valley as the most prone area to desertification [3].

The northern Bekaa Valley is generally considered as a marginal agriculture production area. Sanlaville (1963) described the Northern Bekaa as bare and sterile. Low precipitation rates because those rainfed cultivations succeed in only one out of five years. While the central Bekaa Valley was densely cultivated, in the northern part of the Bekaa cultivation was limited to small patches of land. Northern Bekaa was cultivated in only 10% of its cultivable area [4]. In the 1960s, irrigation in the Northern Bekaa was located in oases. At that time, three oases were found in the area: Laboue and El-Kaa-oases, feed from the Laboue water source, and Hermel oasis that gets water from nearby springs. Next-by populated areas, Hermel and El-Kaa oases were mainly cultivated by fruit trees and orchards (**Figure 1**) [4]. Poorly populated, in the sixties, villages of the Orontes River Watershed were considered as summer homes with only the main villages remaining populated during wintertime.

Since the 1980s, the Northern Bekaa is witnessing a rapid expansion of irrigated farming because of many local and regional factors. Urbanization increased because of the Lebanese internal unstable situation from the mid-seventies to late nineties. Regionally, the hard episodes in Iraq of the early 1990s stimulated agricultural exports from Lebanon as a response to the increasing demand.

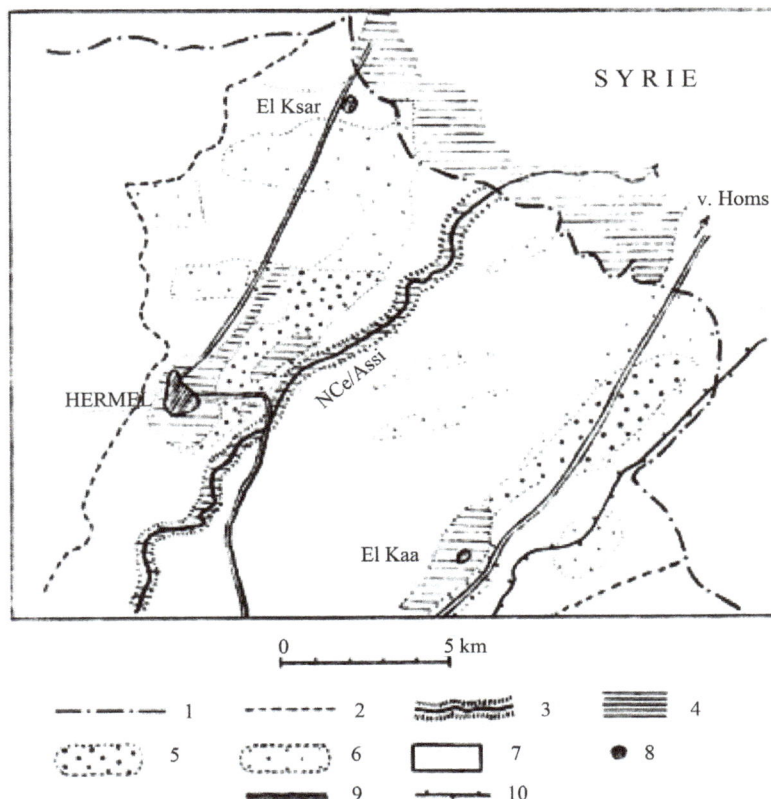

Figure 1. Agricultural map of 1963 for Hermel-El Kaa area [4]. 1: Lebanese-Syrian border; 2: External limits of the plain; 3: Steep valley; 4: Irrigated annual crops and orchards; 5: Poorly irrigated annual crops; 6: Rainfed crops and fallow; 7: Rangelands/Extensive pastures; 8: Localities; 9: Roads; 10: Railways.

In the 1980s, farmers were spotting land suitable for agriculture-production in the Northern Bekaa where groundwater is easily accessible for irrigation purposes. New irrigated plots were brought under cultivation year after year even if sometimes-hard land-reclamations and rehabilitations were first required. Vegetables and new fruit trees varieties were introduced for the local and regional market. Watermelon was among the first vegetables that has proclaimed the area as potentially capable for large agriculture investments. The high land productivity was a major characteristic that it was added to the possibility to produce crops earlier in the season because of the specific weather conditions of the Northern Bekaa. The main elements of land-productivity were the climate, high fertilizer application rates and water availability. Although soils have in most cases a petrocalcic horizon, farmers broke through this horizon and rock pieces were excavated outside to the border of the farmland [5]. Once cleared, these soils became highly productive putting soils however under intensive cultivation regimes.

Groundwater is the main source of water for irrigation in the majority of the newly cultivated land-patches. In few places located close to the river, farmers were using water directly from the Assy River. The amount of groundwater surface water used for irrigation remains among the farmers' secrets where the water use efficiency is considered relatively low. Although drip irrigation is the most widely used water saving system in groundwater pumping areas, the farmers' knowledge of water requirements is considered way off the actual crop water requirement.

Although future scenarios consider the Northern Bekaa an area prone to desertification, it is attracting large agriculture investments at the risk of overexploitation ground water resources. The yearly expansion of irrigation-needs to be monitored for the sake of future water and land management in the Northern Bekaa. The present study provides a temporal and spatial analysis of the expansion of irrigated lands in the Northern Bekaa.

2. Materials and Methods

In Lebanon, the Orontes River Watershed of the Northern Bekaa has witnessed major expansion in irrigated agriculture lands and important changes in crop patterns. The watershed covers 13% of the total Lebanese territory or about 138,400 ha and it is located at the far north eastern part of the country to the border with Syria between 36°10'E to 36°36'E and 34°0'N to 34°29'N (**Figure 2**). Precipitation decreases gradually northward, reaching a rate of less than 200 mm/year at Hermel-El Kaa zone [6]. The population of the watershed reached about 58,000 inhabitants in 1994 (Shema de direction de la territoire Libanais), where it was about 8000 inhabitants in the 1960s [4]. Following the 1/20,000 scale topographic map of 1960s, the cultivated area was about 4600 ha, matching what it was mentioned by Sanlaville. The majority of the cultivated area was dedicated to fruit trees he nearby of the populated agglomerations with a large area of old traditional vineyards agricultural systems.

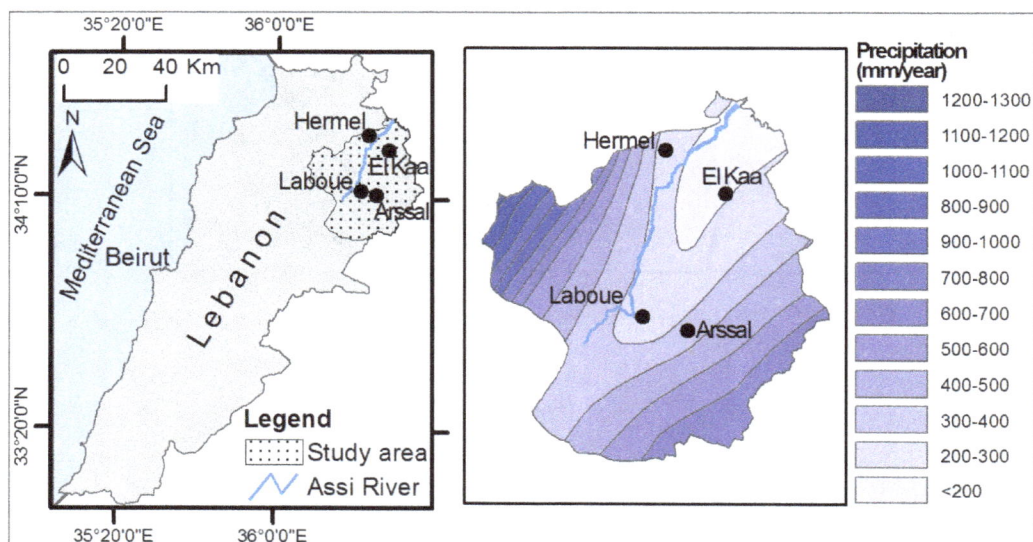

Figure 2. Study area and precipitation distribution.

Data Preparation and Extraction

The topographic maps of 1/20,000 scale were used to extract cultivated land spots of the year 1962 (**Figure 3**). These maps were first geo referenced, according to Lambert Conformal Conic projection that appears on the maps. The cultivated areas were then digitized in a GIS environment into vector formats. The maps describe the land cover of the area, where the vineyard and fruit-orchards were visually spotted and delineated. The areas cultivated lands in 1962 were compared to another source of data of the same year, which is the agriculture areas map of Lebanon [7].

Landsat images were used for monitoring purposes and to determine the evolution of agricultural lands. The multispectral satellite images-data set were of Landsat Multispectral Scanner (MSS), Thematic Mapper (TM), and the Enhanced Thematic Mapper (ETM). The images acquisition date starts from 1972 and continue following the data availability and their importance for the purpose of the study analysis (**Table 1**). The satellite images were cropped fitting the watershed boarders that it helps in comparing results on a time series analysis. NDVI (Normalized Difference Vegetation Index) was generated for each image using GIS tools of ArcGIS software.

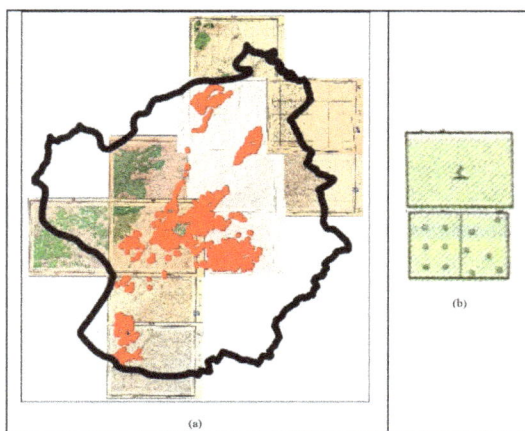

Figure 3. Topographic maps of the Orontes River Watershed; the legend of the vineyards on top and other fruit orchard in bottom.

Table 1. Satellite images acquisition dates.

Image/year	Month
Landsat MSS, 1972	Sept, 15
Landsat 5, 1984	Aug, 5
Landsat 5, 1987	May, 26
Landsat 4, 1988	May, 20
Landsat 4, 1989	Aug, 27
Landsat 4, 1990	Aug, 30
Landsat 5, 1991	May, 21
Landsat 4, 1992	Aug, 3
Landsat 5, 1997	Sept
Landsat 5, 2000	Oct, 04
Landsat 5, 2002	June, 20
Landsat 5, 2003	June, 23
Landsat 5, 2006	Sept, 19
Landsat 5, 2007	Sept, 07
Landsat 5, 2009	Sept, 27

The MSS Landsat satellite image of 1972 has a spatial resolution of 80×80 m whereas all other images are of 30 m resolution. Other Landsat images, from 1984 to 2009, are of 30×30 m spatial resolution. NDVI images were differentiated on a year-to-year sequence. The NDVI image of the year 1972 was directly subtracted from the NDVI image of the year 1984, *i.e.*, the available consecutive year. Whenever vegetation is found on the image of the year 1972, the number will be between 0.2 and 0.45 (highest positive number obtained on 1972 NDVI image). If the vegetation spot was faded out during the available consecutive year, the number will be less than -0.1, *i.e.*, highly negative value. Therefore, a positive value in the NDVI image of the year 1972 will be subtracted from a negative value at the consecutive year, which will be added as another positive number. The obtained image will have a highly positive value on the places of vegetation degradation and a highly negative value whenever vegetation appears newly (or on the newer image date). In other words, we are searching a year-to-year vegetation changes on the level of cultivated land spots.

3. Results and Discussions

3.1. Agriculture Lands since 1962

The cultivated area in 1962 appeared to cover less than 4% of the Orontes Watershed area. Orchards extended on most of the land nearby the village of El Kaa [8]. Vineyards were concentrated mainly in the area between the villages of Laboue and Aarsal, *i.e.*, to the east of the main road that passes into Laboue (**Figures 4(a)-(b)**). The village of Laboue had a large area of fruit trees orchards, especially toward the western side of the same highway. This area is a quaternary soil deposit of low elevation in comparison to the area located to the east of the road. The difference in elevation exceeds 200 m within a distance of 500 m. The surface water from the nearby water sources is easily delivered by gravity throughout the lowland valley to the western side of the highway, allowing farmers to produce high value crops, mainly apricot trees.

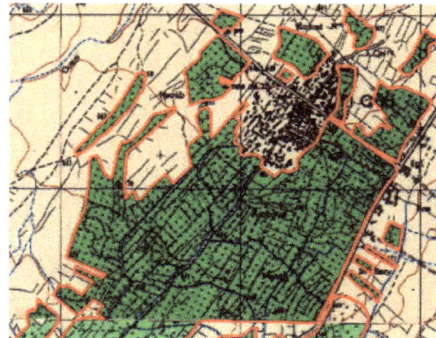

(a)

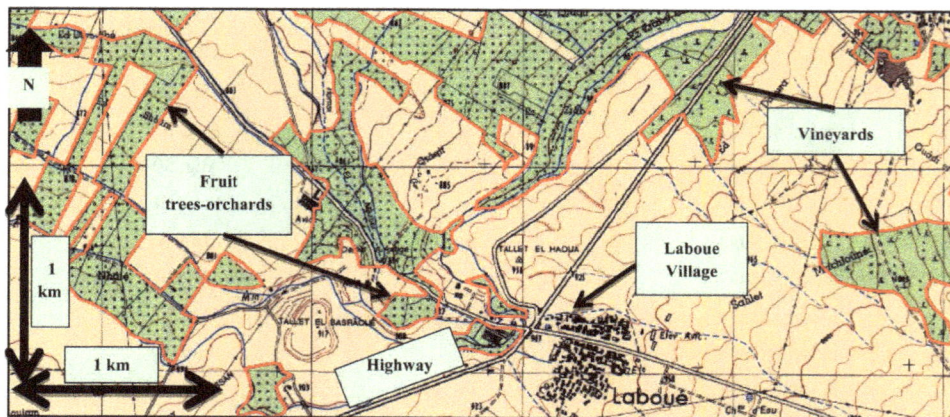

(b)

Figure 4. (a) El Kaa village the orchards at the surrounding; (b) the Laboue village and the nearby cultivated lands of the year 1962.

The Landsat image of the year 1972 showed a similar distribution of cultivated lands in comparison to topographic maps of the year 1962 (**Figure 5**). Although the 1972 satellite image is of coarse spatial resolution, it is visually clear where cultivated lands are concentrated. It is important to note that some cultivation lands have appeared on the 1972 image at the Maharii El Kaa zone, when compared to the topographic maps cultivation places.

3.2. Agriculture Lands after 1972

The NDVIs images of the consecutively available years have demonstrated important changes following the time series (**Figure 6**).

NDVI images show significant changes in comparison to the year 1984 NDVI (**Figure 7**). There was almost a complete abandonment of the cultivated lands nearby El Kaa village. On the other hand, there was a boom of cultivated parcels that appeared at the Masharii El Kaa area. Sanlaville (1963) mentions first deep wells at the central Bekaa by the year 1955, where people of the northern Bekaa mentions that first deep well was drilled around the 1969 in the area of Mashari El Kaa.

The difference between consecutive NDVI images helped in locating important changes in land cover (**Figure 8**). Subtracting NDVI of 1972 out of the NDVI of the year 1984 showed white areas in places where vegetation has receded or disappeared. Areas where vegetation has expanded in 1984 appear in black in the subtracted NDVI image.

The area of El Masharii-El Kaa has demonstrated more than 100% increase in cultivated lands when comparison is made between early eighties and the year 2009 (**Figure 9**). The NDVI images differentiation shows that irrigated plots have expanded on various locations throughout the watershed of the Orontes River. The expansion took place in locations that are far from surface water sources such as between Hermel and El Kaa villages North of Laboue village and to the east of Ras-Baalbeck village, cultivated lands are also expending were there are already important number of parcels appeared in this location. The total irrigated area reached about 19% of the total watershed area in 2009.

Once considered prone to desertification, the semiarid northern part of the Bekaa is witnessing consistent increase in irrigated lands. Development of wells excavation technologies opened the road for large investments in intensive irrigated agriculture. The relatively low cost of water pumps allows having complete set of equipped water-wells with minimum expenses. Adding, the shallow water table of the Masharii El Kaa, as an example, caused farmers to produce with lower costs. In the 1990s, the combination of shallow groundwater, low fuel prices and the highly suitable weather for cultivation attracted investors to implement large farmlands to produce vegetables and new fruit trees varieties. The area has the potential to produce earlier crops in comparison to

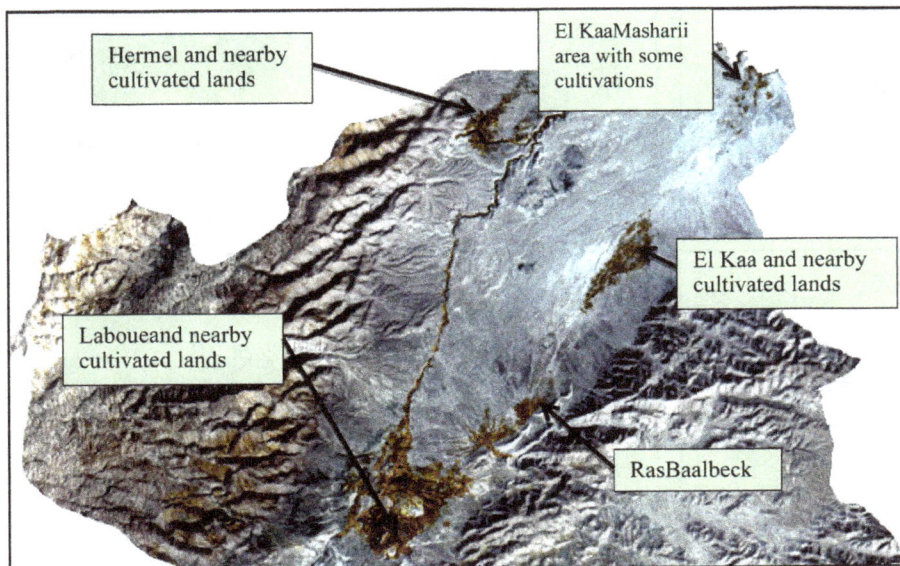

Figure 5. Landsat image of the year 1972 (band combination R = 4, G = 3 and B = 2).

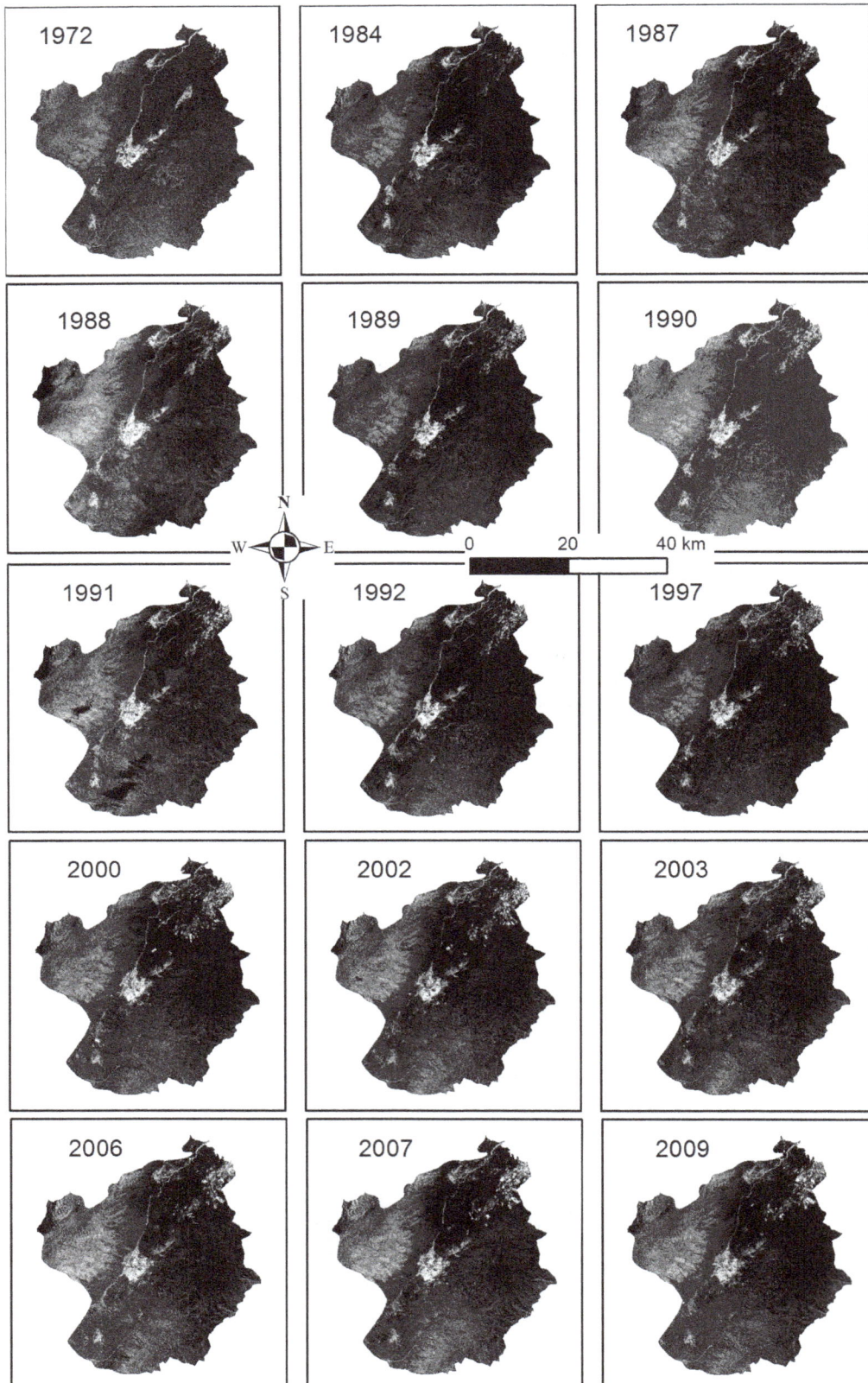

Figure 6. NDVI images of available time series.

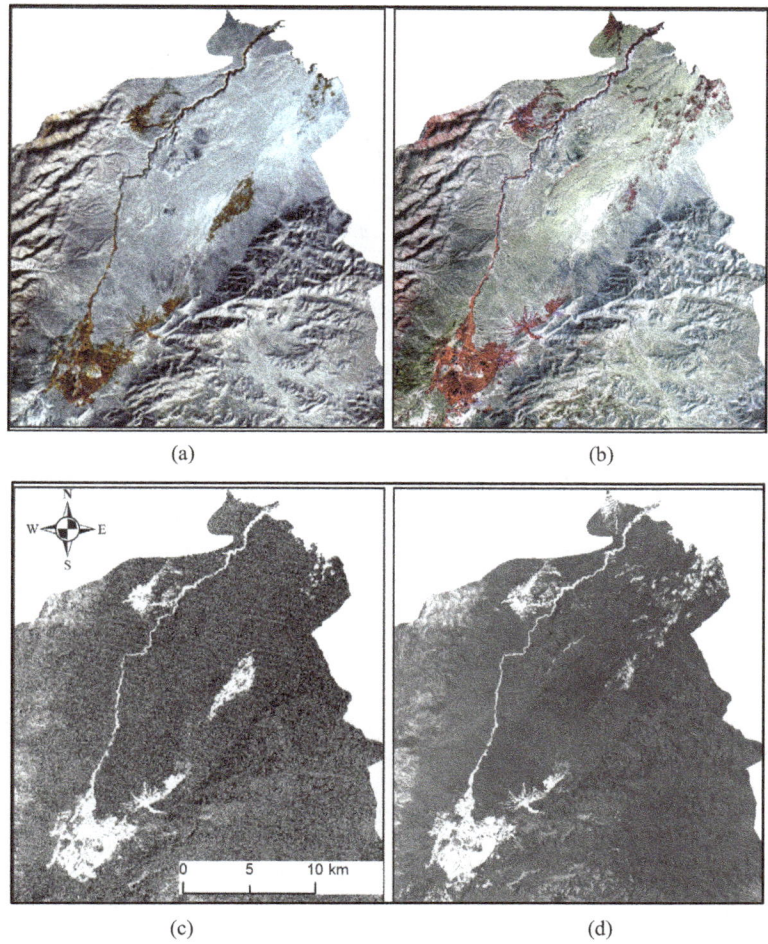

(a)   (b)

(c)   (d)

Figure 7. Landsat image of the year 1972 (a); Landsat image of the year 1984 (b); NDVI 1972 (c); NDVI 1984 (d). Blue circle represents the comparison the El Masharia area of El Kaa; Red circle represents the cultivated lands at the nearby El Kaa village.

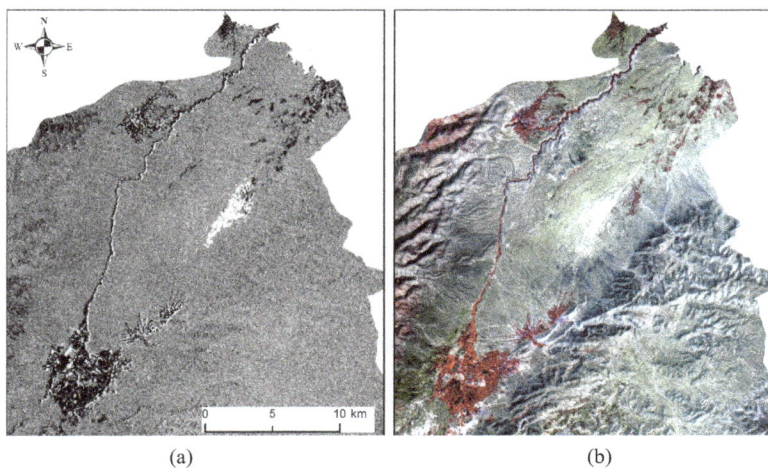

(a)   (b)

Figure 8. NDVI 1972 subtracted from NDVI 1984 (a); Landsat image of the 1984 (b). Blue circle represents the comparison the El Masharia area of El Kaa; Red circle represents the cultivated lands at the nearby El Kaa village.

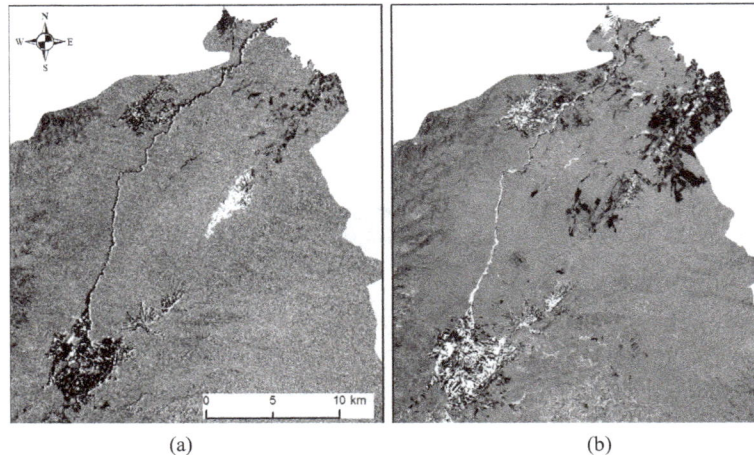

Figure 9. NDVI 1972 subtracted from NDVI 1984 (a); NDVI 1984 subtracted from NDVI 2009 (b). Blue circle represents the comparison the El Masharia area of El Kaa; Red circle represents the cultivated lands at the nearby El Kaa village.

other location in Lebanon, which raise the competiveness of the northern Bekaa. However, it is worth re-noting that intensive irrigation in this semi-arid environment is water demanding and might lead to rapid land degradation and salinization.

At governmental level, the northern Bekaa is still considered as a low productivy area. Although Ministry of Agriculture is calling for rainfed forage crops such as barley and low water requirement fruit tree crops such as olives, farmers are investing in intensive irrigated agriculture. As already mentioned, the local market absorbs the early produced crops of northern Bekaa. Located on the border with Syria, farmers had, before the conflict in the country, an easy to Arab countries like Iraq. In the early 1990s, the boom of intensive cultivation, in the northern Bekaa, coincided with the end of the civil war in Lebanon and the beginning of first gulf war in which severely affected agriculture in Iraq. This was the golden era of intensive agriculture in the northern Bekaa.

Intensive irrigated agriculture is still expanding in the northern Bekaa and future scenarios have yet to be drawn. In comparison to the early days of the expansion era, fuel prices have doubled, groundwater is dropping in depth on a yearly basis, and lands are set under unsustainable management where neighboring Arab countries are becoming producers to supply their domestic markets. All these parameters have to consider in order figuring out possible future scenarios for the area.

4. Conclusions

The early 1970s irrigated lands of the northern Bekaa were mainly concentrated in areas next to open water channels and areas of the surface water availability. Irrigated plots were mainly located near springs and human-made channels. Vineyards were the main crops that were grown in areas far from an existing surface water sources. Fruit trees were largely grown on irrigated surfaces where flood irrigation is being practiced. It should also be noted that urban expansion was limited which might also be reflected on the narrowness in cultivated lands. Petrocalcic soils of the area have added another limiting factor against the expansion of agriculture lands.

The multiplication of wells and the development of groundwater extraction in the 1980's have changed the spatial patterns of water uses and new agricultural practices have immerged. Extensive agriculture areas were changed to intensively irrigated land. Farmers are investing in land that can be irrigated from groundwater for developing intensive agriculture projects.

The northern Bekaa is officially regarded as a dry low agricultural productivity areaprone to desertification. Farmers discovered the agricultural potentiality of the area where they found suitable productive parameters: favorable weather for early production, shallow groundwater table, low fuel cost and access tolocal and international markets. Whenever these producing factors exist, farmers are willing to dig for other land patches for agricultural purposes throughout the northern Bekaa.

Future trends and scenarios have to be built in order to investigate the agriculture potentiality and its negative/

positive trend at the northern Bekaa. In a semiarid environment, intensive agriculture may lead to severe soil degradation. Intensive exploitation could cause relatively rapid depletion groundwater resources. Groundwater is extracted on an individual basis, in absence of monitoring and of an overall water use strategy.

Although the northern Bekaa has demonstrated a large agricultural potential in the short term, the sustainability of the current intensive agricultural production system is questionable. Appropriate water management strategies and higher irrigation water use efficiency will secure the productive sustainability of the northern Bekaa Valley.

Acknowledgements

This paper is made possible as part of the project entitled "Irrigation Water Managements in the Bekaa Valley, Lebanon" supported by *Global Program Water Initiatives of the Swiss Agency for Development and Cooperation.*

References

[1] Yau, S.K., Ryan, J., Pala, M., Nimah, M. and Nassar, A. (2004) Common Vetch in Rotation with Barley: A Sustainable Farming System for a Cool, Semi-Arid Mediterranean Area. *Australian Agronomy Conference* 2004, 12th AAC, 4th ICSC.

[2] Darwish, T., Atallah, T., El Moujabber, M. and Khatib, N. (2005) Salinity Evolution and Crop Response to Secondary Soil Salinity in Two Agro-Climatic Zones in Lebanon. *Agriculture Water Management*, **78**, 152-164. http://dx.doi.org/10.1016/j.agwat.2005.04.020

[3] http://www.codel-lb.org/

[4] Paul. S. (1963) Les régions agricoles du Liban. *Revue de géographie de Lyon*, **38**, 47-90.

[5] Darwish, T., Jomaa, I., Others (2006) Soil Map of Lebanon at 1/50000. CNRS, Lebanon.

[6] Plassard, J. (1971) Carte pluviométrique du Liban à l'échelle de $1/2^{00.000e}$. République Libanaise. Ministère des travaux publics. Direction Générale de l'Aviation Civile.

[7] Boulos, F.B. (1963) Carte agricole du Liban avec divers aspects météorologique, touristique, administratif et social. Conseil National de la Recherche Scientifique, Liban. 3eme Edition 1980.

[8] Jacques, G. Aspect général de l'agriculture libanaise: Tome I: Les circonscriptions foncières et la propriété. Tome II: Les régions agricoles et les principales cultures. Tome III: Enquête sur la bande côtière: Région agrumes-bananes. Rapport présenté au Ministère de l'Agriculture de la République Libanaise, Service Statistique, Beyrouth, 1960-1961.

Permissions

All chapters in this book were first published by Scientific Research Publishing; hereby published with permission under the Creative Commons Attribution License or equivalent. Every chapter published in this book has been scrutinized by our experts. Their significance has been extensively debated. The topics covered herein carry significant findings which will fuel the growth of the discipline. They may even be implemented as practical applications or may be referred to as a beginning point for another development.

The contributors of this book come from diverse backgrounds, making this book a truly international effort. This book will bring forth new frontiers with its revolutionizing research information and detailed analysis of the nascent developments around the world.

We would like to thank all the contributing authors for lending their expertise to make the book truly unique. They have played a crucial role in the development of this book. Without their invaluable contributions this book wouldn't have been possible. They have made vital efforts to compile up to date information on the varied aspects of this subject to make this book a valuable addition to the collection of many professionals and students.

This book was conceptualized with the vision of imparting up-to-date information and advanced data in this field. To ensure the same, a matchless editorial board was set up. Every individual on the board went through rigorous rounds of assessment to prove their worth. After which they invested a large part of their time researching and compiling the most relevant data for our readers.

The editorial board has been involved in producing this book since its inception. They have spent rigorous hours researching and exploring the diverse topics which have resulted in the successful publishing of this book. They have passed on their knowledge of decades through this book. To expedite this challenging task, the publisher supported the team at every step. A small team of assistant editors was also appointed to further simplify the editing procedure and attain best results for the readers.

Apart from the editorial board, the designing team has also invested a significant amount of their time in understanding the subject and creating the most relevant covers. They scrutinized every image to scout for the most suitable representation of the subject and create an appropriate cover for the book.

The publishing team has been an ardent support to the editorial, designing and production team. Their endless efforts to recruit the best for this project, has resulted in the accomplishment of this book. They are a veteran in the field of academics and their pool of knowledge is as vast as their experience in printing. Their expertise and guidance has proved useful at every step. Their uncompromising quality standards have made this book an exceptional effort. Their encouragement from time to time has been an inspiration for everyone.

The publisher and the editorial board hope that this book will prove to be a valuable piece of knowledge for researchers, students, practitioners and scholars across the globe.

List of Contributors

M. Dahiru
Federal University, Kashere, Gombe State, Nigeria

O. I. Enabulele
University of Benin, Benin, Nigeria

Manoel Euzébio de Souza
Department of Agronomy, University of the Mato Grosso State, Nova Xavantina, Brazil

Sarita Leonel and Rafaela Lopes Martin
Department of Plant Production, School of Agronomic Sciences, State University of Sao Paulo, Botucatu, Brazil

Andréa Carvalho da Silva, Adilson Pacheco de Souza and Adriana Aki Tanaka
Institute of Agricultural and Environmental Sciences, Federal University of Mato Grosso, Sinop, Brazil

Hien Thi Thu Le and Thang Nguyen Ngoc
Institute of Geography, Vietnam Academy of Science and Technology (IG/VAST), Hanoi, Vietnam

Luc Hens
Flemish Institute for Technological Research (VITO), Antwerp, Belgium

Asif Nadeem and Zahid Saeed
Agriculture Extension, Panjgoor, Pakistan

Shahabudin Kashani, Nazeer Ahmed, Fateh Mohammad and Shafeeque Ahmed
Livestock Research Institute, Turbat, Pakistan

Mahmooda Buriro
Sindh Agriculture University, Tandojam, Pakistan

Nohar Singh Dahariya, Keshaw Prakash Rajhans, Ankit Yadav, Shobhana Ramteke, Bharat Lal Sahu and Khageshwar Singh Patel
School of Studies in Chemistry/Environmental Science, Pt Ravishankar Shukla University, Raipur, India

Mohammad Tabieh, Emad Al-Karablieh, Amer Salman, Hussein Al-Qudah, Ahmad Al-Rimawi and Tala Qtaishat
Department of Agricultural Economic and Agribusiness, Faculty of Agriculture, The University of Jordan, Amman, Jordan

Shubhra Singh and N. Janardhana Raju
School of Environmental Sciences, Jawaharlal Nehru University, New Delhi, India

Ch. Ramakrishna
Department of Environmental Studies, GITAM University, Visakhapatnam, India

Salam M. Rashrash and Bahia M. Ben Ghawar
Faculty of Engineering, Geological Engineering Department, Tripoli, Libya

Abdelrahim M. Hweesh
General Water Authority, Tripoli, Libya

David Legna-de la Nuez and David Romero-Manrique de Lara
Department of Applied Economics and Qualitative Methods, University of La Laguna, Canary Islands, Spain

Serafín Corral-Quintana
Department of Applied Economics and Qualitative Methods & CIBICAN, University of La Laguna, Canary Islands, Spain

Mahbub Hasan, Tamara Chowdhury and Aschalew Kassu
Department of Engineering, Construction Management and Industrial Technology, Alabama A&M University, Normal, USA

Mebougna Drabo
Department of Mechanical and Civil Engineering, Alabama A&M University, Normal, USA

Chance Glenn
College of Engineering, Technology and Physical Sciences, Alabama A&M University, Normal, USA

M. K. Alla Jabow
Water Management Section, Agricultural Research Corporation (ARC), Hudeiba Research Station (HRS), Ed-Damer, Sudan

O. H. Ibrahim
Crop Agronomy Section, Agricultural Research Corporation (ARC), Hudeiba Research Station (HRS), Ed-Damer, Sudan

H. S. Adam
Graduate Studies and Research Wad Medani Ahlia College, Wad Medani, Sudan

M. A. El-Shirbeny, A. M. Ali, A. Rashash and M. A. Badr
National Authority for Remote Sensing and Space Sciences (NARSS), Cairo, Egypt

Robert J. Lascano, Timothy S. Goebel and Dennis C. Gitz III
Wind Erosion and Water Conservation, USDA-ARS*, Lubbock, TX, USA

Dan R. Krieg
Department of Plant and Soil Science, Texas Tech University, Lubbock, TX, USA

Jeffrey T. Baker
Wind Erosion and Water Conservation, USDA-ARS, Big Spring, TX, USA

Ryan Norman, Kristofor R. Brye and Pengyin Chen
Department of Crop, Soil, and Environmental Sciences, University of Arkansas, Fayetteville, USA

Edward E. Gbur
Agricultural Statistics Laboratory, University of Arkansas, Fayetteville, USA

John Rupe
Department of Plant Pathology, University of Arkansas, Fayetteville, USA

Ramesh P. Tripathi, Isaac Kafil and Woldeselassie Ogbazghi
Department of Land Resources and Environment, Hamelmalo Agricultural College, Keren, Eritrea

Lucia Helena Garófalo Chaves, Danila Lima Araujo and Hugo Orlando Carvallo Guerra
Federal University of Campina Grande, Avenue Aprigio Veloso, Campina Grande, Brazil

Walter Esfrain Pereira
Federal University of Paraiba, Campus II, Areia, Brazil

Alefe Viana Souza Bastos, Renato Campos de Oliveira, Nelmício Furtado da Silva, Marconi Batista Teixeira, Frederico Antonio Loureiro Soares and Edson Cabral da Silva
Instituto Federal Goiano (IFgoiano), Rio Verde, Brazil

Kico Dhima and Stefanos Stefanou
Department of Agricultural Technology, Technological Educational Institute of Thessaloniki, Thessaloniki, Greece

Ioannis Vasilakoglou
Department of Agricultural Technology, Technological Educational Institute of Thessaly, Larissa, Greece

Ilias Eleftherohorinos
School of Agriculture, Faculty of Agriculture, Forestry and Natural Environment, Aristotle University of Thessaloniki, Thessaloniki, Greece

Ynoussa Maiga
Laboratory of Microbiology and Microbial Biotechnology, University of Ouagadougou, Ouagadougou, Burkina Faso

Masahiro Takahashi
Environmental Engineering and Science, Hokkaido University, Sapporo-shi, Japan

Thimotée Yirbour Kpangnane Somda and Amadou Hama Maiga
Laboratory of Water Decontamination, Ecosystems and Health, International Institute for Water and Environmental Engineering, Ouagadougou, Burkina Faso

Daniele De Wrachien
Department of Agricultural Engineering, State University of Milano, Milano, Italy

Mudlagiri B. Goli
Mississippi Valley State University, Itta Bena, USA

Ihab Jomaa
Department of Irrigation and Agrometeorology, Lebanese Agriculture Research Institute, Zahle, Lebanon

Myriam Saadé Sbeih and Ronald Jaubert
Graduate Institute of Geneva, Geneva, Switzerland